Marine Ecology of the Arabian Region

Marine Ecology of the Arabian Region

Patterns and Processes in Extreme Tropical Environments

CHARLES SHEPPARD

Department of Marine Sciences and Coastal Management,
The University, Newcastle upon Tyne, United Kingdom

ANDREW PRICE

Coastal and Marine Programme, IUCN (World Conservation Union),
Gland, Switzerland
and
Department of Marine Sciences and Coastal Management,
The University, Newcastle upon Tyne, United Kingdom

CALLUM ROBERTS

Department of Marine Sciences and Coastal Management,
The University, Newcastle upon Tyne, United Kingdom

ACADEMIC PRESS
Harcourt Brace Jovanovich, Publishers

London · San Diego · New York · Boston · Sydney · Tokyo · Toronto

This book is printed on acid-free paper

ACADEMIC PRESS LIMITED
24–28 Oval Road, London NW1 7DX

United States Edition published by
ACADEMIC PRESS, INC.
San Diego, CA 92101

A catalogue record for this book
is available from The British Library

ISBN 0–12–639490–3

Typeset by Columns Design and Production Services Ltd, Reading

Printed and bound in Great Britain by
The University Press, Cambridge

Contents

Introduction   1

SECTION I   ORIGINS AND THE MARINE CLIMATE

Chapter 1   **Origins, Geography and Substrates of the Arabian Area**   13
Chapter 2   **The Marine Climate**   36

SECTION II   MARINE ECOSYSTEMS

Chapter 3   **Reefs and Coral Communities**   63
Chapter 4   **Coral Reef Fish Assemblages**   87
Chapter 5   **Other Arabian Reef Components and Processes**   108
Chapter 6   **Seaweeds and Seasonality in Arabian Seas**   121
Chapter 7   **Seagrasses and Other Dynamic Substrates**   141
Chapter 8   **Intertidal Areas 1. Mangal Associated Ecosystems**   161
Chapter 9   **Intertidal Areas 2. Marshes, Sabkha and Beaches**   175
Chapter 10 **The Pelagic System**   196

SECTION III   SYNTHESIS

Chapter 11 **Marine Biogeography of the Arabian Region**   221
Chapter 12 **Ecosystem Responses to Extreme Natural Stresses**   242

SECTION IV   USE AND MANAGEMENT

Chapter 13 **Arabian Fisheries**   261
Chapter 14 **Human Uses and Environmental Pressures**   288
Chapter 15 **Coastal Zone Management**   307

References   317
Index   347

To: Anne, Sylvia and Julie

Introduction

Contents

A. The distinctive Arabian region   1
B. Objectives and themes of this book   3
C. Outline of this book   5
D. Place names   5
Authors' note   6

A. The distinctive Arabian region

The tropical Indo-Pacific ocean could be considered the largest ecological system on earth. It is the most diverse marine province also, and the many superlatives which are commonly applied to it are usually directed at the shallow water habitats — especially coral reefs and mangrove systems — which line thousands of kilometres of coast. This immensely complex biological system offers some of the greatest challenges to marine ecologists today.

The Arabian region occupies an extreme corner of this vast biological realm. In one sense it has only recently rejoined it; the low sea levels of the Pleistocene left most of its main components either completely dried out or very saline, and only the Holocene transgression restored Indian Ocean water and rich Indo-Pacific ecosystems to most parts. The transgression began about 17,000 years ago and marks a fast-changing sea level which rose about 130 metres over the following 10,000 years, so that only for the last 7000 years has sea level been more or less stable at its present location. This has allowed the present major marine ecosystems to become re-established. All are represented in the Arabian region. Some, like coral reefs, flourish widely while others, like mangroves, are very much more restricted. Their establishment and degree of development today are not only the result of present conditions but also relate to pre-Holocene reef growth, to vast alluvial deposits and to subsequent erosion patterns in a fairly complex way. It must also be remembered when assessing this region that the duration of the Holocene has possibly not yet been sufficient for many initial, random events of species resettlement and community re-establishment to have become stabilized or even completed.

The Arabian region has several distinctive features. Firstly, parts of it are not especially diverse in Indian Ocean terms. This may be because of their peripheral location or recent re-connection to the Indian Ocean, because of environmental unsuitability, or combinations of all these factors. These themes are explored in this book. However, diversity is a relative consideration; for some groups of organisms the Red Sea at least seems on present evidence to be the richest marine area west of Indonesia, and is certainly far richer than some comparably-sized areas in the eastern Pacific and Atlantic.

Secondly, the Arabian region is unusual geographically in that it is the only part of the Indo-Pacific realm which includes very large, semi-enclosed bodies of water. Further, these bodies differ greatly from each other in their geological origins and physical character. The different physical frameworks that this has led to, often in close proximity to each other (e.g. Gulfs of Aqaba and Suez), has resulted in markedly different communities or species assemblages occurring closely adjacent to each other. Also the semi-enclosed nature has caused increased isolation of some areas which has biogeographical consequences, leading to high levels of endemism in some groups.

Thirdly, the Arabian region is unusual climatically in that it is the most arid part of the Indo-Pacific, with the greatest year-round insolation and the greatest seasonal fluctuations of air and water temperatures. These features combine to create some of the most extreme marine climatic regimes to be found anywhere in the tropics. Consequently there is a greater range of environments providing biota with both optimal and highly stressed conditions in this corner of the Indo-Pacific, than anywhere else. It is possible to criticize the use of the terms "extreme" and "stressed" and suggest that they mean no more than a set of conditions which we imagine species are uncomfortable with. Yet there are absolute values in this case, based soundly on physiology. For example, it is not unreasonable to suppose that the greater the osmotic differential becomes between body tissues and increasingly saline water, the more "stress" the organism experiences because, in general, more energy is required to resist it. Similarly, the more that temperature deviates from the mean values that a group of organisms (e.g. corals, mangroves) are known to live in, the more likely it is that periodic, greater extremes will cause mortality. In the Arabian region, considerable gradients of several natural environmental variables occur, and many of them will be shown to correlate with several aspects of species and community distributions. For these reasons, the Arabian region offers much that is of interest and importance to our understanding of diverse marine habitats and processes in general, and of responses of ecosystems to natural environmental stress.

Finally, the region has been one of the least well known. Despite a long scientific tradition extending back to Forskål in the 1770s, only the last 15–20 years has seen much marine research there, and much of that has been of an exploratory or expeditionary nature. Indeed, much of the region is still inaccessible to outside scientific visitors, so that a considerable proportion of the work reported here was performed on an opportunistic basis by scientists contracted to various engineering or fisheries organizations, or performed by persons engaged on other environmental impact assessments or resource inventories. Due mainly to such initiatives, many of the previously unknown coastal areas have been subjected to at least preliminary studies, and in a few cases to detailed experimental work as well. There remain many hundreds of kilometres, however, which have not been examined scientifically at all, particularly parts of the southern coasts of the Red Sea and Gulf of Aden, much of the Arabian Sea coast of Yemen, the central Red Sea between Egypt and

Sudan, and most of the northern coasts of the Gulf of Oman and "Persian" Gulf. Almost certainly, new and important surprises await, perhaps even of the magnitude of the discoveries of the southern Oman kelp beds or of the central Red Sea barrier reef, which were both studied in detail for the first time only in the 1980s.

B. Objectives and themes of this book

In this book there are three themes. Firstly, we collate as much as possible of the information of the region into one source, and this is intended to facilitate future work because of the difficulty of obtaining literature in the region. Not only was this area, until recently, one of the least known scientifically in the world, but a very highly developed sense of privacy in the region means that official agencies and organizations which commission research commonly see no need to publish their data or make it available to other agencies. Consequently the limited amount of work which is done is very difficult to locate and use, and this has led to much duplication of effort by different agencies who remain unaware of what work has already been completed. This book then, collates much of the "grey" literature which is available only with prolonged searching through local libraries, or by personal contacts, but which nevertheless has often been based on research done to a high standard by many marine scientists.

There are secrets on the surface as well as within the bosom of the ocean, which lie shrouded from human observation and research . . . Where there is mystery there will always be interest, and the greater the one, the more intense the other.

Wellsted 1840

To some extent this first objective has already been partly achieved. The volume *Red Sea* (Edwards and Head (eds.) 1987) is a comprehensive compilation of work undertaken in parts of that sea up to about 1983, and is invaluable for researchers in the region. The multi-author volume covers the main physical systems and biological groups. However, no such treatment exists for the other parts of the Arabian region, and *Red Sea* does not include work done after 1983 which is the period when much of the coastline was explored scientifically for the first time. Much of the material in *Red Sea* was obtained from two locations: the small northerly Gulf of Aqaba from both the Jordanian laboratory at Aqaba and Israeli work in the Sinai, and from the Sudan. Unfortunately scientific output from both areas has declined in quantity in the past few years, but this has been countered to some extent by the increased coverage of other parts of the Arabian coast.

The bibliography in the present book includes a high proportion of local reports and documents, and work reported in local journals. The emphasis is kept regional, and the practice is strongly resisted of including many citations of a general nature for every point or detail.

The second theme is to review marine systems and processes in the intertidal and shallow sublittoral parts of Arabian seas. As already remarked, the area is proving to be particularly interesting because it includes so many environmental extremes and biogeographical gradients. Extreme low temperatures, for example, occur in close temporal and spatial location to high temperatures, while ideal reef conditions and extreme salinity may also be found nearby. Coral reef growth, for which the Red Sea is famous above all, is actually totally suppressed in several parts of the Arabian region including the Red Sea itself, by two or three interacting environmental parameters. Here, every opportunity is taken to focus on marine systems and processes, using the newly researched habitats as material. Gradients and trends, particularly ecological patterns, form the greater part of the book.

The third theme is one of human use and its environmental consequences. The region includes some of the poorest countries in the world as well as rich oil-producing states, and both groups of nations put the coast to different use. Poor nations, for example Yemen and African states bordering the Red Sea, retain a very important artisanal use of the coast, though their low populations have allowed them to avoid the coastal catastrophes seen in many populous areas of poverty. Oil-producing nations have moved toward other uses of the coast; they commonly retain an artisanal use for social reasons rather than for current financial need, and use the coastal zone much more intensively for servicing and producing oil, and for residential development. This reaches an extreme condition on the Gulf coast of Saudi Arabia where 40% of the coastline is now either filled in or else greatly affected by adjacent landfill or oil contamination. Because of widespread oil pollution, several areas (e.g. parts of the Gulf of Suez) have an intertidal region which is now effectively covered in tarmac despite the very low population density. Positive events have occurred in several places, however, and some agencies are taking action before gross, irreparable damage is done. Stimulus for this comes partly from present projections of sea level rise which could have interesting and costly implications in this region where some areas with enormously valuable infrastructures are very low lying. The 1991 Gulf war also had considerable environmental impact, and the effects of this are covered as much as is possible.

In general, shallow benthic habitats rather than pelagic ones are the subject of this book. In part this is of necessity. Visiting researchers have been able to undertake benthic studies which are not hardware-intensive with some frequency, while pelagic work requiring large vessels and local logistics has proved to be more of a problem. Added to the latter difficulty, fishery statistics, which can provide a good source of data, are not collected systematically by most countries even where such data were collected in the past. Therefore, throughout the book emphasis is on benthic and shallow habitats, usually those accessible on foot or by scuba equipment. Available pelagic data for the region are assembled, however, and for those interested in wider fisheries information the volume by Longhurst and Pauly (1987) is almost exclusively concerned with fisheries.

C. Outline of this book

The book has four broad sections, each including several chapters. The first presents the geological, geographical, climatic and oceanographic background to the area. This does not include a detailed review of these subjects but rather gives those data and aspects important to understanding the main, biological themes. The second section includes the bulk of the ecological material. Some parts are descriptive, bringing together what is known of the region's marine communities, and this section begins the interpretation of the relationships between the marine systems and the physical conditions. The third section synthesizes and concludes the biogeographical material and interprets the effects of natural stress on the biota. The final section describes and discusses the human use and management of the region, including fisheries.

D. Place names

Included in the region for the present purposes are: the Red Sea with its two northern Gulfs of Aqaba and Suez, the Gulf of Aden and Arabian Sea, the Gulf of Oman and the body of water formerly called the Persian Gulf. All have been fairly extensively visited by the present authors.

Traditional uses of some names, even when these follow established Atlases or Pilots, are not always appropriate now. Some names are transitional until all concerned can reach agreement.

Nowhere is the problem greater than in the Persian Gulf. The name of the nation and people from which the gulf was named has changed, but "Iranian Gulf" was never universally approved, and Arab states preferred "Arabian Gulf", which is in turn not favoured by the Iranians, who at least have convention on their side. The Regional Organisation for the Protection of the Marine Environment, which comprises all the states bordering that sea including Iran, tends to use "ROPME Sea Area" which, however, lacks a certain crispness. The name "Inner Gulf" is encountered in official documents but has not been widely successful either. With the latter system, "Outer Gulf" refers to the Gulf of Oman. The latter is in any case not challenged, so Outer Gulf is redundant. It should be noted that in earlier literature the name Arabic Gulf referred to the Red Sea (Niebuhr 1792). Finally (so far as we know) "Arabo-Persian Gulf" has sometimes appeared, and while it is a heroic compromise, the name has fallen the way of "ROPME Sea Area". In the present book "The Gulf" is used. It is emerging as the favoured name by default, although its lack of descriptor can be very inconvenient. It is, however, neutral and is also used in an increasing number of maps of the region.

Arabia, properly so called, is that great peninsula formed by the Arabic Gulf, the Indian Ocean, and the Persian Gulf.   Niebuhr 1792

Similar disagreement befalls the Gulf of Aqaba or Eilat/Elat. Gulf of Aqaba has tradition and Atlas use, although Eilat possibly has precedence by being an older city, and it was used to name the body of water first. Most States in the region prefer Aqaba politically, but by far the most references in the scientific literature use Gulf of Eilat because most marine research publications have been done by scientists from Israel. Consistency here has proved to be difficult too; one publisher produced a valuable volume on sabkah ecosystems which names the Gulf of Eilat numerous times (Friedman and Krumbein 1985) only one year after producing another with "Gulf of Aqaba" as part of its title (Reiss and Hottinger 1984). Here, we use Gulf of Aqaba because, unlike "Persian Gulf", the source name (Aqaba City) has not itself had its name changed by its residents who do not, therefore, object to it.

Gulf of Suez and Red Sea are secure names, as is the Gulf of Aden; A'Den itself being an Arabic name. The Arabian Sea here refers to the Indian Ocean from the Gulf of Aden to Ras al Hadd, which is the easternmost point of the Arabian peninsula. The Gulf of Oman continues from the latter to the Strait of Hormuz. Other major bodies of water named here are the Gulf of Salwah which is the southern embayment of the Gulf between Saudi Arabia and Qatar, and the Gulf of Kutch which is an Arabian Sea embayment in the mainland of India. Indian Ocean is of course used also, secure in the knowledge that a claim once made that it be renamed the Indonesian Ocean was rejected.

Following this Introduction are several maps, which include place names mentioned in the text.

Authors' note

All three authors of this book have worked on all of the major ecosystems in the region, in various capacities of ecologist or coastal manager over a combined total period of nearly 40 years. However, there has of necessity been a division of authorship. The chapters on fish and fisheries were principally authored by C. Roberts, while those on mangroves, seagrasses, human uses and coastal management were by A. Price. The remaining nine chapters were principally written by C. Sheppard. There has of course been crossover of material because we intended to provide an integrated account of the marine ecology of this region. This approach, in our view, better reflects the closely integrated nature of the region's biological systems than does the easier alternative of a collection of discrete papers. We hope that the extremely interesting character of the marine part of the Arabian region is reflected in the book.

Arabia, showing locations of larger scale maps.

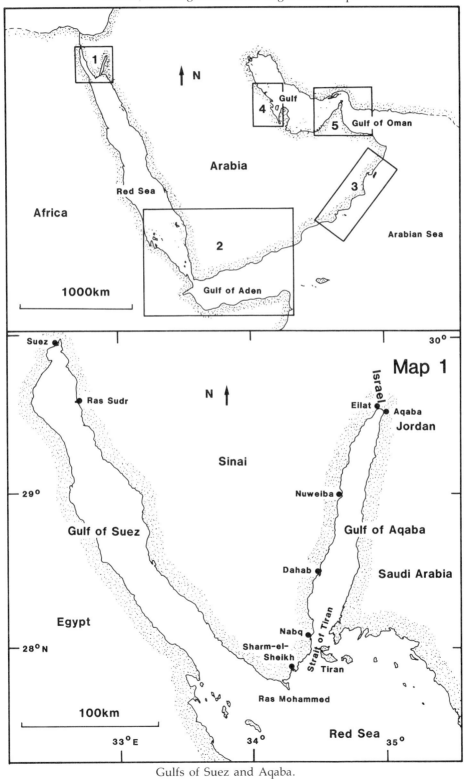

Gulfs of Suez and Aqaba.

Gulf of Aden and southern Red Sea.

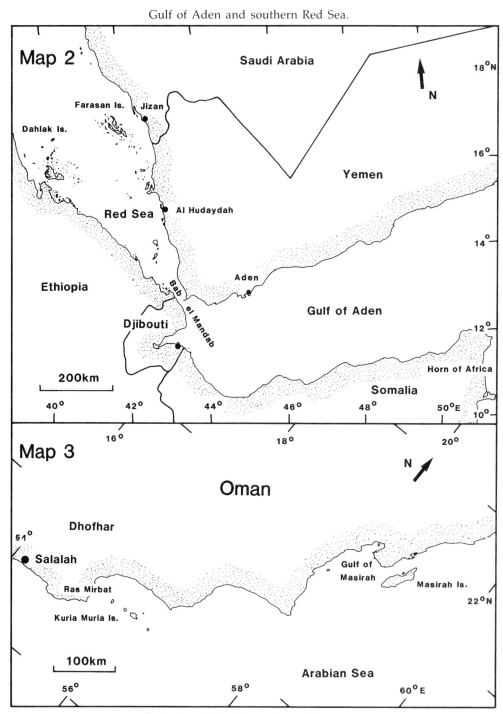

Central and Dhofhar region of Oman.

Central Gulf showing Saudi Arabian Islands, Gulf of Salwah and Hawar Archipelago.

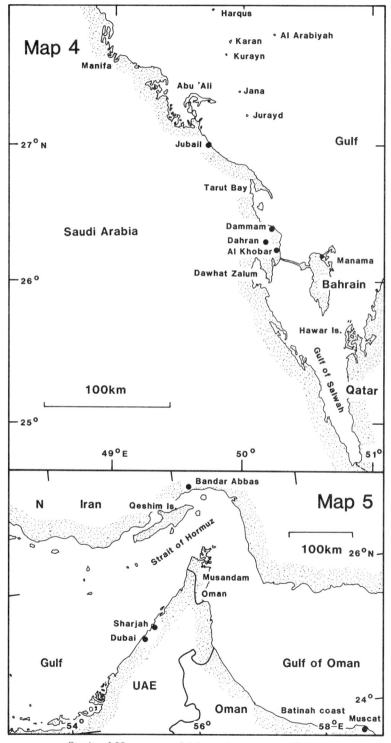

Strait of Hormuz and Musandam peninsula.

Origins and the Marine Climate

Raised fossil reef and fringing reef, North Red Sea.

CHAPTER 1

Origins, Geography and Substrates of the Arabian Area

Contents

A. Introduction 13
B. The coastal structure and formations 14
    (1) Geological history and the formation of the present seas 15
    (2) Substrate precursors and Quaternary sea level change 19
    (3) Shoreline terraces and layers of reefs 24
    (4) The Gulf and Gulf of Suez 28
    (5) The Oman and Yemen coasts 30
    (6) Islands and archipelagoes 31
Summary 34

A. Introduction

This chapter and the next describe the physical foundations to the shallow marine ecosystems and the main hydrological controls (the marine climate) which act on them. Geological processes and events are treated in a manner subordinate to the ecological theme, so only immediate geological foundations and structures in the region which have an important and direct bearing on the present marine life are discussed.

In all coral seas, much of the immediate substrate is itself biogenic, and limestone rock produced by biota is relatively soft and prone to sub-aerial and biological erosion and solution. Quaternary sea level changes have been marked, and have changed the relative vertical and horizontal locations of contemporary biogenesis many times. For these reasons, Pleistocene and Holocene events are seen as most important in the present context. Even older processes may be equally important in determining present day distribution of some habitats, and such events include major features such as the rift formation of the Red Sea as well as more immediately relevant, localized features such as volcanic foundations, upward moving salt domes laid down in periods of extensive evaporation, and Tertiary limestone foundations.

Such features not only explain some of the present biotic distributions in the most obvious sense, but understanding some of the geological processes is the only route to understanding some of the zoogeographical and ecological phenomena which will be described in this book. While the past 17,000 years (the Holocene) are the most important in many respects, several much older events such as earlier episodes of reef building are as important because it is these pre-

Holocene reefs that provide most of the extensive limestone foundations for the present structures, and also the limestone barriers to waves which permit establishment of soft substrate habitats.

In the region as a whole, the total area of soft substrate habitats (whether associated with reefs or not) greatly exceeds that of the reefs, despite the region's fame for the latter. Much of the soft substrate is of biogenic carbonates, several of whose aspects are covered in later chapters. In this region, sub-surface features which support present reef growth, or which form many embayments or gently sloping shelves which contain sediments, have been regarded as having complex and confusing origins, and some of the few recent descriptions of them have themselves been complex and confusing. However, although the detail may be complex it is fairly simple in essence. Some of the difficulties may have arisen because in this region, the processes of reef formation and foundation greatly stretch the classic Darwinian model which is ingrained in students of reefal areas in several respects, as has been recognized for a long time. Some of the most recent information on the foundations and origins of reefs here comes from important engineering and drilling reports. Once the high degree of local variation is recognized, and once it is recognized that hitherto poorly known processes such as establishment of reefs on soft substrate not only occur in this region but are very important, then the geological foundations to the present day marine environments become fairly straightforward.

In the Red Sea accordingly, the reef formation is not brought about, as Darwin and Dana claim to have shown in the case of the Pacific, by the sinking, but on the contrary by the elevation of the sea bottom.   Klunzinger 1878

Darwin's subsidence theory cannot be called into account for the existence of the barrier reefs at a distance from land and separated from it by deep water, or of the atoll forms of certain reefs.   Crossland 1938

B. The coastal structure and formations

The Arabian peninsula occupies, more or less, its own tectonic plate (Figure 1.1). Deduced from lines of earthquake activity, locations of spreading sea floor centres and ongoing crumpling of crustal material, the Arabian plate is a newly separating fragment of the very large African plate. The southern and western boundary of the Arabian plate is marked by the line of sea floor spreading which passes down the Gulf of Aqaba and Red Sea, and through the Gulf of Aden to a point in the Arabian Sea at the Carlsberg ridge. The latter feature, named after the sponsors of the *Dana* expedition which discovered it, not only stamped the name of a famous beer on the oceanographic map, but itself became an important piece in the development of ideas on the origin of oceans, including the formation of the Red Sea. Because the central line of the Red Sea is a spreading centre, the Arabian plate is thus moving slowly away from Africa. In addition, it is rotating slowly anti-clockwise relative to Africa as well as sliding northwards, causing

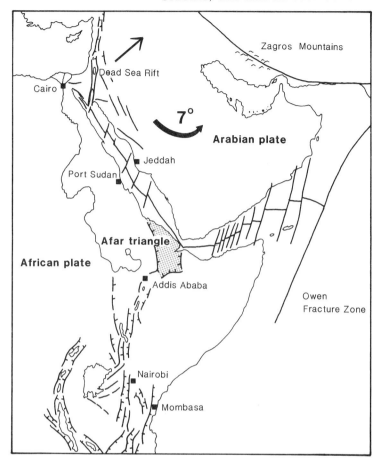

Figure 1.1 *Arabian plate, showing main geological setting, East African Rift system and areas of spreading and rifting. Afar triangle is dotted. Main movement of the plate is shown by arrow; value gives angle of rotation in degrees.*

crumpling in Iran and the formation of the Zagros mountains of Iran and of the Musandam mountains of northern Oman. The latter form a fjord-like marine environment unique in Arabia. The coastal region of Iran along the Gulf marks the northern plate boundary. The rotation of the Arabian plate has been about 7°, so that the part of the Arabian plate along its northern boundary which is moving the greatest amount is the Musandam peninsula which projects into the Strait of Hormuz. The thrusting northwards at this point has resulted in some extensive hard substrate habitat of regional biological importance.

(1) Geological history and the formation of the present seas

The Red Sea is the place where the earth's largest geological feature, the mid-ocean rift system, strikes a continental platform, and splits it. As Braithwaite

(1987) puts it: "For the geologist, the Red Sea is an ocean. This disregard for the realities of scale is rooted in our understanding of the nature of the earth's crust and of what oceans are." On the largest scale, ocean floors are made of dense rock which appears at intervals on the surface as basalt, while continents are composed of lighter, silica rich rocks which float on the basalt and which are pushed about by movements, possibly convective, of the underlying basalt. On a few parts of the earth's surface the oceanic floor is exposed on land, and such examples include those which are important in this account of the Arabian region: near the Capital Area of Oman, and in the southern Red Sea and Gulf of Aden.

In the 1930s, it was discovered that the crest of the Carlsberg ridge, and the later discovered Murray ridge in the Arabian Sea, were split by steep sided gullies which extended a few hundred metres deep. It was noted that these rifts are similar in form to that which extends up the length of the Red Sea and Gulf of Aqaba to Jordan, and similar also to the African Rift Valley. These clefts along the tops of the ridges are zones where the crust is thinned as the two sides move apart. It is now fairly well established that the Red Sea rift has been separating Arabia from Africa for about 70 million years. The Red Sea with its present "oceanic" character thus formed about this time (although the general Red Sea depression was flooded at times well before this). Rifting has not been continuous, and pauses in the first half of the Tertiary saw important episodes of volcanic activity which led to extrusion of the existing volcanic islands and other coastal reef foundations in the south, and to especially well known basaltic larvas in Sinai. Rifting recommenced between 2 and 5 million years ago (Ma BP), at rates averaging up to 2 cm per year. The Gulf of Aqaba seems not to be a spreading centre at present, but instead the Arabian side is sliding northwards relative to Sinai, leading to faulting. These faults have deepened the series of three major depressions in that Gulf (Friedman 1985). It is this tectonic movement which accounts for the great depth of the Red Sea and Gulf of Aqaba, as well as of several other features noted later.

The tectonic theory of Wegener is said to have arisen partly because of the observed fit of Africa and the Americas across the Atlantic Ocean, which is itself another area of ocean spreading. That after such a large distance the fit is imperfect is not at all surprising but the lack of a perfect conformity was the basis of a lot of early dissent from the theory. In the smaller and younger Red Sea, the fit should be much better, and indeed from Sinai south to about 15° N, or the latitude of the Farasan and Dahlak Archipelagoes, the fit is almost perfect. From this point the shores of the Red Sea converge and, at first sight, the fit is destroyed.

However, Wegener (1929) considered this too. The Afar triangle in Ethiopia occurs at the point where the rift system diverges in three directions: along the length of the Red Sea, into the Arabian Sea, and down the African mainland as the Rift Valley. The latter shows little or no spreading at present, but has the same features. The Afar triangle is for the most part below sea level. It is volcanic, and has major basaltic intrusions; it has been called a section of oceanic floor elevated enough to become dry land, and its geologically different nature, graphically described by Sullivan (1974), is clearly distinguished in satellite photographs. Wegener drew attention to the fact that if this triangle of dry ocean floor was flooded, then the fit between Arabian and African sides of the Red Sea would be complete.

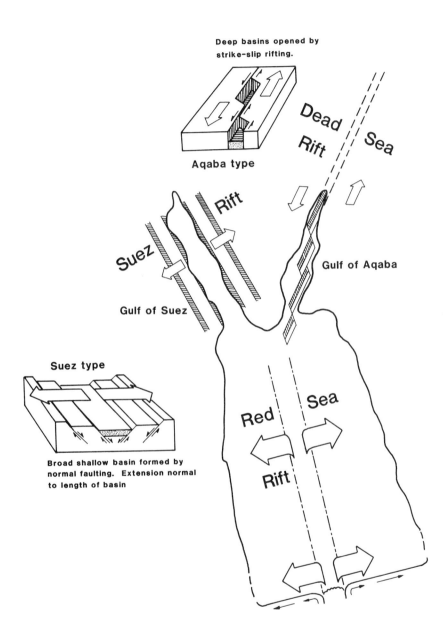

Figure 1.2 *Tectonic framework of the northern Red Sea Gulfs of Aqaba and Suez, showing different rifting and formation. Northern Red Sea is a spreading centre. Gulf of Aqaba is formed by strike-slip faulting and its depth is attributed to basins formed by grabens (inset). Gulf of Suez is formed by normal faulting. From Friedman (1985).*

> *The otherwise accurate parallelism is spoilt by (the Afar triangle) projection. If one cuts this triangle out, the opposite corner of Arabia fits perfectly into the gap.*
>
> Wegener 1929

The Gulf of Aqaba is a continuation of the Red Sea rift, which continues almost to Syria. Its maximum depth is of the order of 2 km. The spreading centre character of the Red Sea is believed to change to strike-slip faulting in the southern third of the Gulf of Aqaba as the Arabian plate slides more in parallel with the Sinai land mass (Friedman 1985), and this has led to three deep basins in the Gulf of Aqaba (Figure 1.2). The total lateral displacement in this gulf is now approximately 105 km, the last phase of which took place in the early Pliocene, amounted to about 45 km of movement, and formed the present shape of the Gulf of Aqaba. The body of water is bordered by mountains and has pronounced underwater relief. It is joined to the Red Sea via the Straits of Tiran at a relatively shallow (250 m) sill.

The Gulf of Suez has a completely different character. It occupies a wide valley and is bordered by wide plains of low relief. It appears to be spreading (Figure 1.2) and exhibits normal faulting. Water depths are mostly less than 50 m, though this reaches 70 m at its southern end. It has no sill at its connection with the Red Sea.

The present, narrow straits at the south of the Red Sea are the Bab el Mandeb, literally "Gate of Lamentations". They are 29 km wide and only about 130 m deep at their shallowest point. Somali folklore has a tradition that their ancestors crossed these straits from Arabia on a land bridge, and indeed an answer required of the Forskal expedition in the 1770s was whether a land bridge was possible (it was answered in the negative). While our present understanding of the time scale of when those straits were closed (over 10,000 years ago at least) makes the Somali tradition surprising, it is still the case that it is by far the narrowest and shallowest part of the Red Sea, and it is an inherited migration path for many birds for this reason. Whenever flooding of the straits occurred, however, it seems clear that it did do so in Palaeolithic times, and the result is of present biological significance.

The Gulf of Aden was never closed. Whereas partial or complete closure of the Red Sea led to increased salinity and build-up of salt deposits, drilling in the Gulf of Aden has failed to demonstrate similar salt deposits there.

The Gulf by contrast is a sedimentary basin, measuring about 1000 km by 200–300 km. It has an average depth of about 35 m, dipping downwards towards the north to a maximum of about 60 m along the Iranian coast, and inclined downwards to about 100 m deep at its entrance in the Strait of Hormuz. Its northwest–southeast axis separates a stable Arabian shoreline on one side from a geologically unstable Asian or Iranian shoreline (Purser and Siebold 1973). This results in the deepest region being very close to the Iranian shore. No part of the Gulf has a continental shelf edge, and there are no large-scale bathymetric changes as seen in the Red Sea, except at the entrance in the east. The floor of probably all of the Gulf lies within the photic zone. As a consequence of the gradual topography and of the favourable environment to carbonate producing

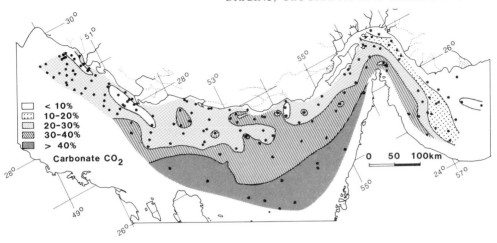

Figure 1.3 *Map of carbonate-CO_2 in surface sediments in the Gulf. From Siebold (1973). Dots indicate the locations of sampling sites.*

biota, the Gulf is a strongly sedimentary province with a dominating soft substrate benthos. Sediments of biogenic carbonates exist over much of the Gulf floor (Figure 1.3), with strong terriginous influences limited to the northwest end where the waterway of the Shatt al Arab discharges into the Gulf.

The Gulf is commonly divided for descriptive purposes into eastern and western sections by the Qatar peninsula although there is in fact a strong similarity of geological character on both sides of the latter. Offshore, underlying salt domes have forced upwards numerous islands and banks of hard substrate which are now colonized by corals. These now provide much of the vertical relief and in this respect the Gulf is similar to parts of the southern Red Sea. At the shoreline along the Arabian side, there is a very gradual slope and a gradual blending of marine conditions with terrestrial, sometimes extending across a band of several kilometres, especially in the United Arab Emirates (UAE). This contrasts with the Iranian side where the Zagros mountains exceed 1000 m elevation close to the shoreline.

In the Pleistocene, complete evaporation is thought to have occurred except for a narrow strip along the northern edge which conducted the fresh water of the Tigris, Euphrates and several smaller streams to a coastline which was located in the Strait of Hormuz.

(2) Substrate precursors and Quaternary sea level change

The long history of formation, flooding, evaporation and alternate connection of the Red Sea with the Mediterranean and Indian Ocean is pieced together clearly by Braithwaite (1987). The basin was at times dry and at times part of the Tethian Sea. It is not necessary to repeat the sequences of events here, and many in any case have very little bearing on the present day marine ecology of the region. The important events for present day ecology derive partly from the eventual but not continuous connection with the Indian Ocean which began 2–5 Ma BP, and partly from the heavy erosion which occurred in the Pliocene and especially in the Pleistocene. The former period brought the modern Indian Ocean fauna and flora, including reef-forming and reef-dwelling organisms, into the Red Sea and Gulf. The erosion later in the Pleistocene was caused by heavy rain, probably in two phases, the later phase between 35 and 17 thousand years ago (Ka BP) (Jado and Zoetl 1984), in other words during the lowest sea water stand (sea level is considered later). During both phases of erosion, sheet flow of alluvial material caused the formation of the alluvial fans and outwash systems which form important foundations to shallow benthic biota today. The material of the fans themselves derived from the mountains on either side of the Red Sea and is of much older, even pre-Cambrian age in the case of the central Red Sea and Sinai.

Two aspects of this erosion are important. Firstly, it cut through some of the early, developing fringing reefs repeatedly, commonly down to around the contemporary sea level or possibly slightly deeper. This appears to be the case especially in the early Pleistocene rainy period where sea level was 30 to 60 m below present. At the same time as the erosion occurred, more alluvial material simply covered extensive areas of fringing reef. Solution from rain also may have added karst relief directly, but this is probably minor in comparison to the erosion from sediments (except on series of coral reefs over 10 km from the mainland areas which are too distant from land for alluvial fans to have reached effectively). Secondly, large amounts of terrestrial material ended up offshore in shallow water where it formed important new substrates which were colonized directly in many cases by new coral reef growth, and they also formed extensive areas of sheltered habitat which retained a soft substrate character.

The extent of the influence of the erosion can be seen by the major wadi systems (called variously sharms, khawrs or mersas in different countries) which developed during the Pleistocene, and which were essentially cut by abrasion into the existing substrate. Examples are shown in Figure 1.4. The depths of the wadis show considerable uniformity at 60 to 65 m deep, though some are deeper such as the wadi at Port Sudan, the bottom of which is over 90 m below present sea level. The excavation of the wadis of the Gulf of Aqaba and Red Sea are discussed and more examples illustrated in Gvirtzman et al. (1977). Similar processes occurred also along much of the Oman coastline where wadis are excavated to similar or even greater depths into coastal cliffs. In the Gulf by contrast, the overall low lying terrain precludes wadi formation, though terriginous sediments grade to seaward for a considerable distance. The importance of the alluvial fans is discussed in a later section.

The cutting of the wadis to 60–90 m below present sea level was possible because of greatly lowered sea levels. From numerous parts of the world the pattern has been deduced of eustatic sea level rise and fall during the late Pleistocene to the present, and a summary of evidence is given by Hopley (1982). Sea level was last at a level similar to the present level about 110–140 Ka BP. This was followed by a long period from 110 to about 30 Ka BP where sea levels fluctuated considerably, but remained between about 30 to 60 m below present levels (Figure 1.5). In this time there were at least eight still-stands lasting for over

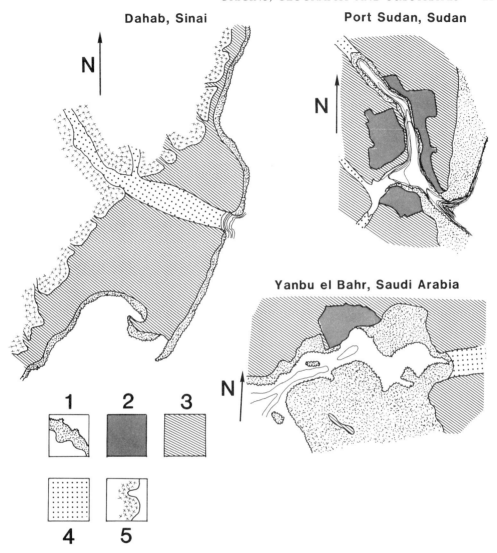

Dahab, Sinai

Port Sudan, Sudan

Yanbu el Bahr, Saudi Arabia

Figure 1.4 *Sketches of three wadis illustrating their major features. Water depth contours are in intervals of 10 m. From Gvirtzman* et al. *(1985) and Admiralty charts. (1) Holocene fringing reefs, (2) built-up area, (3) elevated coastal terraces of fossil reefs and alluvial fans, (4) wadi, and (5) mountains.*

1–2 Ka at intervals between −30 and −60 m. This period has considerable immediate significance in the history of reef growth in the Arabian region. Reefs constructed prior to this were subjected to severe aerial erosion and scouring. Note that the level of the sea at −30 to −60 m corresponds to the depths of major wadis as mentioned above, and it also coincides with important submerged features discussed later. This period lasted about 80 Ka, during which time substantial reef growth took place at the various levels of the still-stands between

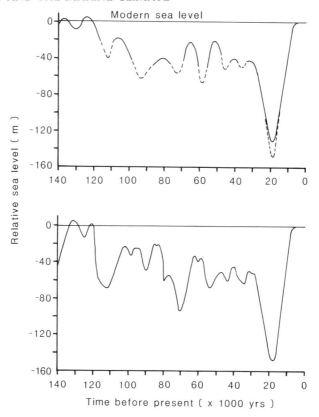

Figure 1.5 *Sea level in relation to present time. From Hopley (1982) (top) and Potts (1983) (bottom).*

−30 and −60 m. It has been suggested that the sea may have remained at present levels until only 30 Ka BP (Milliman and Emery 1968) which would cause various difficulties in interpreting coastal reef sequences in several areas, but this evidence has more recently been discounted (Hopley 1982).

From 30 Ka BP sea levels fell rapidly to a minimum at about 17 Ka BP, at the boundary between Pleistocene and Holocene. The envelope of values surrounding the sea level curve suggests that it probably fell to somewhere between 120 and 150 m below present level, and that this was maintained for perhaps 1–2 Ka. About 15 Ka BP global surface temperatures increased markedly (Milliman and Emery 1968), which led to the Holocene transgression. This rise in sea level commenced about 14–15 Ka BP and proceeded rapidly to near present levels about 7 Ka BP after which movement slowed or stopped. It may for a period have settled at a level slightly higher than at present, though this is arguable.

Three important features therefore are (1) the extensive period when sea levels fluctuated about 30–60 m below present levels, (2) the deepest point to which sea level fell after this period, and (3) the length of time and exact heights of sea level in the period 5 Ka BP to the present.

The first of these three conditions is well established.

Of the second major Recent event, some discussion has centred on the exact level to which the sea level fell. At first sight this is extremely important to the Arabian region. At the entrance to the Red Sea, the Bab el Mandeb is about 130 m deep, so whether the sea level fall was to the lower or upper limits of the deduced envelope might seem to make all the difference as to whether or not the Red Sea was isolated biologically and dried up. Chapter 2 discusses the biological consequences of this. It has sometimes been assumed that complete closure of the Red Sea would be necessary for highly saline conditions to have occurred, but this is not the case. Whether the Bab el Mandeb completely dried or not, evidence shows that the Red Sea became a very saline environment in the period of about 17 Ka BP. Braithwaite (1986) presents the evidence for "The Great Evaporation", and Gvirtzman *et al.* (1977) refer to the Red Sea as a hypersaline lake, as do Sheppard and Sheppard (1991). From the biological viewpoint the result would be similar, namely extinction of marine fauna and flora with the exception of resistant, mainly prokaryotic species such as those flourishing in present day saline pools which are common along shores of both the Red Sea and Gulf.

In the Gulf, lowered levels to any point in the envelope of possible sea level fall values would cause drying out of most of the Gulf floor, leaving only a finger of sea water extending inward from the Gulf of Oman, along the Iranian coast. The lowered sea level, to whatever point in the range of possible values, thus marks a significant and relatively recent point when the marine ecosystems in both bodies of water were effectively "reset".

Geologically, and to students of eustatic changes, the question of complete closure or not is more important than it is to biologists. If complete closure did not occur then water in the Red Sea would have become highly saline but levels would have remained near to that of the Indian Ocean at about 130 m below present. Contemporary sea level erosion features would then be at around that depth. However, if complete closure did occur, then the thousand years or more of intense evaporation following closure (present evaporation is 1–2 m per year) would drop the level in the Red Sea saline lake much more. This would lead to contemporary sea level erosion features being much lower still. The point is not established, though Jado and Zoetl (1984, cited in Crossland *et al.* (1987)) give evidence for widespread lows at −130 m which might indicate that complete closure did not occur.

The third important point concerns whether or not sea level rose slightly higher than present levels, during the last 5000 years. Geological literature in the past has often commented that it is hotly debated whether or not sea levels at various parts of the world were raised above present levels, and strong evidence from several areas suggests a level of 1 m above present about 6 Ka BP and a subsequent smooth fall (Chappell 1983). In fact this too is not likely to be especially important in the Red Sea where local changes in the level of the shoreline relative to land appear to have been of greater magnitude, both from local tectonic uplift activity and from localized but large movement of the relatively mobile alluvial terrain that surrounds the seas of the region. Small vertical changes are also probably of little consequence in the context of present day ecological conditions. The question arises partly because of the existence of elevated reefs which line the sides of the Red Sea and which are visible on the Arabian shores of the Gulf. However, any higher point of sea level may have been only one or two metres, which, as the next section shows, is much smaller than tectonic uplift which has continued in the region.

(3) Shoreline terraces and layers of reefs

During the time of considerable alluvial activity in the last 140,000 years, reefs developed along the Red Sea coast. Their growth was irregular and discontinuous for two main reasons: firstly changing levels of sea relative to shoreline (whether eustatic or caused by uplift or both) caused discontinuities or terraces, and secondly because of alluvial activity.

> *The shore on which we are wandering is under the influence of puzzling forces; it is rising while the sea is withdrawing . . . The geologist calls this "the secular elevation of the land".*   Klunzinger 1878

The fluctuation in sea level according to Hopley's (1982) summary, resulted in at least eight periods of about two thousand years when the sea level was stable at various levels 20 to 60 m below present. This length of time is more than sufficient for fringing reefs to develop on the contemporary shoreline, and for patch reefs to develop on any older limestone (or other) platforms. At the same time, during every drop in level, erosion of the previous reef growth would occur. Also at the same time, sheets of alluvial material deriving from the coastal mountain ranges spread into the coastal zone and sea, filling eroded channels and smothering sections of growing reef. This alluvial activity was particularly marked in the well studied Gulf of Aqaba, partly because the mountains of Sinai are steep and located very close to shore.

Where these alluvial fans ended up in water between the contemporary surface and about 40 m deep, they smothered the fringing and nearshore patch reefs, but they also became consolidated fairly rapidly to the point where they themselves became substrate for new reef growth. Even sediments which are not at all strongly consolidated may become suitable substrate for coral growth, as has been demonstrated in several areas including the Red Sea (Hayward 1982, Hopley *et al.* 1983, Sheppard 1981). The effect of this in areas where alluvial flow was considerable was to increase the observed interruptions in smoothly continuous reef growth. Figure 1.6 shows a section revealed by core drilling in the central region of the Arabian side. In this area, a simple model of alternating alluvial fan material and coral reef is observed. Based on Miocene raghama, there was successively: an alluvial fan, coral reef, more alluvial fan, and then another layer of reef which includes present reef growth. The latter encompasses both fringing reef, which is well developed in this area, and some offshore patch reefs which are supported by older limestone of unknown age but which would appear could be the results of erosional patterns which were sculpted during the Pleistocene. The depth of the deepest alluvial fan in this area extends to about 60 m below the

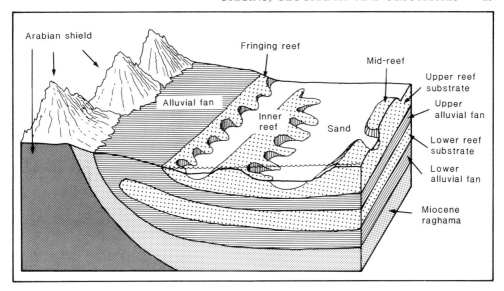

Figure 1.6 *Sketch of section through alternating reef and alluvial material near Yanbu, central Red Sea coast of Saudi Arabia. From Fugro (1977) and Sheppard and Sheppard (1985).*

present fringing reef and slopes down to 120 m below the offshore patch reefs. Geologists at Fugro Inc (1977) interpret this to mean that the lower alluvial fan was deposited at the lowest still-stand at about 17 Ka BP and that all the subsequent layering followed successively from that time. However, no dates are known.

Such layered arrangements are common elsewhere in Arabia. In Oman, the "MAM reef", a Miocene or possibly Oligocene structure now a few kilometres inland, is described by Green (1983) as being a stacked bioherm rather than a true fossil reef. In this example, several layered coral communities with a composition similar to that of adjacent, presently living coral communities, alternate with coarse wadi deposits and storm debris. The coral communities were probably not fully consolidated reef; in this part of Arabia, coral communities commonly develop flourishing communities yet fail to develop into true reefs, and are discussed later. Here in the coast facing the Gulf of Oman, as well as further south in the Arabian Sea near Masirah Island (personal observation), the same conditions of alluvial smothering as seen in the Red Sea appear to occur.

In the Gulf less is known despite the substantial amount of work done on Gulf sediments (Purser 1973). On the Arabian side of the Gulf, carbonate sands predominate, though on the Iranian side these are mixed with a much stronger terrestrial influence due to both wind and small and numerous riverine influences from the Zagros mountains (Purser and Siebold 1973). Coastal formations along the Iranian side are unfortunately possibly the least well known in the world. More is known about the Arabian side, however, where there exist vast areas of hard limestone substrate which are sporadically covered by thin layers of carbonate sand. This type of sea floor substrate has been recorded to about 35 m deep in the Gulf, and Purser and Siebold (1973) attribute its formation to submarine lithification. However, extensive examples of this type of substrate have been studied by direct observation through diving by the present authors, to about 15 m deep. These shallower limestone platforms commonly extend as flat, featureless expanses with remarkably little relief, but in many cases they exactly resemble patch reefs or series of reefs. Possibly the largest is a chain of "patch reefs" in the Gulf of Salwah running north–south, to the east of Bahrain. The summits of these are near low water level, but they have no corals at present due to the severe environmental conditions and support large brown algae instead (they are a non-calcifying community). Yet the appearance of this series is in every way one of a small chain of offshore reefs such as is very common in the central Red Sea. While the suggestion by Purser and Siebold (1973) of submarine lithification is not doubted, there may also be many examples of past reef deposition in areas which can no longer support coral communities.

Below sea level in the Red Sea, Braithwaite (1982) and Sheppard and Sheppard (1985) describe various bench marks in Sudan and Yanbu, central Saudi Arabian coast, respectively. In both areas, enclosed depressions occur in offshore limestone platforms, which descend to about 50 or 60 m deep. Also, in Yanbu at least, many patch reefs form steep sided blocks resting on vast limestone platforms which together extend as a linear feature offshore for about 400 km. This platform is located from 10 to 40 km off the Arabian coast, and is several kilometres wide, with an upper surface (except where living reefs reach the surface of the water) about 30 to 60 m deep. This level matches the sea level during the Pleistocene when sea level varied from 30 to 60 m below present. The limestone series in the central Red Sea was certainly an extensive reef system. Irregularities noted on its surface suggest erosion during low sea level periods, and the distribution of currently growing reefs which reach the surface is likely to be determined by the location of high points left by erosion processes in the Pleistocene. Present reefs are probably therefore a blanket of recent growth (7 Ka or so) covering the Pleistocene highs.

Series of raised fossil coral reefs higher than current sea level are one of the most striking features of the Red Sea coastline. Dullo (1984) notes four levels in the Sinai, a number found in several parts of the Arabian coast further south as well (Table 1.1). Coastal uplift occurs along both shores of the Red Sea and as noted by Braithwaite (1987) the mountain ranges lining its shores derive from faulting and uplift. Figure 1.7 illustrates one series, and the successive growth of reef over successive layers of sand and gravel is clear here too. Tectonic activity continues; in the Sinai the famous cracks containing a pan-tropical cave fauna (see Chapter 9) appeared following an earthquake in 1968 (Holthuis 1973). It is clear from Table 1.1 that even on the same shoreline, the levels of the series vary considerably. In addition to those tabulated, fossil reef elevations of 2 m and about 10 m were observed at Yanbu, central Red Sea and at 0.5 m, 2 m and about 3 m in the southern Red Sea by the present authors. These were not levelled accurately and so are not included in the table. Guilcher (1988) refers to series of 11 uplifted reefs on the Tiran Islands extending vertically to 320 m. Other raised reefs in the Dahlak Islands in the south have been dated as 120 Ka and 170 Ka, while raised reefs in Afar and Djibouti are 80 to 125 Ka old (Angelucci et al. 1981).

Some writers have suggested that the lack of an exact conformity between

Table 1.1 Elevations and dates of series of emerged fringing reefs. Ka = thousand years. Heights in metres above sea level. Series 1, Behairy (1983), Series 2, Dullo (1984), Series 3, Klein *et al.* (1990), Series 4, Al-Sayari *et al.* (1984).

1. Central Red Sea		2. Gulf of Aqaba	3. Gulf of Aqaba		4. Gulf of Aqaba
Height	*Age* (Ka)	*Height*	*Height*	*Age* (Ka)	*Height*
1	9.98				
		2.5			
3	18.1				
3	16.6				
			5–10	108–140	8–10
10	31				
		12.5			
					12–13
			15	140–200	
					15–17
		17			
					25
			30–35	>250	

heights in various parts of the Red Sea makes it difficult to piece together the sequence, but given the 1000 km separation between some of the sites, it is difficult to see why such minor differences should in any way be surprising. There is other evidence, after all, to show that the north of the Red Sea has exhibited greater emergence than the southern part, and the latter may even have suffered some subsidence in Recent times (Crossland *et al.* 1987) a feature difficult to reconcile with tectonics. There is much closer agreement in heights amongst the Gulf of Aqaba series. What does appear to be clear is that there is a series of several elevated reefs, whose height correlates with age (see Table 1.1). To

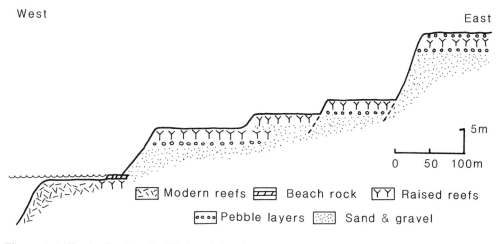

Figure 1.7 *Sketch of series of uplifted reefs (marine terraces) in the Gulf of Aqaba. From Al-Sayari et al. (1984).*

explain his Saudi series, Behairy (1983) remarks on several transgressions, each with a still-stand which presumably allowed a reef to develop. But given that from Behairy's (1983) dates, the highest fossil reef predates both the regression and transgression and grew during the extended Pleistocene period when sea level was lower than today, that the middle reef grew at the time of minimum sea level and that the lowest fossil reef grew during the transgression, the more parsimonious argument is to invoke uplift. Over such lateral distances, slightly different amounts of uplift would account for the different elevations seen at different ends of the Red Sea. There are differences also in the horizontal distances between the platforms of each series, though these are fairly easily explained by reference to different slopes and to horizontal movements from block faulting. In addition to this, the vertical upward movement by ancient salt evaporite deposits described in Guilcher (1988) can be added.

History also records proofs of this gradual recession of the waters; and mentions, as seaports, several places which are at present inland, without noticing the present maritime towns, which must undoubtedly be of later origin than the formation of the land on which they stand.   Niebuhr 1792

In the Gulf too there are examples of shoreline terraces of reefs. The Gulf is a Miocene basin, most of whose mainland shore on the Arabian side is very flat and low lying, but numerous elevations of limestone occur, some, for example, forming the highest parts of the island of Bahrain. As noted earlier, the elevation in many cases is caused by upward forcing by underlying salt domes. Along the shore, the tallest studied limestone cliffs are those which reach 10–15 m in the Hawar Archipelago between Bahrain and Qatar. These drop vertically into sea and are undercut by wave action. Limestone then extends out to some 20 m deep, shelving steeply, and forming limestone platforms to at least 7 m deep (personal observations).

(4) The Gulf and Gulf of Suez

The Gulf of Suez and the Gulf share many geological and physiographic similarities. As will be seen, they share many similarities in biotic composition also. Detailed geological origins of the Gulf of Suez are summarized in Braithwaite (1986) and those of the Gulf may be derived from numerous papers in Purser (1973). Both are shallow, sedimented basins with most of their floor lying within the well illuminated zone. The Gulf of Suez measures 280 km long and 20–40 km wide. It has an average depth of less than 30 m but deepens suddenly to over 100 m at its entrance into the Red Sea. The Gulf measures about 1000 km long and is 200–300 km wide and has an average depth of 35 m, deepening to over 60 m towards the northern side and 100 m deep at its entrance into the Strait of Hormuz.

Both bodies almost certainly dried out completely in the low sea level period of 18 Ka BP and so were recolonized only with the influx of water during the

Holocene transgression. The Gulf of Suez probably had no significant terrestrial input of water (though there is a vast quantity of artesian water of Miocene age under the Sinai and there are some minor springs, mostly sulphurous and of biblical significance along its shores). Its condition therefore was much like the present surrounding desert during the low sea level stand, and much of it was in similar condition for most of the Pleistocene as well.

The Gulf in contrast is supplied by the Tigris, Euphrates and Karun rivers which all discharge into the Shatt al Arab waterway at the northern end of the Gulf. The latter is itself a slowly subsiding region. In addition, numerous small and now quantitatively insignificant rivers descend from the Zagros mountains. The Gulf is shaped to a great extent by a later Tertiary fold system which causes its deepest depression to run along the northern Iranian side. This depression continues through the Strait of Hormuz and is believed to have acted as a conduit or extension of the rivers in the low still-stand. Bathymetry of the Gulf floor is such, however, that the area covered by or affected by the river flow in the time of the lowered sea level would have been small. Today, rivers have only a localized effect on the hydrology of the Gulf, and their sediments are mostly shed before their water enters the Gulf (Purser and Siebold 1973). In this area, greater terriginous input comes from flash floods and minor rivers descending from the Zagros mountains, but the effects of these disappear well before the larger, southern shallow basins of the Gulf.

The hot, arid climate experienced by both areas stimulates formation of evaporitic minerals including dolomite. Carbonate sediments dominate, however, and the general pattern is of increasing carbonate sediments southward. Figure 1.3 illustrated the pattern, and while the entire Gulf was not included in the latter study, it seems clear that the most southerly embayments of the Gulf are also dominated on the surface by carbonate sediments. The sediments are derived, in the Gulf at least, mainly from microfauna, especially foraminifera. In both bodies of water, carbonates from corals and coralline algae are relatively minor, in contrast to the condition in the Red Sea, though reefs are important biologically.

Details of specific categories of biota such as reefs, mud flats, seagrass beds and mangroves are given in the relevant chapters. Here it is sufficient to draw attention to three main parts of the Gulf which provide exceptional conditions. First is the Shatt al Arab waterway which is the only part of the Arabian region which is estuarine and which supports marsh plants in abundance (though limited parts of the Gulf of Suez do also and new colonization in artificial "estuaries" derived from city sewage outfalls is increasing). The second set of exceptional conditions is typified by the Gulf of Salwah which extends down the west side of Qatar south of Bahrain. This Gulf develops most extreme conditions with respect to several hydrological parameters, leading ultimately in the south to expanses of hypersaline algal mats. Third is the barrier complex of islands and shallows lying off the coast of the UAE. This area has restricted water exchange, but its shallow nature makes it rich and productive. These last two areas of the Gulf, and the Gulf of Suez, provide numerous opportunities for examining responses of several community types to increasing stress.

(5) The Oman and Yemen coasts

The coastline of Oman and Yemen has a total length, measured on a large scale map, of about 3500 km, and has several major biogeographical components.

The first is the Musandam peninsula in the far north, a 90 km spur of mountainous limestone cliffs of Permian through to Cretaceous origin (Map 5, see Introduction). This region's fjord-like valleys and mountains have long been recognized as very unusual structures in Arabia: it is the Mons Asabo of Ptolomy, the Maceta of Arrian and the Maka of Erathosthenes (Admiralty 1944). During the Permian to Cretaceous period, much of Oman was covered in a series of shallow seas. Reef forming organisms included algae, rugose and scleractinian corals which formed reefs and other layered deposits during that time. The limestone mountains of Musandam are folded and tilted approximately 10° from horizontal as the result of rotation of the Arabian plate which causes uplift and crumpling as it impacts the Asian plate. Added to uplift from crumpling there is in addition at least 60 m of localized subsidence (Vita-Frinzi and Phethean 1980). Thus Musandam juts into the Strait of Hormuz from Arabia, and while it is contiguous with Arabia it is part of the Iranian mountain system (Green and Keech 1986).

As a result of folding, this 90 km long spur of extremely remote, convoluted and inaccessible rock has a steep sided coastline, most of which is accessible only by sea. The fjords, called khawrs, greatly increase the actual length of the coast. As Mark Twain remarked about Switzerland — that it would be quite a sizeable country if it was flattened out — so Musandam has a much longer coastline than its small area suggests. Even without descending to unrealistic fractal scales, this small peninsula and its countless small islands provides a shallow sublittoral zone probably over 2000 km long. This is of considerable biological importance. The area is extremely spectacular, beautiful and desolate. Villagers in the interior farm on terraces when rain falls and occupy picturesque homes carved into the mountainsides, and leopards still exist in the wild. The peninsula's desolate nature is well illustrated from the fact that the phrase "going round the bend" originated here; the steep sided khawrs shut off breeze, so that officials posted to a telegraph relay station in Khor ash Sham between 1864 and 1865 were driven to distraction by boredom and heat. Missing the marvellous opportunities which would be seized by any naturalist, they gave this area the reputation of being the most oppressive place in the world.

Musandam's limestone coastline has a structure which is exceptional in Arabia in any case, but it is the only rocky sublittoral for a considerable distance on either side. To its west is the Gulf, dominated by soft substrates for considerable distance, while east and south of Musandam lies the exposed, sandy Battinah coast for 300 km. The infolded, fjordic structure of Musandam provides substantial rocky substrate in the intertidal and sublittoral zones. These features ensure that Musandam, at the entrance to the Gulf, has a particularly great biological importance by providing both a relatively enormous habitat as well as a zoogeographical stepping stone for hard substrate species, not least of which are corals and reef organisms. While the Iranian coastline on the opposite shore is rocky too, the limited evidence which is available suggests that it is like much of the Arabian coast and that its sublittoral is mainly exposed, coarse grained soft substrate.

The second zoogeographic zone is the Battinah coast, a gently sloping sandy bay over 300 km long encompassing part of the UAE as well as Oman. Here, rock outcrops terminate at low water. Further south is the Capital Area of Oman, which stretches from Muscat for 250 km to the eastern tip of the Arabian peninsula, Ras al Hadd. This area is mainly rocky, with sandy bays. In the Pleistocene, the coast followed roughly the same line as today, but a few kilometres seaward. Various sandstones and dolomite outcrop, along with some limestones and the green coloured ophiolite, a metamorphosed basalt. At elevations of 2 to 10 m along this section of coast, fossil scleractinian corals exist which Green and Keech (1986) suggest were deposited in the early Pleistocene. This relatively short region of the Arabian Sea is by far the best known ecologically.

The latter faces the Gulf of Oman. The much longer Arabian Sea coast is relatively poorly known with the exception of the island of Masirah, and the Dhofar region of southern Oman. Rocky cliffs alternate with long stretches of littoral and sublittoral sand dune, but in the former case, cliffs generally terminate near low water level and are heavily scoured. As a result, except in sheltered bays or in the lee of Masirah Island, this coastline is predominantly a soft substrate environment. It is mostly a high energy environment also, though there are enough sheltered areas to support Arabia's largest populations of seabirds. The coastal geology here is a complex mix of tertiary rocks including calcareous shale, limestone and gypsum with chert and marly bands interbedded with limestone, and some igneous rock. The Yemen coastline facing the Arabian Sea is almost completely unknown but may be assumed to be very similar to the Dhofar coast. In its far northern extent at least this is certainly true, and Abd el Kuri Island in the mouth of the Gulf of Aden is also similar (Scheer 1971).

(6) Islands and archipelagoes

Archipelagoes and large individual islands are of obvious biological importance, together with their associated vast shallows.

In the Red Sea, the total length of shoreline is substantially increased and probably more than doubled by islands, and thus they have a marked biological importance, especially in the provision of very sheltered habitat. An interesting group are volcanic remnants which occur in a chain near the central axis (Figure 1.8). These appear to be a consequence of the spreading sea floor centre, and occur mainly in the south and central section of the Red Sea. On the mainland, volcanic cones also occur in the south of the Red Sea on the Arabian side as well as in Ethiopia and Afar triangle region. Some lava flows clearly interrupt fringing reef development. Where lava rock flows over the shoreline and into the water there is sometimes a complete break in the otherwise continuous fringing reef, and corals have not formed new reef deposits on the lava, though the surfaces are colonized by corals and macroalgae (Sheppard 1985a).

Other islands support extensive limestones and sedimentary rocks, and by virtue of being remote and inaccessible provide refuges for some of the Red Sea's most important bird colonies. Important groups in the Red Sea include the

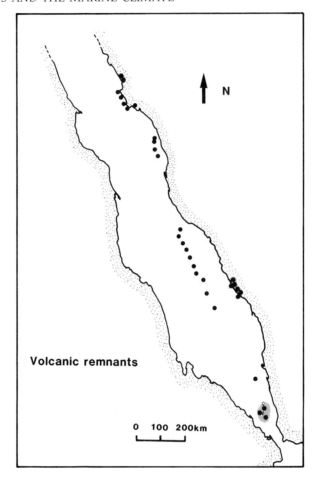

Figure 1.8 *Location of volcanic islands and volcanic coastal structures in the Red Sea. There are no other volcanic areas north of the area shown. From Crossland* et al. *(1987).*

Ashrafi Islands in the mouth of the Gulf of Suez and Tiran Islands in the mouth of the Gulf of Aqaba. Both arise from shallow shelves, and support rich soft substrate habitats. Completely unexplored areas on the border between Egypt and Sudan appear to be limestone (Pleistocene reef) and support mixtures of soft substrate and reef, as do the islands of the Wedj Bank on the Arabian shore, and the Suakin Archipelago of Sudan.

The two largest archipelagoes in the Red Sea, however, are the Farasans and Dahlak Islands in the south, together with a group aligned on the central axis south of these to the Bab el Mandeb. The geology of the Dahlak Archipelago has been reported by Angelucci *et al.* (1981, 1982a, 1985). This group and the Farasans are derived from the relict of a large Pliocene–Pleistocene carbonate platform of up to a few hundred metres thick, which grew on evaporitic salt deposits up to 3 km thick and of Miocene age. Development of the Miocene evaporite deposits continued until halted by the influx of Indian Ocean water about 8 Ma BP. The overlying carbonate platform subsequently developed, which then underwent a large tectonic displacement from the spreading centre of the Red Sea, and was also modified by the rising of the underlying salt domes. These latter, local events, together with Pleistocene aerial erosion, have resulted in a varied topography and hence a very varied combination of marine habitats, including hard Pleistocene substrata for modern reef development, extensive shallow sedimented areas, and large gullies or erosion channels reaching to 150 m deep which cut far into the archipelagoes. Raised reefs exist, of ages 120 Ka and 170 Ka. Modern reefs are mainly fringing coral reefs, and are thin. Those reported by Angelucci *et al.* (1981) resemble those of the mainland (Sheppard 1985a). In addition, mainland shores and the Farasan and Dahlak Islands contain numerous sandbanks and shoals said to support modern coral reefs as well. Although most carbonate production here may come from foraminifera, coral reefs remain very important. Also the southern Red Sea supports a unique (for the Arabian region) development of coralline algal reefs: structures which may reach several tens or even a hundred metres across and formed entirely from calcareous red algae (see Chapter 3).

The geological formation of the southern archipelagoes is complicated, as suggested by the range of topography and sediments contained within them. They include deposits of fresh water species found in deposits in Afar also, suggesting a past physical connection of islands and mainland (Angelucci *et al.* 1981). The southern part of the Red Sea is currently geologically active, experiencing frequent earthquakes.

In the Gulf of Aden the small island of Abd el Kuri near Socotra has been examined briefly by Scheer (1971). No geological details are available, though the substrate supports a few corals and abundant macroalgae in a manner apparently similar to the Dhofar coast. Off Dhofar, the rocky Kuria Muria Islands have a similar character in the sublittoral to the mainland coast. Masirah Island further north differs, being generally low lying and surrounded by a broad, shallow continental shelf, supporting carbonate accumulations and coral patch reefs.

Three groups of islands in the Gulf are both noteworthy and of considerable biological importance. The UAE coast has an archipelago of barrier islands and tidal deltas containing a very complex mixture of coral patches and sedimented areas colonized by seagrasses (Evans *et al.* 1964, 1969). Sediments are almost pure carbonates, produced in great excess to form banks and shoals, and driven on to the mainland shores by northerly winds and waves. Cementation of limestone sediment proceeds rapidly, causing extensive Holocene crusts in shallow water and intertidal regions. Also the extremely hot and arid climate has led to the formation of dolomite and other evaporite deposits in intertidal and sabkah environments. The marine biota is highly productive, though it is an impoverished one in terms of diversity due to the high salinity; open Gulf water here has a salinity of about 43 ppt (parts per thousand) while values of 50–60 ppt are usual even in open bays, and this reaches over 60 ppt in lagoons with more restricted circulation.

The Hawar Archipelago located in the Gulf of Salwah, south of Bahrain and adjacent to the Qatar coast, has a similar mixture of limestone rock and soft carbonate sediments, mixed with a greater amount of terriginous sediment; some of the islands are within wading distance of the mainland of Qatar. In addition, the archipelago appears to have significant raised limestone platforms of unknown age. Coastal cliffs over 10 m tall face the coast of some of the islands, while several islands are simple sandy cays. Underwater, more limestone platforms appear as gently sloping hard surfaces or as vertical steps two or three metres high on the eastern side, facing the main part of the Gulf of Salwah. These have the appearance of old coral reefs, and are in fact near to the limestone structures shaped like patch reefs mentioned earlier. Today, salinities and temperatures preclude all coral growth, and it may be assumed, though data are not available to us, that local carbonate production comes from microfauna in the manner of the UAE coast barrier islands, though much may also arrive from the coral reef complexes to the north. Some undoubtedly derives from erosion of the limestone cliffs which are etched in some places to over 2 m deep at sea level. The biota of these extremely stressed habitats are discussed in later chapters.

Finally, a group of islands extends along the Saudi Arabian waters and into the extreme west of the Gulf off Kuwait. These are coral cays: sandy islands ringed by recent fringing reef growth. They, along with numerous patch reefs without sand cay formations, arise from a sea floor of 10–25 m deep. Purely of coral and coralline algal construction in modern times, these provide the most diverse hard substrate habitats known in the Gulf.

Along the Iranian side of the Gulf large islands derive from offshoots of the land mass, while smaller islands resemble the limestone structures found elsewhere in the Gulf. Although these are undoubtedly important, and the former arise from near the deepest water to be found in the Gulf, almost nothing has been published about them which is of relevance to this book.

Over most of the region, biogenic rock is quantitatively the most important on the surface, that is, in providing a habitat today for both hard substrate and sediment colonists. Only some volcanic islands and parts of the Arabian Sea coast are devoid of recent carbonates, and even the considerable input of terriginous sand in the Gulf is mixed with much continuously produced carbonate material. The marine climatic conditions which cause this and which interact and disperse it are considered next.

Summary

In the shallow sublittoral of the Arabian region, important Tertiary or older processes include the spreading Red Sea centre, anti-clockwise rotation of the entire Arabian plate, widespread volcanic intrusions, upward moving salt domes created in periods of extensive evaporation, and Tertiary limestone foundations. These profoundly influence present habitat type and distribution. However, Quaternary climate and sea level changes are more important to present biological patterns.

Between 30–110 Ka BP, sea levels fluctuated between 30 to 60 m below present levels, moving the sites of contemporary biogenesis and erosion many times. Heavy rain, notably 35–17 Ka BP, led to sheet flow of alluvial material, and to huge fans and outwash systems which are important foundations to shallow benthic biota today. Reefs formed before this were either smothered or severely scoured; levels at −30 to −60 m correspond to depths of major wadis and to surfaces of offshore platforms. Terrestrial material ending up offshore formed important new substrata which were colonized directly in many cases by new coral reef growth, and repeated phases of alluvial flow and reef growth have led to alternating layers in several places. Rain added karst relief directly, but effects of this are probably only important on reefs over 10 km offshore.

A sill at the entrance to the Red Sea is 130 m deep, similar to the maximum sea level fall at the start of the Holocene. The Red Sea became very saline, while in the Gulf evaporation left only a narrow strip along the northern edge which conducted riverine water to the Strait of Hormuz. In both seas, the early Holocene marks a point when the marine ecosystem was effectively "reset".

The question of whether or not sea level subsequently rose above present levels is not especially important in this region. In the Red Sea, vertical changes from local uplift and movement of mobile alluvial terrain are much greater. Striking consequences are series of raised fossil reefs, extending from 2 m to over 30 m above present sea level.

The Red Sea Barrier Reef located 10–40 km off the Saudi Arabian coast is about 400 km long and several kilometres wide. The platform surface is 30–60 m deep, corresponding to Pleistocene sea levels, on which sit many steep-sided patch reefs which reach the sea surface. Present reefs are probably a blanket of recent growth on Pleistocene highs.

In the southern Red Sea, the mainland coast has had less uplift, and even some subsidence in recent times. The Farasan and Dahlak Archipelagoes are the relict of a large Pliocene–Pleistocene carbonate platform a few hundred metres thick, which grew on Miocene evaporitic salt deposits up to 3 km thick. Development of these deposits was halted by the influx of Indian Ocean water about 8 Ma BP. The platform was split by the spreading centre of the Red Sea, and further modified by rising of the underlying salt domes and by Pleistocene aerial erosion. These have resulted in very varied marine habitats.

In the Gulf, countless shallow limestone platforms exist. These flat, featureless or macroalgal covered expanses resemble patch reefs without much coral. With summits near low water level, they appear to be caused by similar upward forcing from underlying salt domes, though several seem to be examples of past reef growth in areas which no longer support corals.

The Musandam peninsula is unique. Although contiguous with Arabia, it is part of the Iranian mountain system, formed by Permian to Cretaceous limestone, crumpled as the rotating Arabian plate impacts the Asian plate. This 90 km long, highly indented spur has a rocky coastline over 2000 km long, in an area which is predominantly of soft substrate, and is thus of considerable biological significance.

Sediments in the Gulf are almost pure carbonates, produced in great excess to form banks and shoals. Cementation of these proceeds rapidly, leading to complex mixes of reef and soft substrate habitats. Extremely hot and arid conditions have led also to dolomite and other evaporite deposits in intertidal and sabkha environments. Some of these have been uplifted as well, and are distinct from the few true coral cays of the Gulf which lie in the centre of that sea.

CHAPTER 2

The Marine Climate

Contents

A. **Seasonal atmospheric cycles** 36
   (1) The main Indian Ocean climatic cycles 37
   (2) Local wind systems 40
B. **Hydrographical influences** 42
   (1) The major Indian Ocean currents 42
   (2) Currents in Arabian seas and their driving mechanisms 43
   (3) Wave energy and tidal patterns 52
   (4) Dissolved oxygen and nutrients 54
C. **Early Holocene marine climate** 56
   (1) Sea level fall and Red Sea isolation 56
   (2) Strength of monsoon and Arabian Sea upwelling 57
Summary 59

This chapter provides an overview of the main climatic features which drive the oceanic, and hence marine biological systems of Arabia. We describe the main features in only as much detail as is necessary to set the context of the physical constraints on ecosystems. Details of many interlocking climatic and physical systems are given in subsequent chapters where the ecology of the various marine ecosystems is discussed.

A. Seasonal atmospheric cycles

The extremely arid nature of the region, high temperatures and constant and intense sunshine, especially along coastal areas, gives the overall impression of a lack of seasonal variability. This is indeed the case in some areas. In much of the Red Sea, especially in the south for example, there is a marked constancy of many important parameters, though because spawning of many marine groups is seasonal there is clearly sufficient seasonal water temperature difference to be biologically important. There is an interesting seasonal change of wind direction, and moderate to marked differences in temperature between winter and summer, but nevertheless many parts have a rather constant marine environment throughout the year. With any wind direction, the exposure of the reefs, mangroves and other main habitats to prevailing energy remains much the same, and even the existence of rain on a few occasions in winter does not affect the marine biota to any great extent.

The Gulf and Arabian Sea are more variable. Although some characteristics such as rainfall do remain constantly low with only rare exceptions, the region lies at the edge of two or more global weather systems which subject these seas to major changes in direction and force of wind and ocean current. This in turn results in some of the most remarkable effects on marine conditions seen in tropical waters. For example, in the Gulf, northerly Shamal winds in winter blow over the shallow water and cause water temperatures to fall to values more usually associated with temperate oceans, sometimes causing massive mortality of the tropical biota. Elsewhere, seasonally reversing winds induce upwelling in the Arabian Sea, which causes the remarkable, low sea temperatures off southeast Arabia in the hottest summer months. This in turn has led to the existence of an enclave of southern hemisphere kelp forest in the sublittoral above 16° N.

(1) The main Indian Ocean climatic cycles

The Arabian region is affected by the Asian weather system which causes two main Indian Ocean monsoons, though they are modified by intrusions from weather systems in the Mediterranean or north Africa. The latter are noticeable in the northern half of the Red Sea, especially in winter.

The broad cycle is basically simple. In low latitudes of the western Indian Ocean, trade winds north of the equator blow essentially from the northeast, while south of the equator they blow from the southeast. The winds are drawn in to meet at a belt of low pressure called the Inter-Tropical Convergence Zone (ITCZ) where air rises to over 12 km high, with a tendency to create cloud and rain. Surface winds in the ITCZ are light ("Doldrums") though they may be strong on either side. The Indian Ocean monsoonal pattern, and the seasonal migration of the ITCZ, dominate weather affecting Arabia and its marine climate.

A) WINTER PATTERNS

In the first quarter of each year the Northeast Monsoon is fully developed and the ITCZ lies far from Arabia, south of the equator (Figure 2.1). High pressure exists over Asia and over northern Africa. In central Africa there is low pressure, projections of which encompass the Red Sea, especially its central part (Couper 1983, Edwards 1987). Winds in the region as a whole during this time generally flow from the east. Winds cross the Gulf from the east or northeast and usually remain below 5 m sec^{-1}, but along the eastern edge of Arabia they are deflected by mountains and flow down that coast from the northeast. In the Gulf of Aden, winds are similarly deflected and funnelled up the Gulf of Aden from the Arabian Sea, continuing up the Red Sea as far as the central part. This is possible because of the low pressure in the central Red Sea, mentioned above. The northern Red Sea lies on the far side of the central Red Sea low pressure belt and is not affected by the Northeast Monsoon. In this region, wind flows into the low pressure zone from the northwest; this is essentially Mediterranean weather penetrating down the Red Sea. Winds over most of the Red Sea are generally less than 7 m sec^{-1} though in the Gulf of Aden and Straits of Bab el Mandeb the funnelling effect

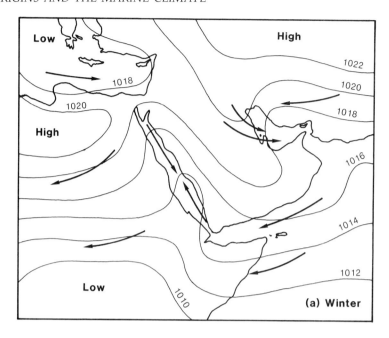

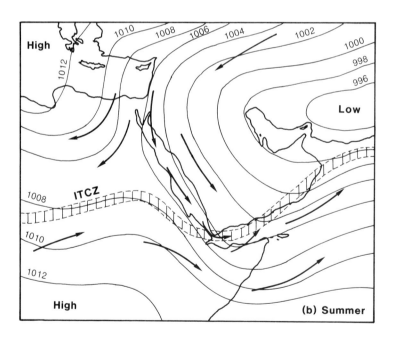

Figure 2.1 *(a) Winter and (b) summer wind patterns. Major features are the summer winds causing upwelling off SE Arabia, the Shamal cold winter winds in the Gulf, the strongly evaporating winter winds over the northern Red Sea, and the seasonal changes of winds in the Gulf of Aden which are largely responsible for a 0.5 m seasonal tidal change in the Red Sea. Modified from Edwards (1987).*

greatly increases the speed, and these areas are notoriously windy. This is important to the marine ecosystems especially in the southern Red Sea, because Indian Ocean water with much higher nutrient levels and a richer planktonic population is forced into the Red Sea.

During the winter, therefore, the Red Sea has two air flows, both travelling along its axis: one from the northwest in the north, and one from the southeast in the south, which meet in a region of low prevailing winds in the middle. It is only during the winter quarter that most of Arabia is crossed by any winds which have passed over water and which become potentially rain-bearing. Thus winter brings occasional heavy rain to parts of the area, and sudden flash floods may be experienced in the Gulf, Gulf of Oman and Red Sea. In the latter area, it is usually very obvious to an observer in the central part, at Jeddah for example, which of the two systems has carried the rain, since the strength and direction of the wind during and immediately before the rain is clear. In the north of the Red Sea, only the Mediterranean system is likely to have an effect, while the reverse is true near the south.

It is very erroneously supposed that the monsoons extend to the Red Sea; but in reality, the wind blows with equal violence from opposite quarters at either extremity, leaving a considerable space between them subjected to light airs and calms.

Wellsted 1838

Several papers list rain data from cities of the region, sometimes tabulating it at great length (e.g. Crossland *et al.* 1987) but all show very low values and these annual totals are of little consequence to the marine environment. The fact that an annual total may be deposited in a few hours is much more significant. The sudden deposition of several centimetres of rain, coupled with the lack of any significant topsoil or vegetation, causes dramatic flash flooding, which causes enormous quantities of terrestrial run-off which can have a locally significant effect on parts of the shallow marine environment. Alluvial fans which extend seaward are maintained by this mechanism, and these are covered in a later chapter.

B) SUMMER PATTERNS

The second quarter is the period when the Southwest Monsoon builds up, which in the third quarter holds sway from the Arabian Sea to southeast Asia (Figure 2.1). The line of the ITCZ migrates northwards, until it touches southern India, passes along the southeast coastline of Arabia, passes up the Gulf of Aden almost into the Red Sea, and then crosses into central or northern Africa. Behind it, the rise in temperature in the northern hemisphere causes strong winds to build up in the Arabian Sea. In the Arabian Sea, now south of the ITCZ, winds are strong and clockwise. Winds over the Arabian peninsula, which is always north of the ITCZ, are variable in strength, but are anti-clockwise, revolving around a low located over Iran and northern India. This causes prevailing winds to flow down the Red Sea for its entire length, so that in summer at the height of this development winds flowing around the Gulf and Iranian low pass down the Red

Sea and continue through the Gulf of Aden, and then are deflected so that they flow to the northeast, parallel to the coast of Arabia. At the same time, the clockwise airflow in the Arabian Sea also causes strong winds to flow towards the northeast on the edge of Arabia, parallel to the Arabian Sea coast. These both reinforce each other, causing sustained and strong southwesterly winds along this coast for about four months, generally exceeding $15\,\mathrm{m\,sec^{-1}}$.

Finally in the fourth quarter, the ITCZ migrates southward again, there is a transitional period of changeable winds, and the winter pattern re-establishes itself. As pointed out by Edwards (1987), there are many additional complexities superimposed onto this basic pattern. In the present context of the effects of the winds on the region's marine biota, however, these are of lesser importance.

The coastlines of Yemen and Oman are the most affected by the Southwest Monsoon in summer. Blowing at over $15\,\mathrm{m\,sec^{-1}}$, it essentially blows warm surface water offshore due to the Ekman effect. This surface water is replaced by deep, colder and nutrient-rich ocean water, whose upwelling is the cause of several remarkable phenomena. These include inhibition of corals and reef growth and the stimulation of large brown algae, including forests of the kelp *Ecklonia*. On a broad scale, effects of the upwelling are greatest along the Arabian shoreline, though upwelling is also marked along the Somali coastline, where it is reflected in the intensity of the fishery. Upwelling effects on the Arabian shore are greater where the continental shelf is narrow. On land, too, effects of this monsoonal·system are equally marked. The terrestrial ecosystem is outside the scope of this book, but it is worth mentioning that in southeast Arabia, the monsoon brings cloud, mist and rain for about four months of the year, resulting in well developed and rich forests on the slopes of the coastal mountains, causing the area to be quite unlike most of the rest of Arabia. This is, for example, the source of Frankincense and Myrrh trees, and is the only part of Arabia which naturally has coconut palms as well as date palms.

(2) Local wind systems

In several parts of the region, greater effects on the marine biota derive from smaller scale wind systems, notably the sea breezes, sometimes called thermionic winds, and from related winds derived from severe temperature differences. The former are diurnal wind changes resulting from differential heating and cooling of the desert and sea, and they are of considerable significance in the Gulf and Red Sea, but less so in the Arabian Sea where the monsoonal system has a major impact. Other thermally forced winds may last for several days, especially in winter where one locally called the Shamal has major biological effect. While average surface water temperatures are of course important and now well documented (e.g. Edwards 1987), extreme values may have a greater limiting effect on the marine biota.

A) SEA BREEZES

Sea breezes in the main body of the Red Sea in summer build up strongly during the afternoon. They are not perpendicular to shore but strike the coast obliquely due to the influence of the prevailing winds, described above. On the African shore the sea breeze is usually from the north or northeast, while on the Arabian side it is generally from the northwest. In the Gulf of Suez and Gulf of Aqaba, the area of water is much smaller and the effects of this wind are less, and in both gulfs there is a predominant northerly wind.

These winds drive important high energy wave conditions. Their effects are most noticeable in the alignment of coral reefs, whose general tendency to grow into prevailing waves makes them angle outward towards the waves wherever there is suitably shallow foundation for them to do so. This also has a marked effect on the distribution of mangrove stands within several embayments, both directly and via the soft substrate distributions. Apart from occasional storms, these afternoon winds in the summer are the strongest experienced in the Red Sea and in the central part they induce a median wave height of nearly 0.6 m along the unprotected outer edges of the barrier reef (Georeda 1982).

On the shores of the Persian Gulf, the south-east wind is accompanied with a degree of moisture which . . . occasions violent sweatings; the north-west, passing over the great desert, is more torrid . . . The Arabians . . . carry with them garlic and dried grapes, for the purpose of reviving such persons as may fall down fainting, from the effect of these hot blasts.   Niebuhr 1792

In the Gulf, similar effects occur in summer, and offshore coral reefs may experience rough conditions. As will be discussed later, these probably have a generally beneficial effect due to mixing and removing water stratification which causes increased stress in summer months; strong winds also keep shallow areas such as reef flats well flushed. Along mainland shores, however, which are shallow and muddy, the effects are important but more complex.

Other thermal winds which have adverse effects on shallow marine life derive from land, and generally occur at night. These are infrequent winds but are intense, and have a powerful desiccating effect on coastal vegetation and, although not investigated, probably on any shallow marine life exposed at low tide as well.

B) THE "SHAMAL": COLD WINTER WINDS

In the Gulf in winter, local effects include the Shamal, which in this season is a cold northerly wind which flows down from the mountains of Iran, bringing severe chilling. One now well known example from 1964 (Shinn 1976) which lasted several days, brought air temperatures of 0.5°C in Qatar, the coldest known from that country. The 65 kph winds built up waves of 6 m, which in turn ensured that chilled sea water was thoroughly mixed. Surface water temperatures fell to 4°C, and even 1 m from the bottom at 18 m deep the water temperature was 14.1°C. This was at a location about 80 km from shore, indicating that severe chilling was extremely widespread. This Shamal caused widespread mortality of corals and other marine groups. While this example is of a particularly severe Shamal, it is not unusual. In Kuwait in 1982 (Downing 1985) cold winds commonly approached 0°C in winter, and during a period of low tides which exposed the reef flats for 2 hours, the recorded minimum was 6.7°C. Low temperatures appear to occur almost annually, and annual, extensive death to fish in the Gulf has been reported to one of us (C.S.) by several fishermen. Mangrove distribution in the Gulf is also limited by cold winter conditions, as indeed it is in the Gulf of Aqaba too, where similar cold northerly winds flow down towards the Red Sea. Even as far south as Jeddah, cool winter winds can have considerable consequences. Morley (1975) describes water temperatures as low as 14.7°C in January 1973 following a period of low air temperature. This lasted several days, after which warmer and stronger winds caused mixing of the surface with deeper water, leading to a rapid temperature increase in shallow water to over 25°C. This rapid change led to mass mortalities of fish.

In the Gulf of Suez, winter winds are also important. In this case, humidity is relatively low and evaporation from the shallow gulf leads to a significant additional drop in water temperature.

One possibly significant effect of these winds is the transfer of large quantities of airborne sand and dust. In the northern Gulf, dust fall-out from southern Iraq amounts to $6.9\,\mathrm{g\,m^{-2}\,y^{-1}}$, and this material also contains adsorbed hydrocarbon pollutants (ROPME 1987, Al-Mudaffar et al. 1990). It appears likely that the recent hostilities will have increased particulate fall-out substantially, especially from burning oil wells. Comparable but slightly higher figures were found on the Red Sea coast (Behairy et al. 1985). Amounts from $8–22.2\,\mathrm{g\,m^{-2}\,d^{-1}}$ were precipitated, and in this case it was noted that heavy metals from industrial facilities adsorbed strongly to the dust which served as a transport mechanism for pollutants into the marine environment. The importance of dust is yet to be evaluated.

The effects on sea surface temperature of local wind systems may be at least as severe as effects brought by the major systems, especially to intertidal and supratidal marine components such as mangroves which are at their environmental limit. Indeed, a large part of the area is scarcely affected by the major atmospheric cycles. Some of the effects of the air patterns are, of course, transferred to marine biota via induced currents and other hydrographic effects.

B. Hydrographical influences

(1) The major Indian Ocean currents

A) SURFACE CURRENTS

Some of the major currents in the Indian Ocean, including those north of the equator which directly impinge on the Arabian region, are seasonal and controlled by the monsoon system described above. In January, the Indian Northeast Monsoon current flows towards Arabia from southern India, and diverges in the Arabian Sea into two main branches. One curves southward, passing the entrance to the Gulf of Aden, returning eastwards after merging with the Equatorial Counter Current, and the other branches northwards towards the Gulf of Oman and dips down to return southward as deeper water. In summer when the wind blows strongly from the southwest along the Arabian coast of the

Arabian Sea, the surface current follows suit to a certain degree in the form of the Somali Current, and this returns eastwards as the Indian Southwest Monsoon Current. However, as will be discussed in more detail, the broad current pattern does not reflect the local condition off Arabia.

Around the equator and to the south of it, the South Equatorial Current flows westwards throughout the year. Although distant from Arabia, this westerly movement of water is probably important to biological dispersal patterns. It is strong, exceeding $2 \, \text{km} \, \text{h}^{-1}$, and interacts in the central Indian Ocean with the easterly currents causing large and important areas of convergence and mixing in regions of major atoll groups and shallow substrates. These areas include the Maldive–Chagos groups and the Seychelles group, and some of the zoogeographical patterns which will be discussed later are based to some extent on the latter "source" areas.

B) DEEP CURRENTS

Deep currents are generally ignored in accounts of tropical, shallow water habitats, because they usually provide only interesting phenomena of marginal or incidental biological significance. In the Arabian Sea, however, they are of considerable importance.

Their pattern in the Indian Ocean is straightforward. From principal water sources in the Antarctic region, a northerly stream flows northward into the Arabian Sea. It is deflected backward toward the central region north of the equator, but some of it is injected well into the Arabian Sea (Figure 2.2). In short, this cold, deep flow carries the water which replaces the surface water forced away from the coastal region in the strong summer Southwest Monsoon. It is thus a major source of nutrients and hence productivity in the Indian Ocean and it is really the only significant source of nutrients in the western Arabian Sea. The Indian Ocean generally is one of the most nutrient poor regions of water in the world (average planktonic productivity is $<150 \, \text{mg} \, \text{C} \, \text{m}^{-2} \, \text{d}^{-1}$, see Chapter 10) so that this injection is particularly important. Also the source of the deep current may be extremely significant. The enclave of the kelp genus *Ecklonia* off the southeast coast of Arabia, a species which otherwise exists only around south Africa, Australia and New Zealand, owes its existence to the lowered temperatures and nutrients brought by the deep upwelling current, but may also have been recruited from it as well.

(2) Currents in Arabian seas and their driving mechanisms

Considerable detail on Red Sea, Gulf and Arabian Sea currents is given in Edwards (1987), Hunter (1986) and Currie *et al.* (1973) respectively. In the former, especially, separate sections deal with currents, salinity and temperature, as well as winds affecting the region. These need not be repeated in detail, as although there have been considerable additional data collected during the 1980s with respect to these parameters, the newer data simply confirm those noted by these earlier reviewers and does not add much new detail.

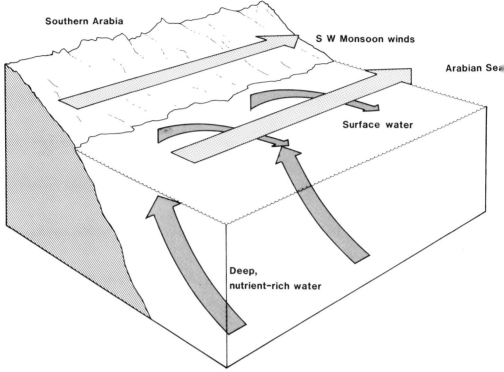

Figure 2.2 *Arabian Sea currents in summer (SW Monsoon). Summer wind blows surface water away from the Arabian shoreline, which is replaced by water from a deep, northward flow. Light arrows are winds, dark arrows are currents.*

The following integrates some known physical parameters, especially currents, from the viewpoint of biologically important effects. Within the two principal, semi-enclosed bodies of water in the Arabian region, surface and near-surface currents are driven to some degree by the prevailing winds, as discussed, but of even greater importance to the hydrology of these bodies are the density currents caused by combinations of salinity and temperature.

A) SURFACE AND DEEP WATER CURRENTS IN THE RED SEA

The Red Sea is unique amongst deep bodies of water for having an extremely stable, warm temperature throughout its deeper waters. Below about 250–300 m, the water maintains a constant temperature of about 21.5 °C, which extends down to the sea floor in all areas except where heated brine pools exist (see later). As noted by Longhurst and Pauly (1987), the meaning of the phrase "tropical oceans" usually relates in a biological sense only to the surface two or three hundred metres of water as below that depth water temperature in all major oceans drops to below 10 °C. The result is that the biological character of all but the surface water layers is not readily distinguishable from that of temperate seas. The Red Sea is different, and in this case the entire volume of water may be regarded as tropical.

i) The main currents and Gulf of Suez density gradient

The mechanism which maintains this situation appears to be driven by a density gradient which causes northerly drifting Red Sea water to become heavier and sink near the mouth and in the Gulf of Suez, and to return south below the thermocline. Morcos (1970) expressed it as: "The northern Red Sea is the natural place where bottom water is formed during winter months." In a sense, the deep water of the Red Sea, and its temperature, is driven by the Gulf of Suez which therefore has a hydrological importance much greater than its relative size.

Surface water temperatures in all parts of the main Red Sea basin vary seasonally between about 22 and 32 °C. At any one time, peak surface temperature is located between 15 and 18° N, and ranges from 27 to 32 °C, according to season. Surface temperature declines slightly towards the Bab el Mandeb because of the influx of cooler water from the Gulf of Aden, and there is a gradual decrease of temperature in a northerly direction also.

As Edwards (1987) points out, provision of a clear and detailed description of Red Sea currents has proved to be difficult, partly because currents generally are weak and variable, both in time and space. Basically it is known that the fundamental movements follow the winds, such that the northerly wind in summer drives surface water south for about 4 months, at a velocity of 12–50 cm sec^{-1}, while in winter the flow is reversed, pushing water into the Red Sea from the Gulf of Aden; the net value of the latter movement is greater than the summer outflow, and the drift continues to the northern end of the Red Sea.

Figure 2.3 shows the driving density gradient in the northern Red Sea. The net annual drift of surface water is therefore to the north. Evaporation in the Red Sea averages 1–2 m y^{-1}, and continues in both summer and winter. This causes a gradual increase in salinity towards the north, from about 36.5 ppt at the Bab el Mandeb to 40.5 ppt at the entrance to the Gulfs of Aqaba and Suez. In addition, surface water temperature falls towards the north such that winter temperatures of about 22 °C prevail in the north, and at the same time evaporation continues strongly. Thus as water drifts northwards its density increases. In the Gulf of Suez, water is cooled more rapidly in winter, to below 18 °C, and evaporation helps to increase the salinity to 42.5 ppt or more in the north. There is thus a steep salinity gradient along the Gulf of Suez, and a marked differential between the latter and the Red Sea. The dense, cool, saline water at the surface of the entrance to the Gulf of Suez pours downwards into the deeper Red Sea, turns under the surface water and returns southward. The turnover occurs near the mouth of the Gulf of Suez; at the entrance to the Gulf of Aqaba there is a sill which precludes similar return. The returning water has the constant temperature of 21.5 °C found throughout the Red Sea below the thermocline at 250–300 m, and it has a constant salinity of 40.5 ppt, in other words close to the value of the water at the southern end of the Gulf of Suez. Thus there is very little difference in either temperature or salinity between the surface and deep layers of water in the north of the Red Sea, and the differences increase between them towards the south.

The density gradient which drives this system derives to a large degree from the raised salinities in the Gulf of Suez. These salinities are themselves important

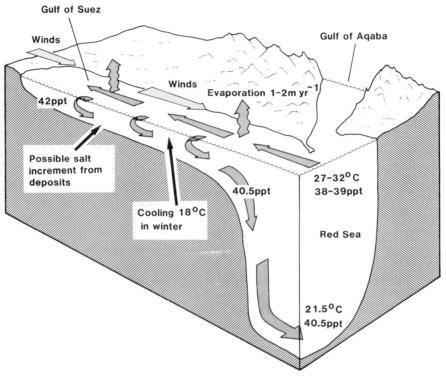

Figure 2.3 *Mechanism of the density gradient in the Gulf of Suez. Northerly flowing water suffers evaporation and cooling, and "turns under" mainly near the mouth of the Gulf of Suez where there is considerable shallowing. Returning water flows south below the thermocline. A sill in the mouth of the Gulf of Aqaba (not shown) precludes a similar effect in that Gulf. Water in the Gulf of Aqaba is also subject to less cooling, since its greater depth buffers it against sharp temperature changes. Light arrows are winds, dark arrows are currents.*

locally to the biota, but affect the pelagic nature of all the Red Sea below the thermocline. Edwards (1987) observes that the increased salinity in the Gulf of Suez appears to be more than would be expected by evaporation alone and has suggested that it may be enhanced by salt from the Suez Canal and Bitter Lakes. The latter indeed derived their name from their very high salinity, but because the main drift in the Canal is towards the north, and because a connection between the Bitter Lakes and the Gulf of Suez is extremely recent in the present context, it does not seem likely that additional salt loading comes from that source. However, as noted in Chapter 1, considerable salt deposits underlie much of this region, not only the Bitter Lakes and parts of the isthmus of Suez, and it would be reasonable to suppose that a "salt increment" comes from such a source, if indeed one is required to accurately explain the density gradient.

ii) The hot brine pools

The southerly deep drift at 21.5 °C contains approximately 90% of the water of the Red Sea. Over 9% of the volume is found in the upper 250 m, and less than 1% is found in the heated, metal-rich deeps. The latter occur below 2 km deep and have elevated temperatures of over 24 °C in almost all cases, often reaching values of over 40 and even 60 °C in the case of the Atlantis, Chain and Discovery deeps (Karbe 1987), though these temperatures vary from year to year. Most of the hot water bodies cover an area of only a few hundred square metres to 1 km², though the two largest are 11 and 55 km². They have a thickness of 1–200 m, and are heated by tectonic activity of the spreading sea floor centre. Their chlorinity is between about 80 and 160 g Cl kg^{-1} (compared with 23 for average mid-depth Red Sea water) and it is therefore dense. This marked vertical stratification in salinity keeps them more or less contained in the central, deep trenches and hydrologically separated, for the most part, from the main body of Red Sea water. Thus they are thought to contribute only little to maintaining the temperature of the main body of water at 21.5 °C.

iii) Red Sea–Gulf of Aden water exchange

The main surface drifts are slow moving and are easily modified and even reversed by local effects and by the small tides. Overall there is a net flow into the Red Sea. The water balance in the Red Sea is negative in the sense that annual precipitation is rarely over 10 mm while evaporation is about 2 m y^{-1}. In addition, there has been a further small net loss northwards into the Mediterranean since the opening of the Suez Canal because of a tidal height difference between the Red Sea and Mediterranean. The balance is made up by an inward flow through the Bab el Mandeb. It is estimated that water renewal time in the upper 200 m (i.e. the water above the thermocline) is in the order of 6 years, while the time for turnover of water for the whole Red Sea is about 200 years.

The flow through the Bab el Mandeb is not a simple one (Figure 2.4). Not only

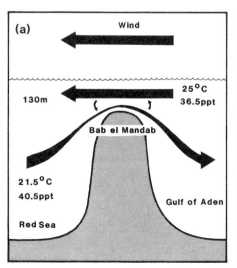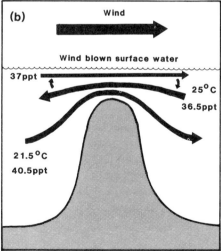

Figure 2.4 *Sketch of (a) winter and (b) summer water flow through the entrance into the Red Sea. Section across the sill at Bab el Mandeb showing the layers of water flow.*

is there an inward flow to make up the evaporation deficit but also a discharge outward of the more saline deeper layer. Both of these processes are affected by the sill of only 130 m deep just north of the Bab el Mandeb, and by the strong winds which drive surface water in the Gulf of Aden.

In the winter, a double layered contra-flow system operates. At the surface, water is driven into the Red Sea from the Gulf of Aden by the prevailing winds; water of this surface flow has physical characteristics of Gulf of Aden water, having a temperature of around 25 °C and a salinity of about 36.5 ppt. Beneath this there is a deeper outward flow which comes from the deeper Red Sea body of water. The latter is both more saline (about 40.5 ppt) and cooler (about 21.5 °C) than the inflowing water, and consequently it is considerably more dense than the surface water. It is also denser than the Gulf of Aden water, and hence the outward flow of deeper water is maintained. According to Edwards (1987), this double layered system is stable from about September to June, or for about three quarters of each year.

There are few portions of the globe where the results of calms and strong currents are felt conjointly with more effect, or in a greater degree, than within the Gulf of Aden
Wellsted 1840

For part of the summer from June to September, however, the prevailing wind through the Bab el Mandeb changes from southeasterly to northwesterly, and therefore opposes the inward, surface flow. During these hottest months, surface salinities in the southern Red Sea increase through evaporation to 38 or 39 ppt, and while this is more than the incoming Gulf of Aden water, its temperature rises to over 30 °C and the water is, despite the raised salinity, less dense. This causes the upper layer to divide vertically into two components, a wind-blown shallowest flow travelling south and out of the Red Sea over the main part which still flows inward. Beneath the latter, the outflow of denser water continues as before.

The observed, broad pattern is not fully explained at present. For example, the sill lies at a depth of 130 m while the deep Red Sea water is considered to exist from 200 or 250 m downwards. Thus there is expected to be considerable vertical mixing at the entrance and just within the Red Sea; much turnover may be expected in the vicinity of Bab el Mandeb whereby water newly arrived into the Red Sea evaporates, warms and becomes denser and sinks before it has time to travel north. The estimates of turnover time of surface water (6 y) and of all Red Sea water (200 y) take this into account. However, an important point biologically is that a large amount of nutrients and plankton is forced into the Red Sea by this process. This has important consequences to local benthic and pelagic biota in the southern region and in the Red Sea as a whole.

B) WATER CURRENTS AND CIRCULATION IN THE GULF

The main, broad scale circulation in the Gulf, or the residual current (current remaining after tidal currents are removed) is an anti-clockwise rotation, affected to some degree by the projection of the Qatar peninsula. This current is also driven by density gradients. While its broad pattern has been known for many years (Emery 1956) it has been pointed out by Hunter (1986) that many details of Gulf circulation are vague or completely lacking, especially along the south and southwestern side, and the latter author has provided several refinements of the original model.

i) The main Gulf circulation

In the Gulf as in the Red Sea, evaporation exceeds combined rainfall and fresh water input, and even though there is a substantial flow into the Gulf from the Shatt al Arab delta, there is annually a net input of water from the Gulf of Oman. The slope of the floor of the Gulf, as described in Chapter 1, is a gradual descent to a trough in the north, which runs roughly parallel to the Iranian coast. Most evaporation in both summer and winter occurs in two extensive and mostly very shallow southern embayments along the Saudi Arabian and UAE coasts. Water enters the Gulf through the Strait of Hormuz at a salinity of 36.5–37 ppt (Figure 2.5). The observations summarized by Hunter (1986) confirm older reports that there is a surface drift towards the west along the Iranian shore, consistent with the anti-clockwise circulation. At all times of the year, the diluting influence of the Shatt al Arab at the northwest corner of the Gulf is evident, especially in winter when flow is greater. In the two large southern embayments, the surface salinity increases to over 40 ppt in open water in both summer and winter (though to much greater values in the semi-enclosed coastal embayments), and this is about 2 ppt greater than values along the Iranian coast. There is a tendency, therefore, for the denser water formed in the southern and

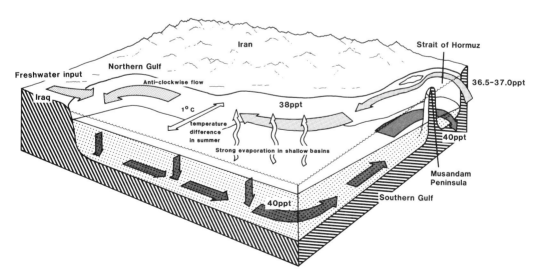

Figure 2.5 *Mechanisms causing the density gradient in the Gulf and the flow through the Strait of Hormuz. Light arrows are incoming surface water from the Gulf of Oman. Dark arrows are a denser, deeper water flow. Light shading in Gulf shows "wedge" of water of increasing density.*

southwestern bays of the Gulf to sink towards the northern trough and towards the Strait of Hormuz which is the deepest part of the Gulf. There is no sill, unlike the situation in the entrance to the Red Sea.

Surface temperatures in the Gulf show greater extremes than those in the Gulf of Oman, being lower in winter and higher in summer. In winter, temperature gradients increase the density flow, since water retained in the south cools more than the inflowing water in the north and becomes denser. In summer, there is greater warming in the southern embayments, and while the presence of warmer and therefore less dense water would tend to counter the density flow, the temperature difference of surface water over the entire Gulf is only 1°C and effects of this are small.

The traditional model of water exchange in the Gulf is similar to that for the Red Sea, namely that denser water flows outward beneath inflowing shallow water. However, Hunter (1986) incorporates the above observations to propose a slightly more elaborate, and more convincing pattern. As shown in Figure 2.5, the water enters not only along the upper part of the entrance into the Gulf but principally along the Iranian side, continuing northwest to the end of the Gulf at the Shatt al Arab. After some dilution, it passes southeast, becoming denser and sinking deeper along the bottom as it does so. It exits the Gulf beneath the inflowing water, but particularly along the southern side of the Straits. In the narrowest part of the Gulf the exact stratification is unknown, but may follow the same pattern.

The surface water flowing into the Gulf has a velocity of 0.1–0.2 m sec^{-1}. This movement is against the prevailing wind which is thus shown to have a less important controlling effect than the changes in density due to evaporation, though it undoubtedly has a retarding effect on the flow. The resulting turnover time of water in the Gulf due to the circulation, defined as the time needed for all Gulf water to come within the influence of the open sea boundary, is estimated by Hunter (1986) to be about 2.4 years. The actual flushing time is estimated to be about 3 to 5.5 years, and is longer because of the effects of vertical mixing and other turbulent processes.

ii) Circulation in the southern embayments

The shallow southern and southwestern Gulf coasts are the principal sites of evaporation which is important because this drives the main Gulf circulation. The bays themselves are indented and include large areas with very restricted circulation, and these increase evaporation further. For this reason, the above estimates of water flushing and turnover can only be averages, and water in the Gulf of Salwah, for example, is likely to be retained for much greater periods. These embayments are important sites biologically for two reasons, firstly because they are highly productive in terms of carbon and nitrogen fixation, and secondly because their conditions of high environmental stress lead to several local extinctions of marine biota which are discussed later.

C) ARABIAN SEA: A "PSEUDO-HIGH LATITUDE EFFECT"

The main currents in the Arabian Sea, cold, nutrient-rich upwelling which occurs in the summer, and wind systems which drive them have been described earlier. Here remarks are limited to components of those hydrographical effects which have important consequences for the shallow, benthic marine biota near shore. Chapters 10 and 13 discuss this with relation to pelagic effects and offshore fisheries.

The strength of the summer Southwest Monsoon which forces surface water away from the coast has maximum effect on temperature depression and nutrient elevation in July and August. The effect varies greatly with local topography. As summarized in Savidge *et al.* (1988) an important consideration is the width of the continental shelf. Upwelling water is approximately 16–17 °C or colder, so impacts on the tropical coastline depend entirely on the relative exposure of the section of coast, and whether there is a buffer to temperature drop provided by retention of localized pockets of coastal water. Two locations studied in detail recently are near the capital of the region, Salalah, and the Mirbat outcrop about 100 km north. Earlier studies of Currie *et al.* (1973) cover a similar, though broader, area. At Salalah, the continental shelf extends about 50 km out and reaches 500 m deep about 70 km offshore. In the Mirbat area, the edge of the shelf is only 10 km out and the 500 m contour located 20 km offshore. The consequence of this is that upwelling water affects the rocky Mirbat promontory to a much greater extent. At the height of the upwelling, surface water temperature off the rocky promontory falls to 17 °C or lower, while it remains over 24 °C where the continental shelf is broad.

Effects of localized topography are similarly reflected in nutrient levels: nitrogen as nitrate being 5–10 μg-at l^{-1} on the broad shelf, rising to over 15 or 20 μg-at l^{-1} 100 km north on the coast where the shelf is only 10 km offshore. Similarly, phosphate concentrations were twice as high in water at the narrow shelf than at the broad shelf. Currie *et al.* (1973) show that phosphate, zooplankton and chlorophyll all decline from their high coastal values to one-half or less 10–50 km from the coast.

Upwelling occurs in summer, so that in addition to greater retention of shallow water where there is a broad shelf, strong solar heating also helps to elevate water temperatures on the broad shelf areas. Indeed profiles of temperature off the Mirbat promontory where some of the most extensive kelp grows shows that in the top 20 m of water there is a rise of 3 °C, and this occurs despite the considerable evaporation caused by the strong winds (as evidenced by a rise in salinity from 35.6 to 35.9 ppt over the same upper 20 m depth). As reported by Savidge *et al.* (1988), cold surface water in areas where the continental shelf is narrow also appears to moderate atmospheric conditions, changing the clear sky to cloudy.

The growth of *Ecklonia* kelp beds for which the region is now renowned occurs best for these reasons on sections of coast where there is greatest effect of upwelling, cold, nutrient rich water. It is also evident that the few, rich growths of corals occur only in semi-enclosed bays and similar small locations where water retention is considerably increased and where solar warming is greatest.

The Kuria Muria Islands also lie close to upwelling water and experience a strong effect from it. Since the area further north remains completely unsurveyed, other areas of local effect may be expected. Generally, therefore, although local effects (in this case local meaning distances of perhaps 100 km) caused by the proximity of the coastline to upwelling water are extremely important, there is an increased overall effect of upwelling in a southerly direction, towards the equator. The entire area under consideration here lies south of the Tropic of Cancer, and the biota gradually assumes an increasing temperate character towards the equator.

This situation extends south to about $8°30'$ N (Currie *et al.* 1973), which is $1.5°$ of latitude south of the northeast tip of Somalia. This is the point where the thermocline of the Somali current, which usually lies at about 90 m deep in the southern part, finally reaches the surface after shoaling steadily in a northerly direction. $8°30'$ N is also the point where the main stream of the Somali current sweeps away from the coast, leaving on its northern boundary an area of cool surface water. South of $8°30'$, typically tropical conditions return over a short distance to the coastal region.

(3) Wave energy and tidal patterns

Tides of the Indian Ocean only correlate closely with tides along the Arabian Sea; those in the semi-enclosed Red Sea follow a completely different pattern. They are summarized in Morley (1975). Tides of the Gulf are similarly independent of those of the Arabian Sea, although they are driven to some extent by the tidal forces propagating through the Strait of Hormuz. Tides within embayments in the Gulf and Red Sea are different again. Generally, tidal height is not very marked anywhere in the region, and ranges of 0.25 to 0.75 m are most common.

In the Red Sea, summaries by Morcos (1970) and Edwards (1987) describe the now well known oscillatory condition, whereby the central part at 20–21°N has almost no daily difference in tidal height but where the northern and southern ends show daily tidal differences whose range increases with distance from the central region. Tidal ranges reach about 0.6 m in the north and up to 0.9 m in the south of the Red Sea, but these are spring tide ranges and greater than the average. They are referred to here as daily tides because they are a complex composite of both diurnal and semi-diurnal tides (details are given in the above references but are not especially relevant here). The Gulfs of Aqaba and Suez continue the northward trend of increasing tidal range, reaching maxima of 1.2 and 1.5 m respectively.

Superimposed onto this in the Red Sea is a seasonal tide of at least equal importance. In winter, mean sea level is over 0.5 m higher than in summer. In the central region where there is no daily tide, this is therefore the only true tide. The change from summer low to winter high occurs fairly abruptly, over less than a month in the spring and in early winter. The mechanism causing the seasonal tide is driven partly by the greater evaporation in the summer, but is mainly the result of wind-driven currents in the entrance to the Red Sea. As described earlier, the surface current in the Bab el Mandeb flows into the Red Sea in the winter, while in the summer strong winds blow the surface part of this outward, and this appears to be the main balancing factor determining the seasonal rise and fall (D. Dixon, pers. comm.). Evaporation is not thought to be the main controlling factor determining mean sea level (Morley 1975) though it must have some effect, and overall, mean sea level is lower by about 50 cm in the north of the Red Sea compared to that in the south.

In the northern part of the (Red) sea during the warm season, from May to October, the reefs are observed to have about two feet less water on them than in the remaining months of the year. The cause originates in the influence of northerly winds at that season . . .   Wellsted 1838

The biological importance of the annual tide is considerable. Added to this is a considerable water level change which periodically occurs as a result of thermally driven winds which blow towards land in daytime. This may add an additional 0.5 m water height, which exceeds the daily tide in the central Red Sea. Water forced shoreward commonly leads to reef flats being submerged in summer when they are usually dry, but the reverse may occur with opposite winds in winter when water is forced away from the shore to cause complete drying of the seasonally submerged reef flats. The latter may lead to widespread mortality and community disturbance in shallow areas, as was first described by Loya (1976a) for the Gulf of Aqaba. An example in which sea level changed locally by 60 cm during one day is also given by Morley (1975) for Jeddah.

In the Gulf, two amphidromic points where tidal range is zero occur off northern Saudi Arabia and off the UAE coast. The tidal regime in the central part and around Bahrain is complex and basically semi diurnal. Over most of the Gulf away from shore, tidal range is <0.6 m, but it rises to 1–2 m near land, especially in the far north and just outside the Strait of Hormuz. Off Kuwait at the northern tip of the Gulf, spring tidal range reaches 2 m in the south and up to 4 m in the north (Jones 1986a), while off Bahrain range is 2 m at extreme springs (Jones 1986b). To some degree the diurnal pattern in both of these locations ameliorates the conditions for shallow and intertidal biota; in summer, high tides cover the shallow reefs in daytime and expose them at night, while in winter when air temperature is low, the high tide occurs at night, thus affording some protection in normal conditions. Hence the conditions noted for Kuwait and Bahrain are different from several other parts of the Gulf; in the central region of Qatar, Saudi Arabia and the UAE, low tides commonly expose the gradually sloping intertidal region during daytime in summer.

Because of barrier effect from a huge shallow reef complex of Fasht Azm between Bahrain and Qatar, water in the large Gulf of Salwah which is located between Qatar and Saudi Arabia is more restricted than the width of its entrance suggests. Tidal ranges which reach 1.2 m to the north of Bahrain are reduced to about 0.5 m in the south of the Gulf of Salwah, and its phase lags considerably. Flushing is reduced and total water retention time increased. Similar constrictions occur near the mainland shore of the UAE where extensive shallows and ponds occur on the particularly gentle transition from land to sea.

In both the Red Sea and Gulf, tidal streams passing through constrictions caused by reefs, current-formed sand bars, and low islands commonly exceed 1–2 m sec^{-1}. They are important mechanisms of water and nutrient movement even where water exchange with the main part of the Gulf is limited. Tidal streams are important in providing the water movement necessary for vigorous benthic biota, even in areas where there is little tide or water exchange other than oscillation of locally confined water. In the Gulf of Salwah in particular, strong currents keep the waters around the Hawar Archipelago and numerous adjacent limestone banks in a constant state of movement, which leads to a considerably richer benthic community than could otherwise be expected.

The Arabian Sea coastline has a tidal range of 1.5–2.5 m over most of its length. Unlike the situation in the Red Sea and Gulf, however, tidal range is of less importance than prevailing, almost continuous, high energy waves.

(4) Dissolved oxygen and nutrients

For the Indian Ocean, Krey (1973) regarded areas of high fertility as those where chlorophyll concentrations in surface water were >0.5 mg m^{-3}, and noted that during summer, only the Arabian Sea and east coast of southern Africa showed such values. In the coastal part of the Arabian Sea, Ryther et al. (1966) had already demonstrated a primary productivity of >1 g C m^{-2}d^{-1}, which is over 10 times the values found in the central part of that sea. Chlorophyll levels are likewise high. In general, productivity and both dissolved and particulate nutrients show very large variation around Arabia.

Most of the Red Sea is fairly poor, though some areas within the Gulf and southern Red Sea may be locally very much more productive. The large amount of data on dissolved oxygen, nitrate and phosphate obtained during part of the International Indian Ocean Expedition are summarized in McGill (1973). The latter also usefully condenses the data into maps showing concentration contours to 4000 m deep, and these place the Arabian Sea and Red Sea in context within the Indian Ocean. Recent summaries for the Red Sea, Gulf of Aden and Gulf are given in IUCN/UNEP (1985a,b) and Edwards (1987).

Some sources state that the Gulf is one of the most productive bodies of water in the world, though there has been confusion between benthic and pelagic production in this shallow body of water, and this statement should really apply only to the total, or benthic production (Sheppard and Price 1991). In the shallow water of the Gulf, even though there is unlikely to be any limitation by light levels, there is evidence of nutrient limitation, with a consequent reduction in true pelagic productivity, as Jones (1985) points out. Primary productivity is greater in mixed central waters and in shallow bays, especially in the influence of the Shatt al Arab estuarine conditions. Jones (1985) reports chlorophyll a values in the Gulf of 0.2–0.86 mg m^{-3}, but as he also notes, this is less than that recorded in the Arabian Sea where upwelling raises concentrations by one or two orders of magnitude.

In the upwelling waters of Oman, Savidge et al. (1988) recorded chlorophyll a concentrations of 5–20 mg m^{-3}. Nutrient levels were also correspondingly high. Nitrate values rise to 5–20 mg at NO$_3$ m^{-3} and phosphate values to 1.5–2.5 mg at

$PO_4 m^{-3}$, which is three to five times greater than the winter, non-upwelling values (Barratt *et al.* 1986).

Nutrient levels in the Red Sea are detailed in Weikert (1987). Water of its two northern Gulfs is poor in nutrients compared with the Indian Ocean, and there is evidence that the production potential of the Red Sea is low. Levels of PO_4 are if anything lower than those shown for the Gulf, namely below 1.5 mg at $PO_4 m^{-3}$. The southern part of the Red Sea has higher levels due to the influx of Gulf of Aden water; there is a sudden drop in nutrients of surface waters around 19°N which corresponds to the limit of direct influence of the northward flow of Gulf of Aden water through the Bab el Mandeb. Values of carbon and its assimilation are discussed in Chapter 10.

Throughout most of the region, dissolved oxygen is near to saturation in surface waters. Exact levels vary according to temperature and salinity, since saturation typically ranges from about 4.8 to 6.5 ml $O_2 l^{-1}$, being lowest in warmer waters of higher salinity. The saturated layer in the Red Sea extends to about 100 m deep, while in the shallower Gulf it extends to the bottom. Below 100 m in the Red Sea, values drop to only 10–25% saturation in the O_2 minimum layer, and then rise to double this in the deep water layer. The Gulf of Suez resembles the Gulf in this as in many other of its characteristics, and has no oxygen minimum, while in the Gulf of Aqaba there is a gradual decline with depth but never to lower than about 50% saturation.

In the Arabian Sea, critically low levels have been reported below 100 m deep. This sea is filled with higher salinity water than the Indian Ocean, caused by surface evaporation and intensified by intrusions of higher salinity water from the Gulf at 300 m deep and from the Red Sea at 800 m deep (Wyrtki 1973). Vertical mixing of these results in a thick layer of high salinity water which extends as far as 2000 m deep. The isolation of this water together with a high pelagic productivity creates a large mass of water with a very low oxygen concentration. This layer extends from 200 to 1200 m deep, and its O_2 values fall within the range of 0.2–1 ml l^{-1} everywhere to 3°N, compared with values two to four times greater elsewhere in the Indian Ocean at this depth. The low levels of oxygen in the deep layers of the Red Sea are of a similar level to those of the Arabian Sea. Conflicting information comes from Rochford (1966), however, who found oxygen rich water in the Arabian Sea under the oxygen poor layer 200–600 m deep. This was identified as coming from the southern Indian Ocean via deep currents, and some water identified as originating from the Antarctic impinges the Yemen–Oman coast. This water was focused onto this coast by the wedges of higher salinity water emerging from the Red Sea and Gulf. The high oxygen-containing stream mixed with high oxygen-containing water exiting from the Gulf, to form the observed layer. In a complementary study, Rochford (1967) determined that the productivity from the Arabian upwelling region can be traced eastwards as far as Sumatra, using dissolved phosphate levels as an indicator.

C. Early Holocene marine climate

Consequences of lower sea levels in the Pleistocene, and the important development of limestone substrate at depths of 30–60 m below present sea level, were considered in Chapter 1. Several aspects of climatic change in the early Holocene are of particular relevance to present day biota, in particular to biogeographical questions. Changes in strength of the monsoon system have occurred over time, and the present aridity of the region is a relatively recent development. Klein *et al.* (1990) demonstrated that earlier in the Holocene, rainfall in the Sinai was considerably greater than it is today. These authors measured fluorescent banding attributable to humic materials in fossils of the coral *Porites* from raised terraces, and showed that it was absent from present fringing reefs. They deduced that heavy summer rains supplied the former. Today the Sinai has rare, winter rains only. Much of the vast quantity of alluvial material swept into the Red Sea in earlier times can be explained only by such rainfall.

(1) Sea level fall and Red Sea isolation

The low sea level at the start of the Holocene was approximately 130 m below present (Hopley 1982). This is very close to the present depth of the sill near Hanish Island just north of the Bab el Mandeb. Questions concerning a possible "drying out of the Red Sea" are not new, and whether sea level fell to just below or just above the level of the sill would seem to be critical to whether or not the Red Sea became isolated. But, whether the exact amount of sea level fall is critical biologically is an important question. The answer is, almost certainly, that it is not.

As described in more detail above in "Red Sea–Gulf of Aden water exchange", for 9 months of the year the entrance of the Red Sea supports a double flow of water, one inward flowing on top of an outward flow of heavier, more saline water. During the summer, the surface layer itself is divided vertically into two, with another outward flow on the surface which again is more saline than the incoming water. Whichever regime is operating, the outward flow or flows are about 10% more saline than inflowing water. Evaporation in the Red Sea basin is about $2 \, m \, y^{-1}$, or ten times greater than the total fresh water input of less than $0.2 \, m \, y^{-1}$. Given an area of $440,000 \, km^2$ and a volume of $240,000 \, km^3$ for the Red Sea, the imbalance of evaporation over fresh water input tends to increase total Red Sea salinity by $0.13 \, ppt \, y^{-1}$, an amount which would not take long to become lethal. That this does not occur is, of course, due to the incoming water from the Gulf of Aden, but the replacement water from the Gulf of Aden is itself not fresh but is about 36.5 ppt. The evaporative increases in salinity are countered principally by the discharge of water with 10% greater salinity out of the Bab el Mandeb (Figure 2.4).

This contra-flow system is maintained with the present depth of 130 m, and indeed, a similar contra-flow system occurs at the entrance to the Gulf at depths which are similar or slightly less (Hunter 1986). However, it seems most unlikely that the double or triple stratifications could be maintained if the depth was reduced to 10 or 20 m, which is the depth which would exist in the Bab el Mandeb if the lower values suggested in the envelope of sea level fall were true. Thus, even if the Bab el Mandeb did not dry out completely, the outward flows of high salinity water could not be supported. Only inward flows to replace water lost from evaporation could possibly take place. Salinity would increase even if the Red Sea remained connected to the Indian Ocean.

There is evidence to support the view that the Red Sea was hypersaline during the glaciation. Reiss *et al.* (1980) determined that salinities of >50 ppt prevailed, enough to kill most plankton and probably all reef based life at least. Data from cores in Reiss and Hottinger (1984) indicate that salinity in the Gulf of Aqaba reached about 70 ppt. According to Friedman (1972, 1985) marine conditions at the height of the glaciation may have resembled those in the Dead Sea at present. Braithwaite (1987) summarizes sedimentary evidence, describing several bands of coccoliths, related to alternating oceanic and hypersaline conditions in the north of the region. In the southern Red Sea too, layers of light sediment with poor plankton content indicate hypersaline conditions, and these alternated with bands of dark sediment and rich plankton, indicating inflows of Indian Ocean water. Braithwaite (1987) remarked that some authors have suggested that even in the lowest sea levels, water might still have entered via an inlet in the Afar triangle, but he notes that this remains disputed.

There is evidence for highly saline conditions in the Red Sea, therefore, but even if the Red Sea was not hydrologically completely cut of by the sill in the early Holocene, the consequences to the water in the Red Sea would be the same, though less dramatic. Salinity would rise at least beyond the point where most eukaryotic life would not survive (see Price 1982a), except perhaps near the southern entrance if water continued to flow in from the Gulf of Aden. It is probable, therefore, that a view on survival of marine life in the Red Sea is not dependent on the precise drop in sea level or whether the Red Sea actually became separated from the Indian Ocean by dry land.

In the Gulf, the question is very much simpler, and there has been no doubt that, to whatever level the sea level dropped at the start of the Holocene, the Gulf dried out over most of its extent. In the biogeographical context, recruitment of macroscopic marine life into the Gulf commenced anew in the Holocene, as is probably the case in the Red Sea.

(2) Strength of monsoon and Arabian Sea upwelling

Effects of lower sea levels in the Gulf of Aden and Arabian Sea are not known. The nature of substrata near the coast at depths of 100–150 m is largely unrecorded, though off Oman there are extensive stretches of both sandy and rocky substrate. Nothing of use can yet be added to the obvious comment that the early Holocene (and earlier) shoreline probably contained mixed substrates, as they do today.

As described earlier, Arabian Sea upwelling, which brings cold, nutrient-rich water to the coast in the peak of summer, is an event of considerable biological

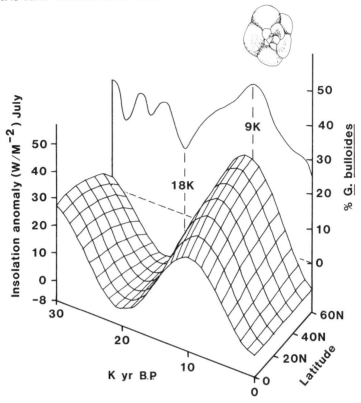

Figure 2.6 *Strength of the Arabian Sea upwelling over the past 30,000 years. From Prell (1984). The percentage abundance of the diatom* Globigerina bulloides *is termed the "upwelling index".*

importance. Of several consequences that derive from this, two are relevant here. Firstly, upwelling permits the existence of kelp beds of a temperate nature, replacing reefs typical of the tropics. Secondly, the upwelling is itself located between the main semi-enclosed water bodies of the Arabian region (Red Sea and Gulf) and the "parent" Indian Ocean. Since the upwelling event is inimical to coral and reef growth, it may act as a barrier to recruitment of tropical species into the Red Sea and Gulf. The latter is important in a biogeographical sense especially if most, if not all, recruitment has been subsequent to the Holocene transgression. The question of interest here is whether the upwelling event, and hence the existence of both the kelp ecosystem and the possible barrier to reef-based life, occurred continuously through the Holocene.

Prell (1984), Prell and Van Campo (1986) and Prell and Kutzbach (1987) demonstrated that the severity of both monsoon and upwelling has not been constant during the Holocene. As described earlier, upwelling is directly associated with the monsoon system, which forces surface water away from shallow, coastal areas of the western Arabian Sea. Prell and his co-authors showed that pollen and upwelling records indicate, firstly, that stronger monsoons occurred in interglacial periods and weaker monsoons in glacials, over the past 140 Ka (Figure 2.6). In addition they examined planktonic foraminifera in sediments to determine temporal changes. The latter organisms respond to both temperature and to changes in nutrients in upwelling water, so Prell used the percent abundance of the common *Globigerina bulloides* as an indicator. Looking first at surface (i.e. present day) deposits, the abundance of *G. bulloides* was shown to correlate negatively, and strongly, with August sea surface temperature (SST); this foraminiferan is a sub-polar species, and provides <10% of the total in areas with an August SST of over about 27 °C, rising to 40–50% in regions where SST is below 24 °C. Nine Arabian Sea cores provided temporal data of abundance over the past 30,000 y. Before the Holocene, *G. bulloides* abundance fluctuated but was slightly higher than today, suggesting a strong monsoon and strong upwelling. Near the start of the Holocene, its abundance dropped sharply and remained low for perhaps a thousand years, then rose steadily to a maximum at 9 Ka BP, after which abundance fell to today's values.

The conclusion is that during the early part of the Holocene transgression, both the monsoon and the Arabian Sea upwelling were considerably reduced, and that this reduction lasted for a period of perhaps 2–3 Ka. About 9 Ka BP, and for 1–2 Ka either side of this, upwelling was considerably stronger than today, and from 9 Ka BP to the present, upwelling gradually declined to its present, moderate state.

By extension, in the early Holocene when sea level rise meant that recruitment into the Red Sea and Gulf could have recommenced, there was probably a much reduced barrier to tropical biota. In the mid Holocene, in contrast, the barrier was probably both severe and broader in geographical extent, providing a greater barrier to tropical fauna and a greater expanse of favourable habitat for the brown algae ecosystem.

Summary

The Arabian region lies at the edge of two global weather systems, whose fluctuations cause varied and severe environmental conditions.

During winter, the Red Sea has two axial air flows converging at the centre. Rain-bearing winds may deposit annual rain totals in a few hours, causing flash flooding which has significant effects on shallow marine biota. In summer, prevailing winds flow down the Red Sea for its entire length, reinforcing the clockwise airflow in the Arabian Sea. This generates strong southwesterly winds there, leading to cool, nutrient-rich upwelling.

Smaller scale wind systems derived from severe temperature differences have equally important effects. In the Red Sea these affect coral reef alignment and control soft substrate distributions. Winter winds in the Gulf may reduce surface water temperatures to 4 °C, with fatal effects on reefs and restricting mangrove distribution.

Water circulations in both the Gulf and Red Sea are driven by density gradients. In the Gulf, evaporation causes a mass flow inwards through the Strait of Hormuz. Greatest evaporation occurs in the two large southern Gulf embayments, where surface salinity increases to 2 ppt greater than values along the Iranian coast. The denser water formed in the southern bays sinks towards the Strait of Hormuz. Thus, newly incoming water enters along the surface and northern edge of the Strait, is driven anti-clockwise, and exits the Gulf in the deeper and southern part of the Strait. Mean turnover time is about 2.4 y, with a complete flushing time of 3–5.5 years because of vertical mixing and other turbulent processes.

In the Red Sea, a northerly fall in temperature, and evaporation of $1–2\,\mathrm{m\,y^{-1}}$ causing an increase in salinity of 4 ppt, both increase surface water density. In the Gulf of Suez, further chilling and evaporation occur. The dense water then turns under into the deep Red Sea, returning southward. Below 250–300 m this water is stable at 21.5 °C and 40.5 ppt, and is a unique Red Sea feature. Mean turnover of water above the thermocline is about 6 years, while that for the whole Red Sea is 200 years.

Flow through the Bab el Mandeb is complex due to its sill. In winter, a double layered contra-flow system prevails in which surface water is driven into the Red Sea by the prevailing winds. Beneath this, a deeper outward flow removes more saline, denser water. This occurs from September to June. During summer the wind divides the upper layer vertically into a wind blown shallow part which exits the Red Sea over the inward flow. Net outward flows are 10% more saline, which balances evaporation in the Red Sea.

Major Arabian Sea water movement comes from removal of surface water in the summer monsoon, and its replacement by upwelling water of 16–17 °C. Biological effects depend on the relative exposure of each section of coast, and its width. Broad areas of shelf buffer the temperature drop, and are affected much less by raised nutrients. *Ecklonia* kelp beds, for which the region is renowned, grow best (and coral growth is worst) where upwelling is greatest.

Tides in the region are generally 0.5–1.5 m. The central Red Sea is almost tideless, and annual water level changes are more important. Winter mean sea level is 0.5 m higher than in summer. Added to this, periodic water level changes driven by thermionic winds may further add or subtract 0.5 m water height. In both Red Sea and Gulf, tidal streams passing through constrictions caused by reefs, sand bars and low islands commonly exceed $1–2\,\mathrm{m\,sec^{-1}}$. They are important mechanisms of nutrient movement, and provide the water movement necessary for vigorous growth of benthic biota.

The Holocene marine climate of the region is important to present patterns. In the entrance to the Red Sea, it is unlikely that the stratifications and outward flow of more saline water could have been maintained during the lowest sea level, even if the entrance did not dry completely. Salinity would increase, and evidence suggests that it reached >50 ppt throughout, and 70 ppt in the Gulf of Aqaba. Thus a view on survival of marine life in the Red Sea is not dependent solely on whether the Red Sea actually became isolated by dry land. In the Gulf, there has been no doubt that it dried out over most of its extent.

In the Arabian Sea, monsoon strength, and hence upwelling, varied considerably during the Holocene. During the early transgression, upwelling was considerably reduced for 2–3 Ka. Then for about 3–4 Ka around 9 Ka BP upwelling was much stronger than today, gradually declining to its present state. Thus, in the early Holocene when rising sea level allowed renewed recruitment into the Red Sea and Gulf, there was probably a much reduced barrier to tropical biota. Later, the barrier was probably both more severe and broader than it is today.

Marine Ecosystems

Fringing reef on alluvial fan sheltering soft substrata, central Red Sea.

CHAPTER 3

Reefs and Coral Communities

Contents

A. Reef distribution and development 64
  (1) Geographical distribution 64
  (2) Spur and groove systems 70
  (3) Southern algal reef constructions 73
  (4) Ridge reefs and atolls 73
  (5) Non-reef coral populations 75
B. Ecological patterns of the coral fauna 76
  (1) Coral distribution and diversity 77
  (2) Zonation in the Red Sea 78
  (3) Gulf coral communities 83
  (4) Arabian Sea coral distribution 84
Summary 85

It is a measure of how recently access to the region has been granted to scientists that until the late 1970s it was assumed that classical fringing reefs known from northern parts extended throughout the Red Sea. In 1987 the comment that "the longest continuous fringing reef . . . lies along the Red Sea coastline, over a total length of 4500 km" (Longhurst and Pauly 1987) was still a guess. The idea has persisted despite hints of muddy shores and difficult mangrove terrain noted by the naturalists of the 18th and 19th centuries, partly because accounts of reefs in both popular and scientific literature led to misunderstanding of the amount of reef, and partly because reefs do indeed dominate most of those parts accessed before the 1970s.

Today, much of the region has been examined, at least briefly. Large areas of the Gulf of Aden and Arabian Sea remain completely unstudied, but enough is known of Oman and parts of Yemen to make some confident predictions of principal habitats and amount of reef development in most areas.

> *. . . the banks of coral are less numerous in the southern, than in the northern part of the Arabic gulf. (=Red Sea)*   Niebuhr 1792

The first part of this chapter briefly outlines the known reef distribution in the region.

A. Reef distribution and development

(1) Geographical distribution

A) GULFS OF SUEZ AND AQABA

The most northerly reefs of the Indian Ocean are those near Suez (Map 1, see Introduction). As noted in Chapter 1, the Gulf of Suez is shallow and sedimented, and is not part of the main, deep Red Sea rift system. On the eastern shore, the most northerly reefs include small coral patches with elevations of 1–3 m, lying on calcareous sandy and silty substrate from 1 to 5 m deep. They experience winter temperatures as low as 18 °C and salinity as high as 41 ppt, but it will be shown later that these levels are not especially limiting to coral or reef growth, so the main constraint on coral reef development probably comes from high sedimentation. The morphology of the corals dominating these reefs supports this view, being stagshorn *Acropora* and an unusual form of similarly-branching *Stylophora*, an ideal shape for passively rejecting sediment in low energy conditions.

Reefs on the western coast of the Gulf of Suez are better developed in general than those on the east. This is particularly so in the far north where relatively well-developed fringing reefs exist to within 50 km of Suez. This marks the beginning of a long stretch of fringing reef running southward, although there are reef patches closer to Suez. At Ein Sukhna, the reef flat extends 30–40 m offshore and is followed by a gentle slope to a sandy bottom at 4–5 m deep. The outer slope supports flourishing coral growth with up to 15 genera present. Communities are dominated by *Porites*, *Acropora* and *Stylophora*. Only patch reefs lying offshore are found at the same latitude on the eastern side, such as at Ras Sudr which itself is a small promontory projecting away from highly mobile beaches.

Extensive reefs first appear in the southern Gulf of Suez, both on Sinai and around the Ashrafi islands on the African shore. This region lies at the point where the northerly surface flow of Red Sea water dips under to return southward at depth (Chapter 2) and, being shallow, has extensive illuminated substrate, something which is greatly limiting in most of the northern Red Sea. As a consequence, the reefs here form some of the largest of all Red Sea reef complexes.

At Ras Mohammed on the southern tip of the Sinai, the bathymetric profile changes abruptly into that of the main Red Sea system which includes the Gulf of Aqaba. Water in most of the latter is over 1800 m deep. The entrance to the Gulf of Aqaba has a broad sill less than 200 m deep where extensive shallows of soft substrate provide a complex, fertile area of reefs and seagrass. In most of the rest of the Gulf of Aqaba only a narrow fringing reef finds a foothold on the steep cliffs. For most of its length, fringing reefs line both shores; they are usually narrow, and have been termed "contour reefs" (Fishelson 1980), but they extend outward to as much as 1 km in embayments and old wadi systems.

B) NORTHERN AND CENTRAL RED SEA

The classical fringing reefs for which the Red Sea is famed extend southward along both east and west sides of the northern Red Sea to 18–20°N. Mostly they are narrow, extending only a few tens of metres from the shore. (Figure 3.1 illustrates profiles of this and several other major forms of reef in the Arabian region.) Shores here, like the Gulf of Aqaba, contain raised fossil reefs. As discussed earlier, they have arisen from tectonic uplift and complex changes of sea level relative to the land. Current fringing reefs are undoubtedly maintaining

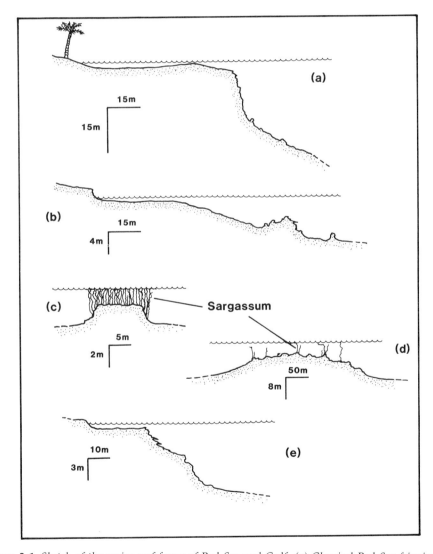

Figure 3.1 *Sketch of the major reef forms of Red Sea and Gulf. (a) Classical Red Sea fringing reef, (b) southern Red Sea fringing reef, (c) Algal reefs of far southern Red Sea, (d) limestone domes of Gulf, (e) profile of coral cay in Gulf.*

themselves at present sea level, but are themselves founded on vertical series of much earlier reef structures.

In areas, especially further south, reefs commonly extend 1 km to seaward. Throughout this region, the alluvial plain is 1 to 7 km wide, and where old wadi systems are present vast amounts of debris form fans up to 1 km from the shore. The larger fans were mainly created during the Pleistocene and may extend 3 to 4 km along the shore, while both large and smaller ones may still be highly active (Hayward 1982). Beaches here are well developed, and many fans are characterized by large back reef lagoons which develop shoals parallel to the shoreline. Fringing reefs are found on the edges of these alluvial fans, which are an important influence on reef growth in such areas.

Offshore in the northern Red Sea lies the Wedj Bank and its islands. These contain extensive seagrass, reefs and mangroves (Ormond et al. 1984a,b). Also offshore, from Wedj to Jeddah, is an extensive series of submerged limestone platforms which form the foundation for a chain termed the "Little Barrier Reef" (Sheppard 1985b). Coral communities on this are the most exposed, varied and vigorously growing known in the Arabian region.

In Egypt, a series of patch reefs and fringing reefs around Hurghada, now an important tourist site, have been studied since Cyril Crossland founded a marine biological station there in 1930. Further south is Quseir, which was Dr Carl Klunzinger's base during two periods between 1863 and 1875, whose studies are a landmark in tropical marine biology generally. These two naturalists revealed the extent of fringing and patch reef systems of the northern part of the Red Sea. Further south is 500 km of little-studied coast. It has been explored by Edwards et al. (1981), Gubbay and Rosenthal (1982) and Barratt (1982) enough to suggest that it is similar to the Arabian side, with probably one remarkable exception. In a section under joint Sudanese and Egyptian control, a mountainous region including Gebel Elba lies close to the coast. The mountains attract unusual quantities of rain, and support a vegetation more like that of central Africa than the desert coast of the Red Sea (Goodman 1985). It contains remarkable refuges for terrestrial fauna, and apparently contains tribes which traditions say are remnants from pharaonic Egypt. The marine environment remains mostly unknown, though the fact that the fringing series of reefs here swing almost 70 km offshore indicates that there are vast expanses of shallow and undoubtedly productive coastal habitats.

The central Red Sea is widest in the region of Jeddah and Sudan, and supports a correspondingly wide range of reef morphologies. The character of the reefs is broadly similar to that of the north, though reef flats begin to broaden in this area and to develop more and larger lagoons, and mangroves begin to assume greater importance along the mainland shoreline. However, winter cooling is not as severe as in the north and salinity less elevated, so reefs still flourish, especially offshore.

The Sudanese coastline covers 750 km between 18° and 22° N. At Dunganab 160 km north of Port Sudan there is a very large sedimented embayment which is considered a unique marine biotope in the country, with characters which may be similar to those of the Gebel Elba region noted above. Sanganeb atoll is well studied, and several similar, circular or annular reef structures form part of the barrier reef system on the Arabian side as well. Most work has, and continues to be done, on the Towartit reefs and in the Suakin Archipelago near and just south of Port Sudan. On the Arabian side, both the offshore and fringing reefs are known in broad detail.

C) SOUTHERN RED SEA

On the Arabian side, fringing reefs become much reduced in size south of about Al Lith at 20° N (Map 2, see Introduction). The continental shelf becomes broader in this area, and the floor slopes more gradually so that the reef base meets soft substrate in increasingly shallow water. The steady increase in muddy substrates and mangroves causes significant reef development to be pushed out further from shore, where there continue to be large expanses of limestone platform. Fringing reefs diminish (Figure 3.1) and have a reduced coral diversity, and in many places there is a complete replacement of fringing reef by broad and thick stands of mangrove. Sand shores and sandy sublittoral habitat with no coral or mangrove growth also increase. These conditions reach their maximum extent opposite the Farasan Islands, and extend into Yemen. There is an unusual development of reefs constructed almost entirely by calcareous red algae on extensive, sublittoral sand (Figure 3.1).

Only rudimentary reefs occur on the continental shore of Ethiopia. The east and west mainland coasts of the Red Sea appear to mirror each other as far as Bab el Mandeb, and this mirroring extends also to the two offshore island groups: the Saudi Arabian Farasan Archipelago and the Dahlak Archipelago of Ethiopia (Map 2). Geological features of the Dahlak group have been well studied (Angelucci *et al.* 1982a, Civitelli and Matteucci 1981) and discussed in Chapter 1, though the marine life of neither archipelago is well known. Fringing and patch reefs are common around both the limestone and volcanic components.

Anecdotal information (Wainwright 1965) and brief observations by us suggest that sublittoral algal and coral populations of the Farasans and offshore reefs are similar to those of the mainland shore, though compared to some of the reefs further towards the central axis of the Red Sea they are possibly richer in corals. Throughout the southern area, fossil reefs occur above sea level along the mainland coast, and these have a higher generic diversity than do the present living reefs, possibly suggesting more favourable conditions in the past.

The shores of the Red Sea converge and shallow at the Bab el Mandeb. About 25% of the Yemen coast in the Red Sea and Gulf of Aden is fringed with shallow reefs or coral communities, found mainly around headlands on remnant fossil reef substrates. Mangroves are less well developed here because of a lack of many sheltered embayments. High energy sandy beaches dominate the coastline on the Arabian side.

D) GULF OF ADEN

Djibouti on the African coast has few reefs (Map 2), but was the site of some notable earlier work by Gravier (1910a,b,c, 1911). Unfortunately little has been done since then. There is even less information for Somalia. It is likely that

Somalia has only weak reefs and scattered corals, due to upwelling water, a condition which continues down the Indian Ocean coast of Somalia for at least 500 km (Angelucci *et al.* 1982b). The upwelling causes large brown algae to dominate hard substrates, while seagrasses are abundant on soft substrates. Darwin (1842) reported a small reef near the tip of the Horn of Africa, but otherwise there is apparently no significant coral growth until the southern part of Somalia, Kenya and beyond. On the north side of the Gulf of Aden, information is also sparse. It is known to be strongly affected by the Arabian Sea upwelling. Scheer (1971) visited the Abd el Kuri Island in the east of the Gulf of Aden and found sublittoral hard substrate dominated by large algae, with few corals. In the Gulf of Aden, therefore, the extremely sparse data suggests that reefs are largely or completely absent, but that scattered coral communities occur amongst the macroalgae.

E) ARABIAN SEA

The same conditions of cold water summer upwelling continue through Yemen to Oman. Although almost nothing is known about the benthic communities of this coast, it is predicted that reef development does not occur but that a low diversity of corals probably occurs amongst the macroalgae in the same manner as for Oman, further north. Upwelling effects are at their most severe here, so coral growth may be much poorer still. The southern region of Oman, Dhofar, has been studied fairly extensively (Barratt *et al.* 1984, 1986, Nizamuddin *et al.* 1986, Anon 1988), though mostly from the viewpoint of its dominant algae rather than corals (Map 3, see Introduction). A low diversity of scattered corals grows beneath the tall and thick algal canopy, which is described in Chapter 6. Bays on the mainland and Kuria Muria Islands do support coral communities with over 75% coral cover, though these are growing on non-reef substrate. There are areas of southern Oman where large coral growths exist, but there seems to be no true reef growth.

North of Dhofar, the only known site where reefs occur is Ras Madrakah, which is a promontory which approaches the 200 m isobath. Otherwise, corals and reefs are abundant only in the shelter of the large island of Masirah at about 20° N.

Most of the 700 km coast of Oman along the Arabian Sea from Dhofar to Ras al Hadd, which is the easternmost point of Arabia, has soft substrate which experiences high energy, even where cliffs descend to water level. Apart from the Masirah region, however, it is extremely poorly known. The southern half is rocky, but the northern half borders the great Wahibah sands desert, whose dunes roll into the sea providing a sandy, very high energy sublittoral. Where sea cliffs exist, scouring is intense and there is no known reef growth.

F) GULF OF OMAN

Where upwelling effects are attenuated by bays, reef growth continues, with typical reef flat and reef slope development. Even where reefs do not develop, prolific coral communities grow on many different types of non-limestone rock, including Semail ophiolite which is uplifted, dark green oceanic crustal material (Green and Keech 1986) as well as on "Oman exotics" (Clarke *et al.* 1986) which are ancient limestones (Chapter 12). Some coral growths develop into vast monospecific beds to a degree seen only in a few other instances in Arabian seas. The reef studded section of coast is limited in extent, however, because to the north of the Capital Area, the extensive sandy Batinah coast sweeps for 200 km towards the Strait of Hormuz. The Iranian coast of the Gulf of Oman similarly consists mainly of sand dunes and sandy beaches, and there appear to be cliffs and some bays containing corals only in eastern Baluchestan.

The unusual geological nature of the Strait of Hormuz at the entrance to the Gulf has been described (Map 5, see Introduction). The clear water in the fjord-like Permian Musandam peninsula provides good habitat for corals and other coelenterates (Cornelius *et al.* 1973). Corals are diverse, with several species not otherwise found in Arabian seas (Sheppard 1985c), but although some true reefs have developed, mostly the corals do not form reefs (Chapter 12).

G) THE GULF

Reefs are reported from Qeshim Island in the Strait of Hormuz, but the only other reported coral reefs from the entire Asian coast of the Gulf are at Shotur Island. This probably reflects only the inability to collect data from Iran, as this northern part of the Gulf is the steepest and deepest of all, and will probably eventually be shown to support the most developed fringing reefs in the Gulf.

The southern Gulf coast along the UAE is low-lying, often swampy and rich in seagrasses. Offshore the water is very shallow but while it is generally muddy and unsuitable for most corals, there are numerous patch reefs dominated by *Acropora*. Fringing reefs grow around numerous low islands, as well as along the east and north coasts of Qatar. These areas tend to have a high coral cover, but a low diversity of perhaps less than 20 species. Their lack of success is probably because of high sedimentation and periodic decimation in near freezing winter air temperatures (Shinn 1976).

Coral reefs in Bahrain and west of Qatar are described by Sheppard (1985d) and, as far as is known, appear to be fairly representative of reefs located in the Gulf near the mainland. Only those well removed from land develop the typical reef profile of a reef flat at or near low water level, and a reef slope which descends relatively steeply. More usually, reef slopes are gentler than is typical for the Arabian region or Indian Ocean (Figure 3.1). Coral reefs nearer land are gentle domes, sometimes of nearly imperceptible slope, whose uppermost extent may or may not reach the low tide mark. This condition is more common and extensive than are the "true" coral reef topographic forms, though the two conditions intergrade. The dome shape, especially where not presently covered by coral, is interpreted to mean that the reefs are not actively growing (Sheppard 1985d, Chapter 12). It is important to distinguish between coral growth and reef growth in areas such as Bahrain where marginal conditions exist. The two are distinct, and there are limestone mounds with extremely gentle slope which do not at present support corals which may have been actively growing in the Pleistocene.

The Gulf of Salwah between Qatar and Saudi Arabia and the attractive Hawar Archipelago within it (Map 4, see Introduction) are too saline for corals. Bahrain has numerous reefs along its northern and northeastern side, and offshore, patch reefs extend down its east coast nearly to a level with its southern tip. These reefs are depauperate (<30 species) but have high coral cover. Throughout this area of the Gulf, corals for the most part merely appear to veneer older limestone structures and only on the Saudi Arabian islands and some patch reefs is there any substantial Recent growth. In most cases, erosion of the limestone platforms is evident, and attachment of corals to the substrate is weak, with the surface of the substrate itself very friable.

The Saudi Arabian islands are coral cays (Burchard 1979, 1983, Basson *et al.* 1977) and are the most developed reefs known in the Gulf (Figure 3.1). Approximately 50 coral species occur. Patch reefs closer to the mainland are much less diverse (McCain *et al.* 1984), and closely resemble those of Bahrain. The most northerly reefs in the Gulf lie around islands off Kuwait (Downing 1985) where around 26 species are present, and like all known reefs in the Gulf, they support insignificant coral growth below 15 m deep. Corals also occur in isolated colonies on rocky outcrops on the southern mainland of Kuwait, but towards the north of the State, the influence of the Shatt al Arab's estuarine conditions precludes corals.

(2) Spur and groove systems

Where wave energy is sufficiently high, spur and groove systems tend to develop. These are reef crest structures whose projections (spurs) point into the prevailing waves, and whose channels (grooves) carry water at rapid velocity. The water oscillates in the grooves such that outgoing swashes collide with incoming waves, resulting in dissipation of energy. The term "spur and groove system" has been applied to structures of this shape on reef crests over a wide range of scales, from large buttresses in the Pacific (Odum and Odum 1955), similar sized structures in the Indian Ocean (Pichon 1978, Sheppard 1981) to small Red Sea structures (Friedman 1985). Those of the Gulf of Aqaba and northern Red Sea have been commonly remarked upon (Figure 3.2) with differing interpretations.

Both karst, erosional foundations and modern growth have been cited as causing spur and groove structure. Until about 10 years ago geologists tended to invoke erosional origins, while biologists invoked growth origins, though there were some exceptions. This was partly because of the differences in scale of the various structures examined, and it is now clear that whereas some of the largest features of this form do follow inherited contours, growth of spurs by calcareous organisms is responsible almost entirely for spur and groove structures up to 2 or 3 m wide.

Spurs are usually constructions of the coralline algal genus *Porolithon* in the Indo-Pacific. In very exposed locations, the algae also develop massive algal ridges from which the spurs project seaward. In the Caribbean *Lithophyllum* joins *Porolithon* (Adey 1975) though there are many examples which result from orientated growth of the large coral *Acropora palmata* alone. In all cases, the series

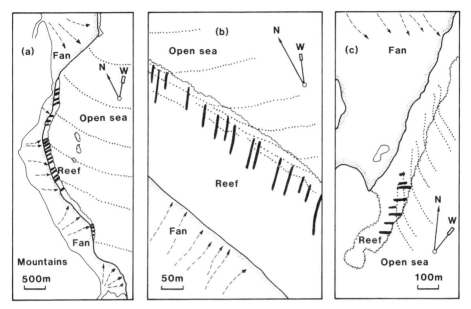

Figure 3.2 *Sketch of spur and groove systems in the Gulf of Aqaba, showing alignment into prevailing waves. From Friedman (1985). (a) El Hibeiq, (b) Dahab (north), and (c) Dahab (south). Spurs are marked by heavy lines, wave fronts by dotted lines and streams on land by dashed arrows. The wind direction (W) is drawn normal to the wave crest and is correct only for the day of the observation.*

obtain their shape from biotic construction rather than erosion in grooves; the latter have a floor usually at or about the elevation of the main base of the reef. Strong water movement and high aeration (Doty 1974, Littler and Doty 1975) are required. Growth is enhanced by sediment forced into cavities by strong wave energy (Ginsburg and Schroeder 1973). The resulting spurs form a strong, dense limestone which may persist for thousands of years; spur and groove structures may persist at depths of 50 m on drowned reef crests (Sheppard 1981), and are known also from elevated fossil reefs.

   It is not known if spur and groove formations of different scales are initiated by the same processes, although because the larger series are found in higher energy conditions it seems likely that the prevalent wavelength is partly responsible (Munk and Sargent 1948). The mechanism regulating the spacing is clearly a correlate of energy too, because the greater the energy the larger are the series. Wave energy transmission at the reef edge is discussed by Dexter (1973) but although this suggests how energy is dissipated, nothing yet explains how the well regulated spacing of spurs is achieved. At all sizes, the structures greatly ameliorate the impact of breaking waves, in a self-maintaining arrangement. Over every scale, whether measured in metres in the Red Sea or tens of metres in oceanic atolls, the tops of spurs are at the approximate low tide level at their landward end. In the Red Sea as elsewhere, spurs dip gradually to seaward, grooves have steep sides, usually with a scoured appearance, the widths of the spurs and grooves in any one series are similar both with each other and throughout their length, and at their seaward ends spurs fade out at a depth where average wave turbulence, but not storm conditions, is diminished.

The fact that they are strong and persistent structures gave weight to the earlier arguments that grooves are inherited from erosional features cut into pre-existing limestone platforms (Hopley 1982). That etching to this degree and greater is possible is clearly shown by wadi systems which are cut into limestone to depths of tens of metres. As pointed out by Hopley, however, Red Sea spur and groove features are aligned at angles of up to 25° from perpendicular to the reef front in response to wave refraction, which by itself supports the hypothesis of spur growth rather than erosion as the dominant process in the Red Sea. It now seems clear that features of this dimension (i.e. 1–2 m wide and about 10 m long) are almost exclusively growth features, and that pre-Holocene erosion explains only the wadi fan formations and associated canyons which are larger by orders of magnitude.

Very large spur and groove structures as seen in the Pacific (Munk and Sargent 1948) and Indian Oceans (Sheppard 1981) or the even larger, possibly erosional features seen in Madagascar (Pichon 1978) do not occur in the Arabian region. The mean wave energy required to cause such structures to develop is not present in the Red Sea or Gulf, which are the only two seas with the necessary reef development. Severe storms occur, but spur and groove formation clearly needs sustained high energy conditions, which is missing in Arabian areas with limestone biogenesis. The Arabian Sea has a hydrological regime commensurate with their formation, but it is not a limestone biogenic area; it is almost all brown-algal dominated, except in sheltered bays.

Series exist in the Gulf of Aqaba (Friedman 1968, 1985, Sneh and Friedman 1980) and on the African side of the Red Sea between the southern Gulf of Suez and Quseir, as well as on the barrier reef which runs down the northern Red Sea. In the Gulf of Aqaba, remarks that the "Spur and grooves dominate the fore reef" (Friedman 1985) show that such structures are important in limited areas, in this case the northern Gulf of Aqaba, but this is far from universal. The Gulf of Aqaba can present some long fetches in northerly winds which are sufficient for spur formation, and these areas add considerably to the range of exposure and habitat experienced in the region.

The true coralline algal ridge elevated about 0.3 to 0.5 m above the level of the reef flat is not seen anywhere in the Arabian region. These structures require sustained, higher wave energy than do spurs (Sheppard 1981). In the Red Sea, exposed patch and barrier reefs have crests covered in crustose, coralline algae, with cover values of 50 to 90%. However, the crests are never raised in true algal ridge form, and it appears that mean or long term energy conditions are insufficient to allow, or force, their development. In addition, the crests are well populated with corals which form typically high energy communities (later section). Significantly, reef crests of the open Indian Ocean which experience high enough energy to develop algal ridges have a minimum, or no, coral in the uppermost 1–3 m. It must be concluded that even the highest energy parts of the Arabian seas which support corals experience lower energy than is common in many oceanic areas (Sheppard and Sheppard 1985).

(3) Southern algal reef constructions

In the southern part of the Red Sea an entirely different reef type is encountered in moderately and very sheltered conditions, which appears to be formed from the same, or very similar, crustose coralline algae that forms spurs in the north. These are termed "algal reefs" (Sheppard 1985a). They appear to be analogues of the crests of true coral reefs, but they exist in the absence of coral reefs. All examples seen are found in low energy conditions in the southern part of the Saudi Arabian Red Sea, where coarse sand extends several kilometres to seaward. They arise from sandy substrata nearshore where water is 2 to 4 m deep, and develop steep sides which reach to the low tide level. They support very few and sparse corals, and are conspicuous from shore because they are covered in dense *Sargassum* whose fronds float in thick mats at the water surface. Notably, the very few corals present include *Siderastrea savignyana* which is extremely tolerant of sedimentation, high temperatures and high salinities (later section).

These reefs provide an important part of the total sublittoral productivity on the mainland shore of the Arabian side at least. Coral reefs are not present in the region (which includes over 100 km of coast), even where fossil coral reefs exist which provide platforms at and near the low tide level. The region appears to be unfavourable for coral reef growth today, although the fossil reefs show that typical, high diversity fringing reefs have occurred before. The algal reefs are more than just a curiosity, however. Outside the Caribbean, living examples of algal reefs are rare, and in this region which includes several hundred square kilometres of inshore, shallow water, they provide most of the limited amount of hard substrate. Their living communities are discussed in the context of responses to increasingly severe environment in Chapter 12.

(4) Ridge reefs and atolls

Guilcher (1988) described a type of reef structure called a ridge reef for the Red Sea, which occurs in both fossil and Recent form. He called it "a heretofore neglected type of coral reef" which should join the classical Darwinian forms such as fringing, atoll and barrier reefs. These structures are described by Guilcher (1988) from the Sinai, Farasans, Jeddah and near Port Sudan, but in fact reef structures with the same appearance are abundant along much of the Saudi Arabian Red Sea coast, although they are not usually referred to by this name. The reefs so defined are longitudinal ridges lying along the axis of the Red Sea, and probably result from a combination of normal faulting from the progressive opening of the Red Sea, and of underlying salt deposits (diapirs) moving upwards along these faults. Those on the Sinai appear to have only a veneer of coral capping the ridges, while others such as off Port Sudan may have thicknesses of tens of metres of coral reef. In cases where the coral cap is presumed to be thick, a later subsidence of the substrate is invoked, and the latter has to occur after the initial formation of the ridge.

It cannot be doubted that a tectonically active coral sea which experiences tectonic uplift, faulting and upward movement of salt ridges will provide complex combinations of substrate type. However, whether a fundamentally new kind of reef type needs to be described for this is more questionable. Indeed, Guilcher (1988) remarks that the only atoll described from the Red Sea, Sanganeb off Port Sudan, rests on a ridge reef. The question must be asked then, can a reef be both a ridge reef and an atoll? There is no reason why it cannot, if the former is described as having a foundation formed from rifting and salt deposit uplift and the latter is defined as a ring of reef resulting from progressive subsidence over a long period. Both events may occur sequentially. Furthermore, almost all the ridge reefs of the Red Sea also support patch reefs, and those off the central Saudi Arabian shore fit the definition of barrier reefs in the classical sense too (Sheppard and Sheppard 1985). Some also have coral islands or at least sand banks as well, and these are commonly rimmed by fringing reefs. There may not be a great advantage to defining a new kind of reef as unique to the Red Sea when there is so much overlap. The new interpretation of the foundation of the "ridge reef", however, is valuable since it illustrates one of the major kinds of reef foundation in the Red Sea, especially in the central and southern parts.

Sanganeb atoll (Figure 3.3) has been described as the only atoll in the Arabian region. However, naming this reef an atoll is largely a question of semantics because numerous similar structures occur off the central Saudi Arabian coast too, but these oval shaped reef structures, enclosing lagoons with depths of tens of metres, are not usually described as atolls. These occur along the "Little Barrier

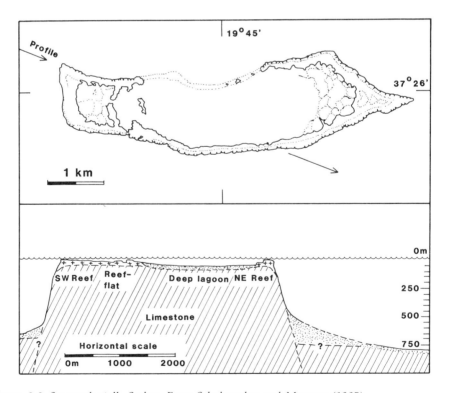

Figure 3.3 *Sanganeb atoll, Sudan. From Schuhmacher and Mergner (1985).*

Reef". If the definition of an atoll requires growth on top of progressive subsidence (as classically it does), then there is little evidence that any of these structures are true atolls; they could equally well be formed from upward growth of reef from the pre-Holocene reef substrates described earlier and, as pointed out by Guilcher (1988), confirmatory drilling into enough representative sites has yet to be done. Figure 3.4 shows two cross-section profiles obtained by echo-sounder across the central Red Sea barrier reef. The section at Yanbu was particularly broad and shows how the outer and inner edges of the platform reach the surface, in a manner identical to Sanganeb atoll. For these, the interpretation was made that the structures grew upwards from a Pleistocene platform (Sheppard and Sheppard 1985).

What is clear is that reef substrates in the Red Sea have moved vertically upwards by both tectonic activity and salt dome movement, and vertically downwards by faulting, by tens or even hundreds of metres. In addition to this, the pre-Holocene low sea levels allowed substantial reef platforms to develop at levels now 30–60 m deep, and these factors together have resulted in a greater complexity of present reef foundation than in many coral seas.

(5) Non-reef coral populations

Coral communities are not limited to growing on coral reefs in the Arabian region. A distinction first drawn by Wainwright (1965) between coral reefs and coral communities on non-reefal substrates has been taken up by Pichon (1971), Heydorn (1972), Hopley (1982) and Sheppard and Salm (1988). These communities range from closely resembling rich coral communities on true reefs, to being scattered colonies coexisting amongst brown or green algae. These areas form a large part of the content of Chapter 12 and are described only briefly here.

Coral communities which survive in stressed conditions can be found on old limestone, algal-derived limestone or even on non-limestone substrates. Several areas demonstrate this. The southern algal reefs, for example, are made almost

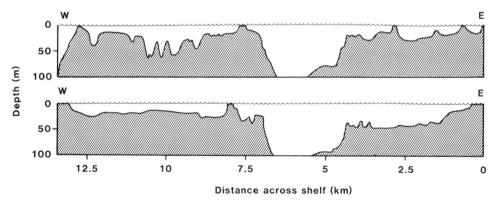

Figure 3.4 *Echo-sounder traces across barrier reef at Yanbu, central Red Sea, showing typical atoll cross-sections. Shore is at east.*

entirely from calcareous red algae and not coral reefs in any sense, though they support scattered corals. In stressed locations of the Gulf near Bahrain, such as the Gulf of Salwah, corals are scattered amongst the brown and green algae, and the latter dominate the pre-Holocene limestone (Sheppard 1985d). Examples are known from the Saudi Arabian Gulf coast also (McCain *et al.* 1984). The best documented examples occur in Oman, where Sheppard and Salm (1988) show a gradient in a southerly direction, from reefs and flourishing communities on non-limestone substrata, to corals widely scattered beneath algal canopies, a condition discussed in the Gulf by Coles (1988) also.

For Oman, three categories of coral population were identified. First, clear coral reef growth occurred where original bedrock is overlaid or obscured by a characteristic reef topography: a horizontal reef flat at low tide level and a steeper reef slope.

This grades into the second category, where coral framework is present at 25–75% cover and sometimes more, but there is no typical reef topography. Instead, gross substrate topography and slope is that of an irregular or continuous descent, commonly identical to that of the adjoining coast where corals may be almost completely absent. The substrate may be nearly obscured by corals which provide the appearance of a true coral reef, and other benthic and fish fauna are also those of a coral reef, but the corals are attached to a wide range of older rock. Both *Porites lutea*, *P. solida* and *Acropora valenciennesi*, *A. valida* communities occur commonly, though *Acropora clathrata*, *Montipora* spp. and *Pocillopora damicornis* may also form substantial and in the last case monospecific stands as well, predominantly in water less than about 5 m deep. Rich coral communities such as these are common from Musandam to the Capital Area of Oman, with others recorded in southern Oman. In deeper water, several areas of Musandam and one location seen in the Daymaniyat Islands, Gulf of Oman, support high diversity faviid and mussid coral communities identical in appearance to the deeper communities of the Red Sea.

In the third category, corals provide <15% cover, and usually considerably less. This is too low to be considered as framework, especially when algal cover is ten times greater than coral cover. Corals remain diverse, but they do not include more than rare examples of the framework genera, *Pocillopora*, *Porites*, or *Acropora*. The change in community dominants is fairly abrupt, and results from a sudden demise of high cover, framework species of corals, leaving hardier species which are ubiquitous but rarely abundant either in these habitats or on reefs. This final category appears to continue, losing species gradually, into remarkably severe environmental conditions.

B. Ecological patterns of the coral fauna

The range of environment, latitude and geological formations combine to produce very varied coral habitats within the region. This results in several different coral communities which are distributed according to geographic location and depth. Recent analyses of several of the communities are summarized, together with a synthesis of data by, in particular, Burchard (1979), Basson *et al.* (1977), McCain *et al.* (1984), Sheppard (1985a,c,d, 1988), Sheppard and Sheppard (1985, 1991), Scheer (pers. comm.) and Wranik (pers. comm.) for the Gulf of Aden.

(1) Coral distribution and diversity

Coral communities are generally defined as a product of the species abundance, diversity, and identity of the dominant forms. The latter usually lends a name to the community as well, though while the species is usually the most conspicuous, it may not be necessarily the most abundant. The term "assemblage" may be more appropriate and is used by some, though the two terms have been used for the same thing in the Red Sea.

Coral diversity throughout the region follows a clearly established pattern: the Gulf has few species, and the Red Sea is particularly rich, even for the Indian Ocean region. Using a division of six large, broad and convenient areas, diversity is shown in Figure 3.5. The dendrogram shows a single, principal division into two groups: the three Red Sea areas fuse together first into a closely related cluster, while the Gulf of Oman and Arabian Sea also fuse together closely. The Gulf then fused with the latter, this cluster being slightly separated from the Red Sea group. Similarity values are high throughout, illustrating the very large overlap in species composition.

I have already had occasion to speak, in the course of my travels, of the astonishing mats of works formed by marine insects; namely the immense banks of coral bordering, and almost filling up, the Arabic Gulf . . . The reader may therefore conceive with himself what a variety of madrepores and millepores are to be met with in these seas.

Niebuhr 1792

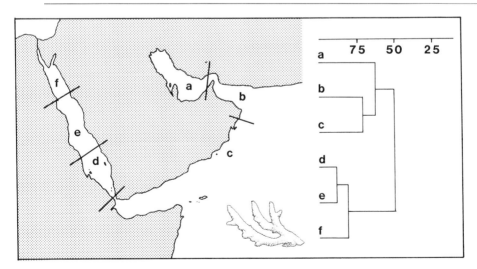

Figure 3.5 *Cluster analysis of corals of Arabian region divided into six major areas. The dendrogram shown is based on a "commonality index" (Sheppard 1987) which takes into account the inequality of diversity between the areas. From Sheppard and Sheppard (1991).*

(2) Zonation in the Red Sea

A) NORTH–SOUTH GRADIENTS

Ormond *et al.* (1984a,b,c) were the first to examine coral reefs from the full length of the Red Sea. Using a mixture of dominant coral species, fish communities and overall appearance or "quality", they described four main zones which were distributed north–south within the Red Sea. While there had been several suggestions that reefs diminished considerably in the south and somehow changed in character, Ormond *et al.* were the first to demonstrate that the Red Sea coral reefs were not homogenous throughout. Their four zones were 1: Gulf of Aqaba; 2: north Red Sea southward to near Jeddah; 3: the mainland coast around Jeddah and some offshore islands in the southern Red Sea; 4: the southern mainland coast and southern Farasan Islands. Not included was the African side of the Red Sea, though the few localities known suggest that the same pattern would be expected there also. Only the Gulf of Suez would be expected to differ from the Gulf of Aqaba.

Later work (Sheppard 1988, Sheppard and Sheppard 1991) analysed coral data from nearly 200 sites from both northern Gulfs and the Arabian mainland coast to 16.5° N, the Yemen border. This showed the existence of 13 principal coral communities. A summary of the groupings is shown in Table 3.1. Several communities could themselves be subdivided further into easily recognized units. The latter in most cases simply represent different depth divisions on one reef type, so the 13 group level was presented as being the one required for most purposes while its extension to the 22 group level allows comparison and compatibility with results of other workers, notably Loya (1972) and Braithwaite (1982).

Figure 3.6 maps the distribution of these communities. This shows a clear north–south trend with most communities showing considerable localization. Eight of the 13 coral communities are limited to either the northern, central, or southern areas of the Red Sea. Only four communities span two areas, and only one (no. 6) is spread throughout. While the distribution appears to be based largely on latitude, gross changes in coastal bathymetry which correlate with latitude are undoubtedly of major importance.

Cluster analysis is an interpretive tool, not generally regarded as a proof, even though statistical limits are occasionally applied to its results. Scale is important, and taking any section of the total picture, further patterns may be extracted. Indeed, one grouping arising from group d of Fig. 3.6 was earlier subjected to the same analysis; this had used 50 southern sites spread over 550 km (Sheppard 1985a), and it showed a finer separation of six types of coral community (Figure 3.7). Another analysis of the central sites (mostly group c in Figure 3.6), again predating the total analysis (Sheppard and Sheppard 1985), showed a division corresponding to four of the main communities, subdivided further into 13 clusters, in this case separated by exposure and depth. It would seem to be possible to define the coral communities of the Red Sea principally by latitude, and then within a smaller area, by exposure and depth. In the Red Sea, exposure relates to site location on fringing reefs, offshore patch reefs and, most exposed, the outer slopes of the barrier reef of the central region.

Table 3.1 Key features of the communities derived from the cluster analysis of corals. Under "Reef type", shallow = crest − 5 m, mid = 5–20 m, deep = 20–40 m. A, B, C, refer to sub-groupings in the relevant clusters. (From Sheppard 1988)

No.	Location	Reef type, exposure and habitat summary	No. of sites	Dominant species (% cover)	Sub-cluster characters
1	N (Suez)	Shallow patch reef. Exposed, but lying in shallow, sandy plain	1	*A. horrida* 50% *A. formosa* 10%	
2	N–C[a]	Shallow, exposed fore-reef slopes	17	*A. hyacinthus* 4% *A. humilis* 3%	
3	N–C	Shallow–mid depths, moderate exposure	18	None	
4	N–C	Sheltered fringing reefs or backs of patch reefs	8	*Porites lutea* 47%	
5	N–C	Shallow–mid in north, mid–deep in central area. All moderate exposure	31	None	A *Millepora* abundant B None abundant C *Goniopora* abundant
6	All	Moderately exposed mid-depths (5–25 m) and moderately turbid sites	37	None, all have generally fairly low cover	A Deep slopes, all areas B Turbid reef in south area C Turbid holes
7	C	Mid depths, patch reefs beside and amongst sand	3	*P. damicornis* 12% *A. eurystoma* 5% *A. clathrata* common	
8	C	Patch and barrier reefs, mid-depths	12	*P. verrucosa* 5% *A. hemprichii* 3%	

9	C	Barrier and very exposed fringing reefs. Shallow areas only. Sub-group A is most exposed, B intermediate, C least exposed within this group	20	*Acropora* zones	A *A. hyacinthus*
					   A. digitifera
					   A. humilis
					B *A. danai*
					   A. hyacinthus
					   A. hemprichii
					C *P. verrucosa*
					   A. hemprichii

10	C–S	Barrier, patch and fringing reefs, mid-depths	14	*Porites* together with:	A *M. circumvalata* B no other C *G. pectinata*
11	S	Shallow reefs near mangroves	3	*Porites* sp. 38%	
12	S[a]	Fringing reefs meeting sand at <5 m. Clear water	22	Stagshorn and *Porites*	A Stagshorn 0–3 m B *Porites* 2–5 m
13	S	Red algal patch reefs	3	None	

[a] Cluster 2 contained one exposed site from the southern area also. Cluster 12 has one site from the central area. Three very low diversity, low cover sites were discarded.

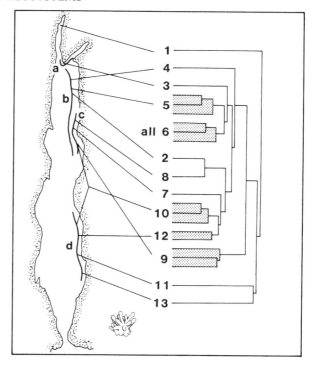

Figure 3.6 *The distribution within the Red Sea of 13 clusters of coral community type. a–d denote major geographical groupings in which each cluster is found. Cluster 6 is found in all four geographical groupings. From Sheppard and Sheppard (1991).*

Several important areas still not surveyed in adequate detail are the Farasan and Dahlak Archipelagoes. Some of the coral communities of the Farasans were photographed by Ormond *et al.* (1986a). These show large areas dominated by *Acropora* species which could be good, if extreme representations of the communities observed so far, but this is uncertain.

B) ZONATION AND PATTERNS ON REEFS

In common with reefs almost everywhere, coral communities on most Red Sea reefs change with depth, exposure and sedimentation. The very few exceptions are the shallow reefs of the far southern region which have very limited depth spans. Zonation in the Red Sea has been particularly well studied, and a very large body of work, quite disproportionate to the short length of available coastline, has been carried out by groups working in Israel and the Sinai coast. Indeed, from the early 1970s, much of the work which brought coral reef study from its early descriptive phase to a modern analytical phase was carried out here. In particular, long series of papers from Fishelson, Loya, Benayahu and Rinkevich (e.g. Fishelson 1971, 1973a, Loya 1972, 1976a,b, Benayahu and Loya 1977a, 1981, Loya and Rinkevich 1980, to name possibly some of the most influential) became much cited in the burgeoning reef literature. This work, and

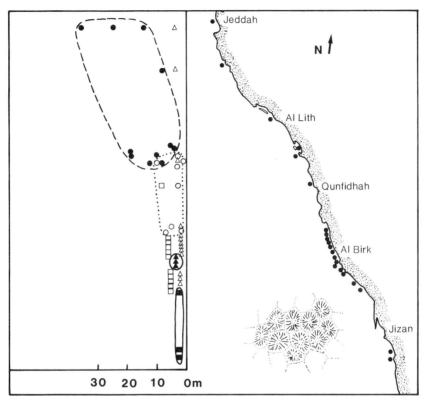

Figure 3.7 *A finer distribution of coral communities within group d of Figure 3.6. Different symbols refer to membership of six different clusters. Four spatially distinctive clusters are outlined. Horizontal axis denotes location of each cluster with depth (m) below sea level. From Sheppard (1985).*

others from the same group, is important material to any present studies of the reefs of the Arabian region.

Many of these studies described the distribution of corals and soft corals, in particular, on reefs and across physical gradients. The work is now well known, has been frequently reviewed (e.g. Sheppard 1982a, Head 1987a) and it is not necessary to elaborate on it again here. Perhaps of particular importance was the development, sometimes confirmation, of ideas on effects of exposure, sedimentation and depth on the coral community, and control by these physical factors on abundance and distribution of many species. Much of it also involved some of the earliest studies of pollution effects and coral physiology, and it related these to ecological distributions. Patterns of corals, or zones, became increasingly apparent during this important phase of reef research, and parallels in the patterns recorded in the Sinai were also seen across the Indo-Pacific in most places where similar work was performed.

However, in the context of the Arabian region as a whole, or even of just the Red Sea, it later became apparent that the Gulf of Aqaba is relatively sheltered. Coral zones recorded for example in Loya (1972) show a pattern which is typical of fringing reefs in rather sheltered situations. Ormond *et al.* (1984c) even suggested, as already noted, that the Gulf of Aqaba was one fairly small, discrete, sub-region within the Red Sea. This became even more apparent when sites further afield were examined, such as in the Sudan (Mergner and Schuhmacher 1981, Schuhmacher and Mergner 1985) and Saudi Arabia (Ormond *et al.* 1984a,b, Sheppard and Sheppard 1985, Sheppard 1985a). These showed a considerably greater range of coral communities, both those which appear to have a geographically distributed pattern in the Red Sea as already noted, and those which are found in very exposed conditions. The latter include the shallow, outer slopes of the offshore patch reefs, the "atoll" at Sanganeb, and the barrier reef which extends for over 200 km off the coast of central Saudi Arabia.

Nowhere can one contemplate the life of the corals, and what belongs to it, more quietly and comfortably than here, although he has to lie on his belly — a trifling matter for the naturalist — and hold his magnifying-glass at the point of his nose above a coral bush.   Klunzinger 1878

The general pattern of coral diversity with depth on Red Sea reefs follows that of most Indo-Pacific reefs. On reef flats diversity is low, and increases sharply near the reef crest. In shallow water on the reef slope diversity generally increases to a maximum at depths variously from 5 to 30 m, below which it goes into a gradual, continuous decline as depth increases and light diminishes (e.g. Loya 1972, Sheppard and Sheppard 1985). This pattern applies, in broad principle, to sheltered and exposed reefs. Coral cover does not follow the same pattern exactly. Over most reef slopes, scleractinian coral cover is usually less than 50%, but depending on degree of exposure, much greater cover is provided by a relatively small sub-set of the total species complement. In sheltered areas, *Porites* may cover over 80% of the substrate, especially on the sheltered slopes of offshore patch reefs. Several species of *Acropora* likewise occasionally show high cover, and there are many instances of particular, turbid areas where *Goniopora* or even *Galaxea* provide very high cover. Even in such areas, the general pattern of diversity remains the same, and is largely unaffected by total coral cover.

A detailed account of coral diversity or zones with depth is given in Loya (1972) for the Sinai, where diversity peaked at a perhaps unusually great depth of nearly 30 m. Also in the Sinai, the distribution of species with depth was quantified (Sheppard and Sheppard 1991). About one-third of species had a depth distribution which was significantly skewed to deep or shallow water, but the interesting result was not that some species are so restricted, which fits the readily observed pattern, but that those which are not form the majority. However, while each of these bathymetrically widespread species may be present at different depths, they are not equally abundant at each depth. Species with a strong skew to shallow water include the species which show high dominance there too, and hence defined the zones which are used by many authors for Red Sea reefs.

The peak in coral diversity at 30 m in the Gulf of Aqaba (Loya 1972) is deeper than elsewhere in the Arabian region, and deeper than most regions of the Indo-

Pacific, where peak diversity is commonly 20 m or less (Sheppard 1982a). This reflects the high water clarity, which has led to several advances in physiological and photosynthetic studies from this area, often utilizing *Stylophora pistillata* which is especially abundant (e.g. Falkowski and Dubinsky 1981, Rinkevich and Loya 1984a,b, Gattuso 1985). This clear water also extends the vertical range of corals to possibly the greatest so far recorded; nine zooxanthellate species have been recorded to 100 m, while the leafy *Leptoseris fragilis* extends to 145 m (Fricke and Schuhmacher 1983, Schlichter *et al.* 1986). While this species is not rare in shallow water in reef recesses, between 110 and 120 m in the Gulf of Aqaba it has colony densities of 13 m^{-2}. This growth is explained by the discovery of a chromatophore system which shifts the wavelength of the very low level of ambient light (0.5–1.7% of surface illumination) to a longer, more photosynthetically useful wavelength (Schlichter *et al.* 1985).

(3) Gulf coral communities

Within the western Gulf, several coral communities are clearly defined, though less work has been done than in the Red Sea. Kuwait reefs are species poor (Downing 1985), and at present it appears that reefs of Qatar and the United Arab Emirates are only a little richer (Shinn 1976), though substantive surveys have not yet been done. However, although these reefs are poor in number of species, coral cover may be substantial. In Kuwait, the most important reef builders are *Porites lutea*, *Acropora eurystoma* and *A. valida* (Downing 1985), which appear to dominate in a way analogous to some Red Sea communities. In Qatar, stagshorn *Acropora* heavily dominate shallow but extensive reefs (Shinn 1976).

In Bahrain, one quantitative study may be representative of many of the Gulf reefs near land (Sheppard 1985d, 1988). Total diversity is poor, with only 28 species of corals recorded. Presence and abundance of corals were examined at 32 sites around the islands and offshore reefs of Bahrain and between Bahrain and Qatar further south in the Gulf, using the same method as was used for the Red Sea sites. Recorded salinity was never lower than 42 ppt, and was often 43–45 ppt, compared with <42 ppt in Kuwait and Saudi Arabia. Despite this, coral cover was high, and did not depend on, or even correlate with, coral diversity. Five coral communities were clearly distinguished, whose dominant species are comparable in morphological type and cover with those from some of the Red Sea zones (Table 3.2). Stagshorn *Acropora* dominate in the least saline and least turbid conditions found in northern parts of the Bahrain archipelago. These two communities correspond to the *Acropora* zones of the Saudi Arabian islands. However, *Porites lutea* communities which are seen in the shallowest parts of reef slopes in Kuwait and Saudi Arabia do not occur in Bahrain, possibly due to the greater salinity in all Bahrain sites. Instead, a deeper community dominated by *Porites compressa* occurred below the stagshorn *Acropora* community in the lower salinity sites of Bahrain, while *Porites nodifera* dominated where salinities reached 43–45 ppt. In higher salinities still, diversity and coral cover fell drastically, and then disappeared. The coral communities of Bahrain are stressed; an aspect discussed in detail in Chapter 12.

Table 3.2 The five main coral communities around Bahrain and in the Gulf of Salwah. "Rich" = >15 spp. per site, "Poor" = <5 spp. per site. Numbers are % cover of total substrate; salinity is in parts per thousand. For algae, + = all but scattered plants, ++ = tall plants over $1 m^{-2}$ and often forming a good canopy. NB. Numbering of clusters is separate from cluster numbers in Red Sea.

	Cluster number				
	1	2	3	4	5
Diversity	Mid	Rich	Mid	Poor	Poor
Total coral cover (%)	30–90	35–70	1–10	12–70	2–5
Stagshorn *Acropora*	20–75	5–15			
Porites compressa		8–20			
Porites nodifera				10–65	
Brown algae			+		++
Salinity	<43	<42	43	43–45	>45
Turbidity			++		++

Iranian reefs are reported to have high coral cover (Harrington 1976, Marini 1985) though only few species are recorded from the region so far. The richest, studied islands are two Saudi Arabian coral cays, Jana and Karan (Basson *et al*. 1977). Massive *Porites* dominates the upper reef slopes, from about 2 m to about 4–5 m deep, providing very high coral cover. Below this is a zone of table or fan shaped *Acropora* species, extending to about 10 m deep. Cover at this depth is high, and diversity is greatest. The community includes a considerable number of faviid corals, as is the case in several Red Sea communities. Below this, Basson *et al*. (1977) define several more small zones: a community of encrusting and foliose *Montipora*, then an intermittent zone of bushy *Acropora* or *Pocillopora*, and finally a zone of encrusting and massive faviids with conspicuous *Turbinaria* which, at about 18 m, forms the deepest coral community known for the Gulf. Although these communities have not been analysed quantitatively, they are well illustrated in Basson *et al*. (1977).

(4) Arabian Sea coral distribution

Glynn (1983), Green (1983, 1984), Green and Keech (1986) and Sheppard and Salm (1988) have described the coral communities of Oman in varying degrees of detail and illustration. This area is especially interesting, not because of the diversity or cover of corals, which is not particularly unusual for the Arabian region, but because many coral species here are living close to their environmental limits. With increasing distance southward along the Arabian Sea there is an increasingly temperate marine climate which leads to the eventual demise of reef construction, and then of the coral communities.

The nature of the coral communities is broadly similar to those of other parts of the Arabian region. Quantitative analysis in the manner described for other

Arabian regions has not been performed, though most of the above authors recorded observations quantitatively, allowing interpretation and some comparison to be made. The Musandam peninsula contains some reefs made from, and heavily dominated by, *Porites*, with various *Acropora* abundant too. The latter generally contain mixed coral communities on Permian to Cretaceous limestones. Their diversity is high, and the mixed communities contain both abundant *Acropora* and faviid zones, and sometimes dense *Pocillopora damicornis*. In this unusual setting (for the Arabian region) of vertical underwater cliffs in extremely sheltered fjords, the mixed communities do not appear to match any community defined elsewhere.

In the Capital Area, *Porites lutea* again forms substantial fringing reefs, while *Pocillopora damicornis* forms monospecific reefs reaching several hundred metres across. These were never seen in the Gulf or Red Sea even though this species is common in both areas. Other mixed and *Acropora*-dominated communities occur too. Further south, true reefs gradually disappear in favour of rock dominated by macroalgae, though Salm (pers. comm.) has subsequently found reefs dominated by *Montipora* on the mainland near Masirah Island, and Glynn (1983) recorded *Acropora* patch reefs near Masirah itself.

In Dhofhar, *Acropora* forms a few accreting reefs reminiscent of some *Acropora* zones of the Red Sea and zone 1 of Bahrain (above), and *Porites lutea* forms some enormous solitary growths, though these cannot really be called small reefs. For the most part, however, coral dominated communities are sparsely scattered in Dhofhar.

Wranik (pers. comm.) collected corals in the Gulf of Aden which increased the known diversity of this area considerably, though it still appears to be much poorer than the Red Sea. As noted earlier, it is predicted that high cover communities, whether forming true reefs or not, occur only very rarely in this area, and that substantial coral communities will not occur until near the entrance to the Red Sea.

Summary

Diverse and spectacular coral reefs for which the Red Sea is renowned occur only in its northern and central half. Greatest development occurs in offshore barrier reefs and in reefs fringing 1–7 km wide alluvial plains on the mainlands. Much thinner reefs veneer the Gulf of Aqaba and other northern shores. Further south, the continental shelf widens and mainland shores are dominated by mangroves and sand beaches. Better developed reefs remain around the Farasan and Dahlak islands, which also support extensive mangroves. Fossil reefs are abundant, and these have a higher generic diversity than living reefs, suggesting more favourable conditions in the past.

The Gulf of Aden has very poor reefs due to upwelling water and sandy shorelines, a condition which continues down the coast of Somalia for 500 km and north to the Capital Area of Oman. Further north, Musandam has the most diverse reefs, while Iran probably has the most developed. The UAE coast is low-lying, mostly swampy and rich in seagrasses. Offshore the water is very shallow, but while it is generally muddy and unsuitable for most corals, there are numerous patch reefs.

Rudimentary spurs and grooves develop on the most exposed reefs in the northern Red Sea. These red algal constructions are aligned to wave refraction. Large series as found in the Pacific and Indian Oceans do not occur; mean wave-energy is too low for their development in those parts which have reefs (Red Sea or Gulf), and where wave energy is highest, reefs are absent (Arabian Sea). Coralline algal ridges elevated 0.3–0.5 m above the reef flat are not seen anywhere in the region although exposed patch and barrier reef crests have a coralline algae cover of up to 90%.

Algal reefs occur in the southern Red Sea in low energy conditions. They are analogues of coral reef crests, but exist in isolation. They support dense brown algal cover and provide important hard substrate in otherwise sandy areas.

Longitudinal series of coral reefs lie along the axis of the Red Sea on ridges resulting from normal faulting and upward movement of underlying salt deposits. These are widespread in the Red Sea. Structures sometimes called atolls are also numerous and mostly are found on the ridges. They are probably not true atolls in classical sense. It is likely that they grew upwards from pre-Holocene reef substrata, which have moved upwards by both tectonic uplift and salt dome movement, and downwards by faulting, by tens and even hundreds of metres.

There is a fairly distinct Arabian coral species grouping. Within it, there is a single, principal division into a Red Sea group, and a Gulf of Oman/Arabian Sea group, which then fuses with the Gulf. In the Red Sea there are 13 principal coral communities, some of which can be subdivided further into a total of 22 recognizable units. Most show considerable localization, correlated with latitude but linked with gross changes in coastal bathymetry and morphology. On any one reef in the Red Sea, the general pattern of coral diversity with depth follows that of most Indo-Pacific reefs, rising to a maximum at 5–20 m deep before declining. About two-thirds of species have a depth distribution which is not significantly skewed to deep or shallow water. Coral cover is usually less than 50%, but in sheltered areas one or two species, especially *Porites*, cover 80% of the substrate.

In the Gulf, fewer coral communities exist; only five are recorded from Bahrain. Kuwait, Qatar, Bahrain and the UAE have 30 species or less. Despite this, coral cover is high. The richest reefs known are Saudi Arabian coral cays. At the entrance to the Gulf, Musandam contains reefs dominated by *Porites*, and *Acropora*. In the Capital Area of Oman, substantial monospecific reefs of *Pocillopora damicornis* occur. Coral dominated communities become rare further south.

Many coral communities occur on non-limestone substrata. These may closely resemble true reef communities, but grade to being scattered colonies under canopies of brown algae in stressed locations. Change in community dominants and cover occurs fairly abruptly in stages as framework species disappear leaving only a few hardy species.

CHAPTER 4

Coral Reef Fish Assemblages

Contents

A. **Regional patterns in assemblage structure**   88
   (1) North–South differences within the Red Sea   88
   (2) Gulf of Aqaba   93
   (3) Gulf of Suez   95
   (4) Djibouti and the Gulf of Aden   96
   (5) Gulf of Oman   96
   (6) The Gulf   97
B. **Vertical distribution of fishes on reefs**   98
   (1) Diversity and abundance: the patterns   98
   (2) Processes   101
C. **The role of fishes in reef processes**   104
   (1) Productivity   104
   (2) Nutrient transfer   106
Summary   106

Fishes constitute a dominant component of the reef fauna, and those inhabiting reefs comprise the most diverse and abundant assemblages of vertebrates found anywhere on the planet, and certainly within the marine realm. Although reefs of the Arabian region support diverse assemblages, they have fewer species than those in much of the Indian Ocean or Indo-Australasian region. About 1000 species of fish are known from the Red Sea (M. Dor 1984) and perhaps 200 from the Gulf (Downing 1985, Kuronuma and Abe 1972, Smith *et al.* 1987), compared with over 3000 in Indonesia and the Philippines, the acknowledged global centre of diversity for fishes (Sale 1980).

Other sections of this book show that many areas of hard benthic substrata occur in the Arabian region which are not coral reefs. However, the amount known of such non-reefal hard substrata is much less than that for reefs, and of necessity this chapter concentrates on reef fish assemblages. As previously noted, the Red Sea contains by far the majority of the region's reefs in terms of area. Consequently, it also supports the most important reef fish assemblages. This chapter focuses on the Red Sea, both for this reason and because its fishes have received substantially greater scientific attention than those of the Gulf or other parts of the Arabian region.

(For the Arabian fisherman) there are below the waters charming genii who are eager to marry human beings, though to be sure, only when the latter have mortified themselves for months previously with unsalted bread and water, so as to give their flesh and blood a half-ethereal character.
   The naturalist, however, cannot allow himself to be allured . . . Klunzinger 1878

A. Regional patterns in assemblage structure

Reef fish assemblages of the Arabian region are as varied as the reefs themselves. There are marked differences among areas in species richness, assemblage composition and species' abundance. When examined in the contexts of the environments in which they exist, fish assemblages are helpful in illuminating many important ecological processes. The following account explores patterns of regional variation in assemblage structure and seeks to explain them in terms of the environmental settings in which they occur.

(1) North–South differences within the Red Sea

Just as there are marked changes in the structure of coral communities moving from north to south within the Red Sea, so there are equally pronounced changes in the structure of fish assemblages. Detailed studies made along the Saudi Arabian coast (Ormond *et al.* 1984a,b,c, 1986a,b, Roberts 1986, 1991, Roberts *et al.* in press) have identified major differences in faunal composition and relative abundance of species in all families investigated. They include butterflyfishes (Chaetodontidae), angelfishes (Pomacanthidae), damselfishes (Pomacentridae), surgeonfishes (Acanthuridae), parrotfishes (Scaridae), wrasse (Labridae), emperors (Lethrinidae) and snappers (Lutjanidae). Almost certainly similar differences may be expected to exist in other families.

The Red Sea (excluding the Gulfs of Aqaba and Suez) divides into two main regions on the basis of fish assemblage structure: a central and northern, and a southern region. The division between them is abrupt, with a major change in the region of 20°N, over a distance of only about 60 km of coast. This boundary has been defined on the basis of cluster analyses of the species composition of and patterns of abundance within four families (Roberts 1986). The results of these analyses are shown in Figure 4.1. They show that in terms of species composition, or on the basis of the abundances of the species present, these two areas are very different.

The differences are immediately recognizable in the field with major shifts in species dominance between regions. Figure 4.2 shows differences in the abundance of six species of butterflyfish which are typical of patterns for the family overall (Roberts *et al.* unpublished). Similar differences were found in the other families studied. At least 14 of the 34 species of damselfish which occur within the Red Sea show marked changes in abundance between north and south of 20°N. These changes occur in both directions, some species becoming more common moving from one region to another, others less so. For example, the schooling planktivorous damselfish *Chromis dimidiata* is very abundant in the central and northern Red Sea whilst virtually absent from the south. It is replaced by abundant *Neopomacentrus xanthurus*, a species only found in small numbers further north.

Why do assemblages differ in this remarkable way? There are a number of alternative causal factors which could lead to the observed patterns. So far the available evidence cannot allow the acceptance or rejection of any of the

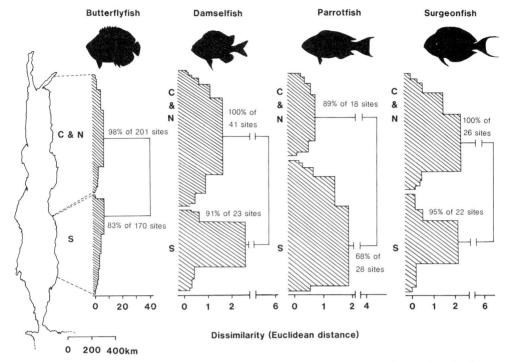

Figure 4.1 *Results of cluster analyses (Ward's method) on species presence/absence data for four families studied at sites on the Red Sea coast of Saudi Arabia. C & N = central and northern region (north of 20° N), and S = southern region (south of 20° N). Percentages show the numbers of sites within each cluster correctly classified on the basis of location north or south of 20° N. Data are from Roberts (1986) and Roberts et al. (in press).*

alternatives. However, some insight as to the likely causes can be gained through examination of the correlative evidence which exists.

A) THE INFLUENCE OF HABITAT

That habitat exerts a major influence on which species are able to live in a particular place, and their patterns of commonness or rarity, is undisputed in ecology. There are striking differences in reef structure and coral assemblages from north to south within the Red Sea. These have been described in detail earlier in Chapter 3. In summary, north of 20° N the continental shelf is narrow, and water depths of several hundred metres are found less than 1 km from shore. By contrast, to the south it broadens, extending between 30 and 130 km offshore. The central and northern region (north of 20° N) is dominated by well-developed fringing, platform and barrier-like reefs which typically drop steeply into clear water. In the south, reefs exist within a shallow environment dominated by large areas of soft substrata. Inshore reefs are less well-developed than to the north, and further south reef structures become shallow, macroalgal dominated frameworks. In the far south there are few areas of hard substrata, and these are

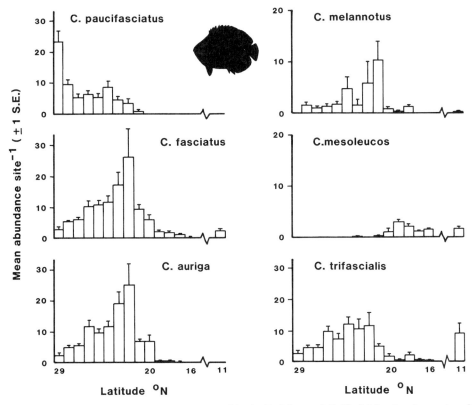

Figure 4.2 *Patterns of abundance distribution within the Red Sea and Gulf of Aden for six species of butterflyfish of the genus* Chaetodon. *Figures show the mean abundance per site for each degree of latitude. Latitudes from 12–15° N were not sampled. Data are from Roberts* et al. *(in press).*

mainly coralline–algal reefs covered with dense growths of the brown algae *Sargassum* and *Turbinaria*. Typical profiles of reefs within the Red Sea are shown in Figure 3.1, indicating some of the differences in structure encountered among regions.

As humming-birds sport around the plants of the tropics, so also small fishes, scarcely an inch in length and never growing larger, but resplendent with gold, silver, purple and azure, sport around the flower-like corals. Ehrenberg 1832

In splendour of colour and diversity of form the fishes of the coral region do not yield to the most brilliant birds. Klunzinger 1878

In the past, many have argued that reef fish assemblages are deterministic, with species adapted and coevolved to live in particular habitats with particular groups of other species (Ehrlich 1975, Anderson *et al.* 1981, Gladfelter and Johnson 1983). Thus the differences in fish assemblages within the Red Sea could be considered to result from differences in the types of habitat present, with reef structure and composition of coral assemblages the causal agents.

This line of reasoning can be illustrated by differences in damselfish (Pomacentridae) assemblages between regions. The distributions of a number of species might be explained on the basis of simple presence or absence of a particular reef habitat. For example, *Chromis pembae* and *C. weberi* are generally restricted to the deeper regions of the fore-reef slope (>15–20 m) in the central and northern region. In the south, reefs generally reach a sandy sea bed at depths of less than 10 m; the zone they inhabit does not exist, so their absence there is not surprising.

However, not all changes in abundance can be so easily explained. The herbivorous damselfish *Plectroglyphidodon lacrymatus* is common on the shallow fore-reef slope of central and northern reefs but rare on southern reefs where suitable habitat appears plentiful. A second form of habitat-based explanation might be invoked to account for the distribution of such species. Differences in habitat are argued to alter the outcome of competition between species. On southern reefs there are large areas dominated by thick algal turfs composed mainly of fleshy red, brown and green algae, in marked contrast to the much thinner turfs of filamentous greens typical of central and northern reefs. Such a difference might favour some species over others. It is perhaps significant that a second species of herbivorous damselfish, *Pomacentrus trilineatus*, becomes very common in the south, but is absent from, or rare, on most reefs further north. A third species, the herbivore *Plectroglyphidodon leucozona*, is also absent from the south but dominates the shallow region of central and northern reefs. This might also be due to a shift in competitive dominance towards *P. trilineatus*. With the available data it is impossible to determine causes.

B) WATER QUALITY EFFECTS

As the Red Sea covers 18° of latitude there are many changes in environmental conditions throughout this region, described in detail in Chapter 2. There are large gradients from north to south in surface temperature (approx. 6–8 °C), salinity (5–7 ppt), nutrient concentration and turbidity (Gordeyeva 1970, Morcos 1970, Morley 1975, Thiel 1980, Edwards 1987, Weikert 1987). Turbidity increases abruptly south of around 20° N, probably largely due to suspension of sediment from the shallow seabed. In the north, sediment can quickly drop out of the photic zone into deep water. 20° N also marks the approximate northernmost point reached by nutrient-rich water flowing in from the Gulf of Aden. Increased turbidity in the south may be due partly to increases in water column productivity.

Latitudinal gradients of water quality may be the cause of north–south changes in fish assemblage structure. Differences in environmental tolerances among species could mean that some are better adapted to conditions prevailing in the south than to those further north or vice-versa. Although the gradual changes in salinity and temperature could lead to abrupt boundaries to species' distributions (Carter and Prince 1981), it has been argued elsewhere (Roberts 1991, and

Roberts *et al.* in press) that it would be very unlikely for the boundaries to so many species' distributions to coincide around 20° N.

The large Aynunah Bay in the far northern Red Sea (28° N) is particularly interesting in the present context. The water within it is shallow and very turbid and its reefs are not unlike those of the southern Red Sea. Assemblages of fishes present were very similar to those of the south, especially in terms of dominant species. Sites sampled here fell into the same clusters on the bases of similarity in species composition and patterns of abundance, as those from the south for all four families of fishes studied in detail (Figure 4.1). This strongly suggests that temperature and salinity differences are not responsible, these being little different in Aynunah Bay from other sites around the same latitude.

Turbidity could be of considerable importance. It may affect assemblage structure either directly, or indirectly through effects on benthic communities. It has been discussed earlier how sedimentation rates may have major influences on coral community composition. These in turn might influence fish assemblages through deterministic habitat effects such as those discussed above.

However, there is little reason to suppose that post-settlement life stages of reef fishes will be the only ones subject to environmental influences. Almost all reef fishes have a pelagic larval dispersal phase. Larvae drift with the plankton for a period of days or weeks before settling onto the reef where they remain for the rest of their lives (Sale 1980). It is entirely plausible that pre-settlement stages may be strongly influenced by the pelagic habitat (Leis 1986, Roberts 1991). Indeed their tiny sizes, small energy reserves and extreme vulnerability to predators make this one of the most decisive stages in the life history (Doherty and Williams 1988). This view is lent strong support by the huge mortalities which occur during the larval stage (Underwood and Fairweather 1989).

How might turbidity exert an influence on fish larvae? A number of fisheries studies have suggested that larval fishes have a critical period early in development, and insufficient food during this period results in poor year-class strength (Lasker 1981, Miller *et al.* 1988). Differential mortality of larvae could result from differences in feeding efficiency, related to turbidity. Since larvae respond to food particles only over short distances (Blaxter 1986) the concentration of food items is critical. Presence of suspended inorganic, particulate matter may reduce feeding efficiency of some species, thereby increasing mortality, particularly if there were a critical period early in development. Vinyard and O'Brian (1976) showed that increased turbidity greatly decreased the distance over which bluegill sunfish (*Lepomis macrochirus*; Centrarchidae) reacted to their prey (zooplankton), thus reducing effective food availability. Differences among species in ability to detect prey under conditions of differing water clarity could thus result in differential survivorship.

Variation in feeding efficiency may arise if some species are better adapted to detect prey over short distances and others over longer distances. The former would feed better than the latter in turbid water and vice-versa in clearer water. Very high levels of mortality during the pelagic dispersal phase mean that only tiny differences in mortality rates between regions could produce large differences in recruit composition. For example, if mortality during dispersal was 99.999%, then an increase of only 0.0005% would reduce numbers of recruits by 50%.

Levels of nutrients and water column productivity cannot be ruled out as possible structuring forces on fish assemblages. As pointed out above, there are substantial differences in both between the two regions distinguished. However, it is difficult to see any simple way in which these might result in the kind of effects observed. Nutrient levels and productivity within Aynunah Bay in the north are unknown and so this site is not at present helpful in determining their importance.

C) OCEANOGRAPHIC ISOLATION OF REGIONS

A further way in which major faunal differences between regions might be sustained is if water bodies were essentially isolated by circulation patterns. Under this hypothesis, larval exchange between regions would be so limited that differences in assemblage composition might develop and be maintained. However, no such barrier to water exchange between the central and northern and the southern regions exists (Patzert 1974, Mustafa et al. 1980). Prevailing currents flow northwards for around seven months of the year and southward for the remainder. There is thus ample opportunity for free exchange of larvae and this hypothesis can be discounted.

(2) Gulf of Aqaba

The Gulf of Aqaba continues the Rift Valley to the north from the Red Sea trough. Consequently, conditions there are rather similar to those prevailing throughout the central and northern region and sites sampled in the Gulf of Aqaba were grouped with them in the classification analyses shown in Figure 4.1. However, there are some distinctive differences in fish fauna between the two areas. The most notable is a small but significant fall in species richness. Figure 4.3 shows detailed clines in numbers of fish species within the northern Red Sea. A number of species present at the entrance to the Gulf of Aqaba appear unable to live within it, for example, the butterflyfish *Chaetodon semilarvatus*. Similarly, there are some species present which have not been recorded in the Red Sea proper, such as the damselfish *Chromis pelloura* and the dottyback *Pseudochromis pesi* (Pseudo-chromidae). Differences such as these led Ormond et al. (1984c) to consider the Gulf of Aqaba a distinctive zoogeographic region within the Red Sea.

Fish assemblages are somewhat distinctive also in that normally deep water species are found closer to the surface than elsewhere in the Red Sea. For example, the flashlight fish *Photobleraphon palpebratus* (Anomalopidae) generally occurs in water deeper than 50–100 m but is found up to the surface in the Gulf of Aqaba and at the southern tip of Sinai. The angelfish *Holacanthus xanthotis* is a deep water species normally found at depths greater than 30 m in the Red Sea. In the northern Gulf of Aqaba it frequently occurs in water less than 5 m deep. The damselfish mentioned above, *Chromis pelloura*, may be restricted to the Gulf of Aqaba where it has only been recorded from depths of 30–50 m off Eilat. However, there remains a possibility that it usually occurs in much deeper water and so has not yet been recorded from elsewhere within its range (Allen and Randall 1980).

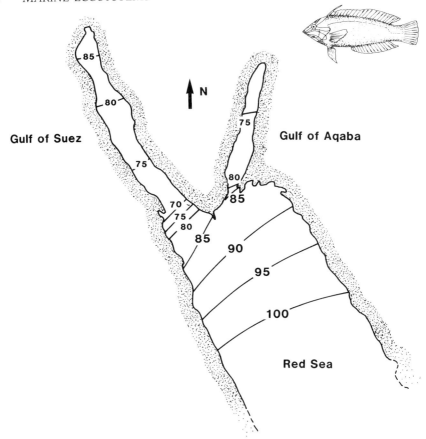

Figure 4.3 *Clines in average number of species per site within the northern Red Sea for the following families combined: butterflyfishes (Chaetodontidae), damselfishes (Pomacentridae), wrasse (Labridae), surgeonfish (Acanthuridae), groupers (Serranidae) and parrotfish (Scaridae). The reduction in species richness in the central Gulf of Suez needs confirmation by further sampling. Previously unpublished data.*

Many shallow water species (typically found between 10 and 50 m deep) occur closer to the surface in the Gulf of Aqaba than further south in the Red Sea. For example, the butterflyfish *Chaetodon paucifasciatus* is usually found deeper than 15 m in the central Red Sea, but occurs right up to the surface in the Gulf of Aqaba. Similarly, the damselfish *Chromis pembae* and the angelfish *Genicanthus caudovittatus* generally occur deeper than 20 m in the central Red Sea but may be found close to the surface around steep drop-offs in the Gulf of Aqaba. Many other species have vertically shifted distributions.

The penetration of deep water fishes into shallower regions, and the shifted distributions described above have been attributed to temperature (Ormond *et al.* 1984a,b). Lower surface temperatures in the north may allow species to penetrate shallower water than areas where they are higher, if their movements are temperature dependent. At present, this hypothesis remains conjectural. Several other explanations could also be offered such as niche expansion in the absence of certain competitor species (Diamond 1984).

(3) *Gulf of Suez*

The difficulty of gaining access to much of the Gulf of Suez coastline due to military restrictions and uncleared minefields has meant that little ichthyological work has been done there. However, reefs of the area have been visited by one of us (CMR), and this account is based largely on unpublished information from three trips there. Despite covering a very similar latitudinal range to the Gulf of Aqaba, and having roughly similar dimensions, the Gulf of Suez contains a very different fish fauna. This is also very distinct from that of the adjacent northern Red Sea. There are differences in species composition and major changes in relative abundance. It was argued above that faunal differences between the central and northern and the southern regions of the Red Sea were due largely to environmental differences between them. Environmental differences are also likely to be responsible for the distinctiveness of the Gulf of Suez.

The very different physical and environmental conditions in the Gulf of Suez have been described in Chapters 1 and 2. Most importantly, it is a shallow basin with an average depth of only around 40 m. Consequently, temperatures fall lower than the deep Red Sea proper, and the water is more turbid due to frequent resuspension of bottom sediment. Such conditions restrict the development of reefs, and those present have a different structure and benthic composition from the rest of the Red Sea, as discussed in Chapter 3.

The fall in species richness of fishes at the entrance to the Gulf of Suez (Figure 4.3) may be an artefact of sampling since it rises again in the far north. Interestingly, limited availability of reef habitat, as well as decreases in degree of reef development and differentiation of reefs into different zones, do not appear to have reduced species richness on far northern reefs. No sites from the area were included in the cluster analyses shown in Figure 4.1. However, in many respects fish assemblages are more like those of the southern Red Sea than those of the north. For example, the groupers *Epinephelus chlorostigma* and *E. summana* (Serranidae) are abundant in the south and in the Gulf of Suez, but very rare in the intervening region. The butterflyfish *Chaetodon larvatus* and the wrasse *Minilabrus striatus* are characteristic of the south whilst very rare in the northern Red Sea and absent from the Gulf of Aqaba. Both are present in the Gulf of Suez, although rare there.

There are, however, many differences from southern Red Sea assemblages. A number of species abundant in the Gulf of Suez are absent from or rare in the south, such as the dottyback *Pseudochromis flavivertex*, and the butterflyfish *Chaetodon paucifasciatus*. Several species appear to be endemic to the Gulf of Suez, such as an undescribed damselfish of the genus *Neopomacentrus* and blenny of the genus *Ecsenius* (Blenniidae) observed in the far north around Ein Sukhna.

(4) Djibouti and the Gulf of Aden

The waters of Djibouti support quite well-developed coral reefs and associated fish fauna. Being located between the Indian Ocean and the Red Sea, it shares faunal affinities with both regions. However, in terms of species composition, the Red Sea influence dominates, especially in areas close to the Bab el Mandeb (Barratt and Medley 1990). There are many species present which are endemic to the Red Sea and Gulf of Aden but do not penetrate to the Indian Ocean. Sites in Djibouti were classified as a sub-cluster of the southern Red Sea for butterflyfishes (Figure 4.1). The generally turbid conditions of Djiboutian reefs are very similar to those prevailing in the southern Red Sea.

On the south coast of Djibouti, close to the border with Somalia, the effects of upwelling nutrient-rich water begin to be discernible in fish assemblages. Water there is very turbid and reefs poorly developed. They support fewer species and lower abundances of reef-associated fishes than reefs further north. However, non-reef species are more productive and this area represents the main artisanal fishing ground in Djibouti (Barratt and Medley 1990).

Across the Gulf of Aden, in Yemen, there are few reefs but little is known of their fauna. Moving east, reefs rapidly disappear under the influence of cold water upwelling and do not occur again until the south coast of Oman.

(5) Gulf of Oman

The reefs of Oman are structurally simple and cover only a small area of coast, mainly around the Capital Area, within the Gulf of Oman and near Masirah Island. Much of the rest of the coastline is dominated by other hard substrata (Barratt *et al.* 1984). The composition of fish assemblages present depends to a great extent on the degree of reef development, with the most diverse assemblages found where reef corals are abundant. Fewer species are present than might be expected for an open Indian Ocean setting, probably partly due to the seasonal upwelling, but mainly to limitation of available habitat. Most families are represented by many fewer species than in adjacent areas, such as the Red Sea. For example, there are only five species of butterflyfish compared to 14 in the Red Sea, and 12 species of damselfish compared to 34. Nevertheless, those reefs which are present support similar abundances of fish to reefs of richer regions (Sheppard and Salm 1988).

Limitation in the total area of reef available will partly depress the species richness of assemblages. However, the diversity of habitats available within reefs is also important, as noted above for the Red Sea. The structural simplicity of Omani reefs means that they are differentiated into few distinctive zones (Green and Keech 1986) and so the between-habitat component of fish diversity is also low.

(6) The Gulf

The main difference between reef fish assemblages of the Gulf and those of the adjacent Indian Ocean is a further drop in species richness. Smith *et al.* (1987) recorded only 72 species on reefs off the coast of Bahrain, Basson *et al.* (1977) and McCain *et al.* (1984) 70 and 106 species respectively off the east coast of Saudi Arabia, and Downing (1985) 85 species on Kuwaiti reefs. In a brief survey of inshore reefs of Qatar one of us (CMR) recorded only 35 species during ten hours of observation. Even fewer appear to be present in the waters of the United Arab Emirates to the south (Dipper and Woodward 1989). These trends in species richness accord well with patterns of reef development in the Gulf, the richest assemblages being found where development is greatest.

The most prolific coral growth of the Arabian Sea shoreline is found at Musandam, Strait of Hormuz (Chapter 3). Coral diversity exceeds that of the Gulf but little is known of the fish assemblages inhabiting this area. However, it is likely that fishes will follow the pattern for coral and that their species richness around Musandam will be high.

Reef development clearly has a strong influence on fish diversity. Many studies have shown that species richness increases with factors such as area of habitat (MacArthur and Wilson 1967), and the differentiation of this into zones (Goldman and Talbot 1976, Roberts 1986). Roberts *et al.* (1988) showed that trends in butterflyfish species richness and abundance in the Red Sea closely matched patterns of reef development. Offshore reefs fringing islands off the central Saudi Gulf coast support the most diverse assemblages and constitute the largest areas of reef in the Gulf. They also have a greater depth range than those closer inshore (Basson *et al.* 1977). Reef development is controlled largely by the physical environment, and extremes of temperature and salinity, coupled with turbid water, make the Gulf a harsh setting for reef growth (Downing 1985, Sheppard 1988). It is very likely that environmental extremes exert direct effects on fishes as well as these indirect effects (Coles and Tarr 1990). As discussed above, the early life stages, especially larval stages, are almost certainly very sensitive to environmental stress. This may contribute to the reduced species richness in the Gulf.

There is a marked difference between assemblages inhabiting inshore and offshore reefs of the western Gulf. Up to 50% of species are restricted to either inshore or offshore reefs (Coles and Tarr, 1990). Offshore reefs have richer assemblages with nearly twice the abundance and around 1.7 times as many species as inshore reefs (Coles and Tarr, 1990). These differences do not appear simply due to the greater depths and habitats present on offshore reefs. They were of similar magnitude when the same habitats were compared between locations. Coles and Tarr argue that offshore reefs support richer assemblages primarily because of greater structural complexity, and biological diversity of the substratum. However, direct physical effects cannot be discounted. Inshore reefs suffer greater extremes of temperature and higher turbidities than those offshore.

Reef fish assemblages of the Gulf are notable for being composed largely by species not entirely dependent on the reef for survival. These include species such as the bream *Diplodus sargus* (Sparidae), the parrotfish *Scarus ghobban* and the goatfish *Parupeneus margaritatus* (Mullidae). Coles and Tarr (1990) found that numbers of fish on Saudi Arabian Gulf reefs fluctuated seasonally, with an average 50% increase in numbers of species and 75% in abundance from spring to summer. This could not be attributed to a summer peak in settlement, but instead appeared due to seasonal off-reef emigration by adult fishes. Similar seasonal differences were reported in McCain *et al.* (1984) in their study of Saudi reefs.

B. Vertical distribution of fishes on reefs

Wherever an environmental gradient exists, the animals and plants which occupy that environment usually have discontinuous ranges, each species inhabiting only part of the total area (Pielou 1975). There are many changes in conditions moving from shallow to deep water on coral reefs. For example, on fringing reefs the shallowest parts are usually tidal and subject to periodic exposure; wave action decreases from shallow to deep water; light intensity decreases with depth and light quality changes with depth; rates of sediment deposition are higher towards the shore and in deeper water than at the edge of the reef, etc. (Sheppard 1982a, Huston 1985).

There are few, if any, cosmopolitan species of animals or plants which can equally thrive throughout the whole of this spectrum of conditions. Coral reefs typically have many recognizable zones with characteristic morphologies and compositions of coral species (Sheppard 1982a, Done 1983; see Chapter 3). Zones may be defined primarily by their structure (e.g. spur and groove zone) or by dominant species (e.g. *Millepora* zone of Red Sea reefs; Loya 1976a).

The physical structure of the substratum has also been shown to have a considerable influence on the diversity and abundance of fishes on coral reefs (Risk 1972, Luckhurst and Luckhurst 1978, Carpenter *et al.* 1981, Roberts and Ormond 1987). However, fishes may in turn affect the structure of the substratum, primarily through algal grazing and patterns of territoriality (see later this chapter). Fishes may also have highly specific habitat requirements which can to a large extent determine their distribution. For example, anemonefishes (Pomacentridae) are dependent on stoichactid anemones for protection (Allen 1972). Likewise, some species of goby (Gobiidae) and damselfish will only inhabit one or a few species of coral (Tyler 1971, Sweatman 1983). As with corals the distribution of fishes is a result of a complex interaction of biotic and abiotic influences.

(1) Diversity and abundance: the patterns

The vertical distribution of diversity and abundance on reefs is far from even. There are marked changes in both numbers of species and overall abundance of fish with increasing depth. Specific examples of such patterns shown here come from studies in the Red Sea. This is because the depth range covered by reefs there is greater than anywhere else in the region, and because relevant studies

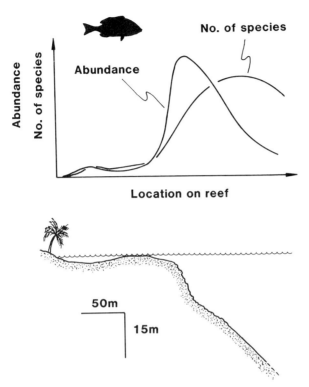

Figure 4.4 *Typical overall patterns of abundance and numbers of species with depth on a Red Sea fringing reef. Based on data from Roberts (1986).*

have been concentrated there. However, the principles illustrated using Red Sea examples may be applied to reefs of any region.

Typical patterns of overall species richness and abundance of fishes on fringing reefs of the central and northern Red Sea are shown in Figure 4.4. Species richness first increases from the shore to the reef-edge and continues to increase to depths of 10–15 m before falling again. Abundance follows a similar pattern but peak abundances are generally found close to the reef-edge, in somewhat shallower water than maximal species richness. Similar patterns have been found on reefs elsewhere in the world, such as in the Pacific and Caribbean (e.g. Goldman and Talbot 1976, Clarke 1977), suggesting the action of general underlying processes.

The overall patterns do not necessarily reflect those present for individual families. In fact these often differ considerably, as illustrated for five different families in Figure 4.5. Groupers tend to be restricted to the fore-reef, for example, whilst surgeonfishes are commonest in shallow regions.

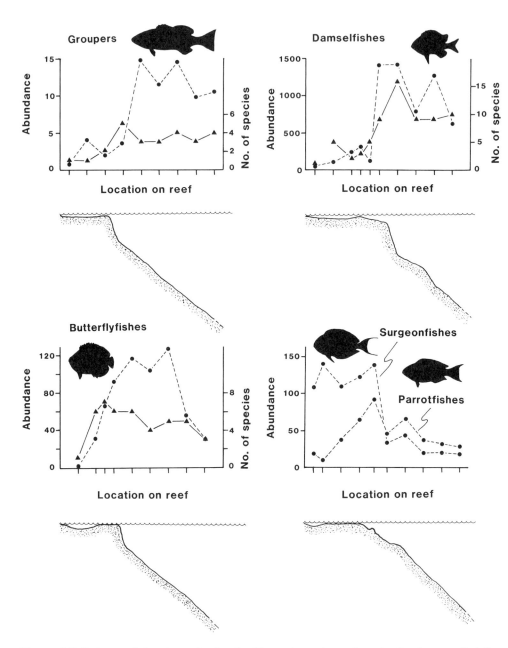

Figure 4.5 *Patterns of abundance for five families and numbers of species for three on Red Sea fringing reefs. Dashed lines = abundance; solid lines = number of species. Data for groupers in Sinai are from Zyadah (1989), for damselfishes in the central Red Sea from Roberts (1986), for butterflyfishes in Jordan from Bouchon-Navaro and Bouchon (1989) and for surgeonfishes and parrotfishes in Jordan from Bouchon-Navaro and Harmelin-Vivien (1981).*

(2) Processes

No single process determines the distribution patterns shown above. Instead, a multiplicity of interacting forces shape the distributions of individual species which combine to result in the characteristic patterns shown in Figures 4.4 and 4.5. However, three factors probably assume the greatest importance: environmental stress, food supply and structural complexity.

A) ENVIRONMENTAL STRESS

The shallowest parts of the reef comprise a stressful environment in which to live. Variations in water level due to tides, temperature fluctuations which may be extreme, and turbulent conditions all make it hard for fishes to inhabit regions less than a few metres deep. This partly explains why species richness and abundance are lowest on these parts of the reef. Conditions are most extreme on the shallow reef flat. Some areas are subject to periodic exposure at low tide, and water temperatures may vary between 16 and 40°C due to evaporative cooling in winter or solar heating in summer. Life is most difficult in flat, featureless areas with few deeper refuges into which fish may retreat during extreme conditions. For example, in the southern Red Sea, the reef flat typically nearly dries at low tide and is composed mainly of shallow sediments over Pleistocene reef rock. Only one or two species of burrowing goby (*Cryptocentrus* spp.; Gobiidae) are able to permanently occupy this habitat.

Towards the edge of the reef flat, turbulence is usually high and it is here that the reef absorbs most of the energy from breaking waves. Only a few species of fish are able to inhabit this part of the reef, such as the surge wrasse *Thalassoma purpureum*. Although most of the reef flat and lagoonal habitats are protected from the worst effects of breaking waves, they are still affected by them. Even low levels of turbulence may cause heavy sand-scour in shallow water, reducing growth of corals and other benthic organisms on which many fishes depend. It is widely held that shallow reef environments represent a spatial refuge from grazing for macroalgae, such as *Sargassum*, because turbulent conditions prevent the main herbivorous fishes from feeding there for most of the time.

B) FOOD SUPPLY

All organisms are clearly dependent on their food sources, and therefore will live in areas where they are able to forage effectively. It is thus unsurprising that vertical distribution patterns, especially of abundance, also appear to be substantially determined by the availability of different kinds of food.

Figure 4.6 shows a breakdown of the abundance of damselfishes (Pomacentridae) by their feeding behaviour. Peaks in abundance for herbivores, omnivores and planktivores reflect cross-reef differences in availability of their particular diets. Algae are most productive in shallow water due to higher levels of illumination there (Borowitzka 1981), and herbivorous species are most abundant in the shallows. This is also apparent at the family level. Surgeonfishes and parrotfishes are predominantly herbivorous and show similar patterns (Figure 4.5).

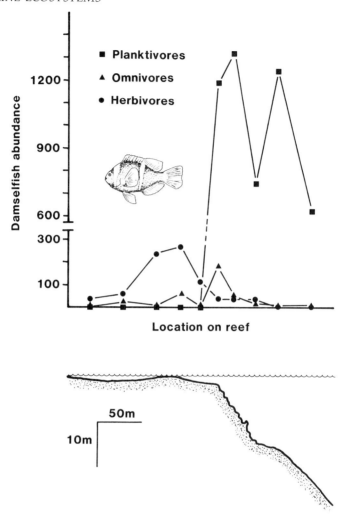

Figure 4.6 *Patterns of abundance by trophic category of damselfishes from a fringing reef in the central Red Sea. Data are from Roberts (1986).*

Plankton is also most abundant in shallow water (<10 m), again primarily due to higher levels of illumination close to the surface. Planktivorous fishes remove this from water as it flows across the reef, and thus they tend to accumulate in upcurrent areas to exploit the resource prior to depletion by others. This effect has been described as the "wall of mouths" by Hamner *et al.* (1988). It leads to high densities of planktivorous fishes in shallow water close to the reef edge, a pattern apparent in Figure 4.6. Virtually no wholly planktivorous fishes occur on the reef flat, although a number of omnivorous species which feed partly on plankton may be found in deeper areas of this zone or in shallow lagoonal habitats.

Piscivores are strongly influenced by prey distribution. The groupers (Serra-

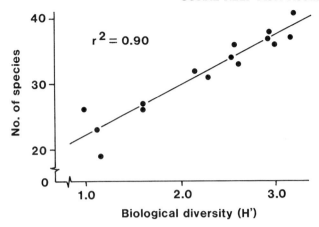

Figure 4.7 *Relationship between biological diversity of the substratum and number of fish species present. Biological diversity was calculated from percentage cover data for different substrate types on 10 × 1 m transects at five depths at four sites using the Shannon–Weiner index (H'). Fishes were counted on 200 × 5 m transects at four depths at the same four sites. Methods are described in Roberts and Ormond (1987) and data are from Roberts (1986).*

nidae) are most abundant on the fore-reef where greatest numbers of other fishes occur. Small planktivorous species form important food sources for many, explaining the peak in grouper abundance coincident with maximum abundance of planktivores.

These simple cases illustrate the importance of food availability to abundance. However, species richness may also depend on distribution of food resources. Roberts and Ormond (1987) measured the diversity of different types of substrata on reefs of the central Red Sea, an index of the variety of different components of the benthos. They found a strong positive correlation between benthic diversity and fish species richness (Figure 4.7), suggesting that resource distribution strongly influences where species are found.

C) STRUCTURAL COMPLEXITY

Reefs are well-known for their high abundances of piscivorous fishes. For example, Goldman and Talbot (1976) found that piscivores comprised the largest proportion of the biomass of fishes of any trophic group at One Tree Reef on the southern Great Barrier Reef. Predation rates have rarely been effectively measured, but where they have appear to be high. Sweatman (1984) found that a lizardfish, *Synodus englemani* (Synodontidae), ate 1.8 fishes per day with an estimated weight of 12% of the predator's body weight.

Whether there are real coral-eating fishes is still doubtful . . . Many fishes eat plants, others subsist on the numerous worms and molluscs that live here, or on decaying animal matter; the greater number are predaceous, and eat other fishes.

Klunzinger 1878

Studies of fishes which have just settled from the plankton show very high mortality rates, presumed due to predation, although they vary between taxa. Shulman and Ogden (1987) estimated that 90% of newly recruited grunts, *Haemulon flavolineatum* (Haemulidae), died within one month of settlement. Similarly, Doherty and Sale (1986) found that over 25% of individually recognizable fishes disappeared within five days of settlement. The majority of reef fishes respond to predators by seeking refuge within the reef framework. Individuals of many species habitually use the same one or two holes located within their territories or home ranges (Roberts 1985). The availability of refuge holes is dependent on the structural complexity of the reef (Roberts and Ormond 1987).

C. The role of fishes in reef processes

Whilst the structure and benthic composition of the reef evidently affects fish assemblages, fishes may in turn affect the structure of reef benthic communities. Such effects are a direct result of their feeding activities. The following account looks at the most important of them in terms of their influence on reef structure and function. Some aspects of grazing and bioerosion are deferred to the following chapter.

(1) Productivity

Many herbivorous fishes, especially damselfishes and surgeonfishes, defend territories interspecifically against other herbivores, particularly parrotfishes (Vine 1974, Roberts 1985). There are around ten species which defend such feeding territories in the Arabian region. Most inhabit the shallow regions of the fore-reef and reef-flat (e.g. Figure 4.6), although one damselfish, *Plectroglyphidodon lacrymatus*, occurs to depths of 17 m in the central Red Sea. Most defend territories interspecifically against a wide range of potential competitors. For example, *P. leucozona* were observed to attack 30 other species on Saudi Arabian reefs (Roberts 1985).

Roberts (1985, 1986) has shown that associations between species can improve the efficacy of territorial defence. There is a symbiotic relationship between the surgeonfish *Acanthurus sohal* and several damselfish species which share territories and the algal food they contain. *A. sohal* provide the majority of defence against larger herbivores such as parrotfishes and other surgeonfish, whilst the damselfish mainly exclude smaller fishes such as juveniles from these families. This combined defence results in almost total exclusion of roving herbivores (Roberts 1985) and leads to the existence of algal lawns, often called algal turfs, which are a characteristic feature of the shallow parts of Red Sea reefs. Whilst *A. sohal* is found around the whole of the Arabian peninsula, most of the damselfishes with which it shares territories occur throughout only part of the range. The composition of the partnership changes but the functional significance of it remains the same.

Herbivorous blennies also share space with damselfish and surgeonfish, but unlike the relationship between species from the latter families, this relationship appears to be 'parasitic'. Blennies do not contribute to interspecific defence of territories, although they defend their individual territories against each other. They are small enough to escape exclusion by damselfish by use of refuge holes inaccessible to the latter.

Large areas of the reef above 3 m deep are dominated by the territories of herbivores. Damselfishes spread over virtually the whole reef flat excluding sandy substrata. *A. sohal* is restricted to a band close to the seaward edge of the reef, usually living in dense colonies. Defence of territories results in a thick turf of filamentous algae developing within them. This contrasts starkly with the thin layer of predominantly coralline algae covering much of the reef outside them. Algal biomass measured in territories at one site in the central Red Sea was on average ten times greater than that of non-territory substrata (Roberts 1986). Such values are typical for central and northern regions. However, not only is the biomass of algae greater but the productivity per unit of that biomass is higher than outside territories (Klumpp *et al.* 1987).

Differences between algal assemblages inside and outside territories were once thought to be due to reduced grazing pressure within territories. However, Russ (1987) has shown that rate of removal of algae is comparable inside and outside them. It is almost certainly the way in which they are grazed that creates the differences. Damselfishes, surgeonfishes and blennies crop algae, removing only part of the tissue with each bite. By contrast, the dominant herbivores outside territories, parrotfishes and in some cases echinoids, scrape the algae to their basal parts. In doing so they may remove some of the inorganic material of the substratum (see Chapter 5).

Klumpp *et al.* (1987) have suggested three possible reasons why algae within territories are more productive per unit biomass than those outside: (1) territorial fishes weed out less productive kinds of algae, (2) algae in territories are grazed so that they remain within an exponential growth phase, and (3) fishes "fertilize" territories by excreting within them close to the algal turf. Although anecdotal accounts suggest some weeding may take place, there is no firm evidence. Better evidence exists to show that excretion within territories can lead to relatively efficient recycling of nutrients within turfs which may enhance productivity (Polunin and Koike 1987). However, the main explanation is probably that heavy grazing by scraping herbivores outside territories keeps plants at such low levels of biomass that their growth rates are reduced. Whatever the mechanisms by which high productivity is achieved, algal lawns, as they are collectively known, contribute significant quantities of organic carbon to the reef system as a whole (Klumpp and McKinnon 1989). This input constitutes an important component of the very high overall productivities achieved by coral reef communities.

Grazing is also important to nitrogen fixation on reefs. Heavy grazing shifts the balance from dominance by larger, fleshy algae, to smaller filamentous and blue-green species (Cyanophyta). Blue-green algae fix nitrogen, and they dominate the very heavily-grazed areas outside algal lawns of territorial herbivores. Experiments on the Great Barrier Reef have shown that rates of nitrogen fixation are highest on unprotected substrata, low on caged algal communities, and intermediate within damselfish territories (Wilkinson and Sammarco 1983). Grazing thus enhances overall levels of nitrogen fixation.

(2) Nutrient transfer

Fishes play an important role in the cycling of nutrients and energy on the reef, and in transferring them among reef zones. Many herbivores for example, particularly parrotfishes and surgeonfishes, feed on the shallow reef-flat but defecate upstream of this, their faeces dropping onto the deeper outer-slope zone. In this way, nutrients are actively transported to parts of the reef they would probably not reach by passive means.

There are also large numbers of species which undertake daily feeding migrations, moving between reef habitats such as the outer-slope and lagoon zones (Ogden and Quinn 1984). Especially notable are the large schools of grunts and snappers which feed over sandy areas or seagrass beds by night and rest over large coral heads by day. Schools habitually rest over the same places, and Meyer *et al.* (1983) showed that corals with schools grew faster than those without at a site in St Croix in the Caribbean. They argued that this was due to fertilization by faeces and suggested that foraging migrations transferred significant quantities of nutrients among habitats, particularly nitrogen and phosphorus.

Similar kinds of foraging migrations are undertaken by a wide variety of other fishes (Ogden and Quinn 1984) and have been claimed to constitute an important link between the ecosystems in which fishes rest and feed (Meyer and Schultz 1985). There are many species in the Arabian region which behave in this way but probably the most important are snappers.

Summary

Fishes are a dominant component of the reef fauna. There are marked differences throughout the Arabian region in the structure and composition of fish assemblages which reflect the heterogeneous nature of the environment.

The most diverse assemblages occur within the Red Sea with a total of approximately 1000 species present (including non-reef species). The Gulf supports only about 200 species in total, of which at least 125 are found on reefs. Within the Red Sea there are major differences in assemblage composition between areas north and south of latitude 20°N. Several hypotheses have been advanced to account for this. The first suggests that differences in reef habitat between areas generate the patterns with species abundance determined by the distribution of their particular habitat requirements. North of 20°N reefs are typically well-developed and drop steeply into clear water. To the south reefs occur within a shallow, turbid environment, and are less well-developed. Habitat differences might also affect the outcome of competition among species, resulting in species replacement between areas. A second hypothesis suggests that differences in water quality between areas can produce differential survival of fish larvae. 20°N represents the approximate northerly limit of penetration by nutrient-rich water from the Gulf of Aden. There are also north–south gradients in temperature, salinity and turbidity. Differences in mortality rates could result from effects of turbidity on feeding efficiency of larvae. Only small differences would be necessary to generate large differences in assemblage structure due to the very high overall mortalities of fishes prior to settlement. A third hypothesis, invoking oceanographic isolation of northern and southern parts of the Red Sea, can be discounted since there is free exchange of water between areas. The data available at present do not allow an evaluation of the validity of the first two hypotheses.

The Gulfs of Aqaba and Suez support distinctive fish faunas and also differ substantially from each other. Those of the Gulf of Suez share greater affinities with southern Red Sea assemblages than with the Gulf of Aqaba, probably due to its shallow, turbid nature. The Gulf of Aden marks a division between a fauna dominated by Red Sea species to one dominated by Indian Ocean species in the Gulf of Oman and Gulf. Upwelling of cold water in the Arabian Sea and off the Horn of Africa appears to provide this major biogeographic barrier. Compared with the Red Sea, reefs of the Gulf and Gulf of Oman support low diversity fish assemblages. This is probably an effect of a scarcity of reef habitat and the rather extreme environmental conditions prevailing.

The vertical distribution of abundance and diversity of fishes on reefs is very uneven. Typically, species richness increases to depths of 10–15 m before falling with increasing depth. Patterns of abundance are similar but peak abundance occurs closer to the surface. These patterns are shaped by many forces including environmental stress (e.g. wave action and temperature fluctuations), food supply, and structural complexity of the reef.

Fishes play an important role in a number of reef processes. Defence of feeding territories by herbivorous species results in increased rates of carbon fixation. Heavy grazing by fish and echinoderms in areas outside territories results in dominance by nitrogen fixing blue-green algae. Foraging migrations by fishes form important pathways of nutrient transfer both within and among reef habitats.

CHAPTER 5

Other Arabian Reef Components and Processes

Contents

A. **Cementation**   109
   (1) Foraminifera, Bryozoa and sponges   109
   (2) Red algae   111
B. **Grazing and bioerosion**   111
   (1) Grazing effects of fish and echinoderms   111
   (2) Effects on benthic community structure   114
   (3) Grazing by other groups   115
C. **Soft corals**   116
Summary   119

It is commonly held that coral reefs support particularly diverse communities of invertebrates. Although at supra-generic levels this is undoubtedly true, the fact that such a high proportion of the benthic fauna is cryptic means that often it has been difficult to obtain data to support the contention. In part, different perspectives affect the judgement on relative diversities. The 1000 or so species of reef fishes currently attributed to the Arabian region certainly seems to be a high number, although whether the 250 species of scleractinian corals, or 180 soft corals, are deemed to be particularly diverse depends on what area is compared to it. For example, the Gulf's 70 species of corals and handful of alcyonarians is almost always referred to as being very poor compared to the Indian Ocean, even though many tropical regions contain fewer. For many other species groups, taxonomic or ecological knowledge is extremely scarce.

From work undertaken elsewhere, it is expected that Arabian reefs also contain relatively high diversities of cryptic, burrowing and both attached and motile invertebrates; Head (1987b) lists numerous taxonomic groups from the Red Sea with brief bibliographies relevant to each. Diversity itself, of course, is not necessarily a firm measure of ecological importance (as opposed to genetic resource importance which is not considered here), but it may at least be regarded as a measure of the complexity of the system.

The Red Sea is very rich in invertebrate animals, but of this class of products it is only the pearl-oyster and the black coral that are made any use of, except now and again by a European naturalist. The natives seldom trouble their heads about such things.

Klunzinger 1878

In this chapter, several functional groups of animals and plants are considered for which there are insufficient data for treatment in separate chapters. In line with the aims of the book, however, they are not treated by group, the practice followed for example by Head (1987b) for the Red Sea, but by their known function in the ecosystem. There is of course a strong correlation between many species and species groups and their function. Several aspects which could appropriately be included in this chapter are only briefly touched on if they are covered in greater detail elsewhere; examples are algae, reef fishes and grazing. For others, the discussion is limited to their known or even presumed importance in the Arabian region. In these, there is no attempt to comprehensively review the subject matter, and reference is made in several cases to existing, recent reviews which already do this. Instead, emphasis is on their importance in Arabian reef systems.

The first two major categories are the approximately balancing factors of reef cementation and erosion by a wide range of biota.

A. Cementation

Cementation of carbonate sands is as important to reef growth and maintenance as is the initial coral growth. There appear to be various mechanisms for cementation, discussed in detail by Hopley (1982), which fall into two main groups, chemical and biological. The net result is consolidation and eventually transformation and solidification of aragonite and calcite particles. It is becoming clearer that the role of biological agents is in part due to the fact that they smother and stabilize sediment so that chemical and biochemical processes may act on and transform the particles. Groups of organisms which are now known to be very important in this include sessile Foraminifera, Bryozoa, some sponges and encrusting algae.

(1) Foraminifera, Bryozoa and sponges

Foraminifera are mentioned elsewhere in connection with their role in Gulf sediments, and the evidence they supply on Pleistocene and early Holocene upwelling severity in the Arabian Sea. They are also important components of coral reefs where they form a significant part of the particulate and detrital recycling processes. One species, however, is sessile and grows into small, bright reddish brown coloured sheets on the undersides of corals and on other concealed limestone substrate. This species is *Homotrema rubrum*, and it is particularly conspicuous in sheltered parts of reefs where large boulders of *Porites* and branching corals with dead basal parts are abundant. In such habitats it commonly covers up to 25% of dead basal parts of coral colonies. It provides a considerable component of the sediment itself in some reef areas in the Red Sea (Head 1987a) though possibly its most important role in reef maintenance lies in providing a living veneer of tissue over dead coral skeleton which is otherwise prone to bioerosion. This species, in conjunction with sessile foraminifera from other parts of the Indo-Pacific and Atlantic, contains photosynthetic symbionts. A recent study of northern Red Sea foraminifera by Reiss (1979) underlines the fact that this group is very widespread in the region even though its members are inconspicuous at first glance.

Bryozoa provide a very similar function in that they also are found in encrusting sheets in cryptic habitats, on dead basal parts of corals or on their undersides in sheltered areas, and on semi-consolidated sediments. Head (1987b) very briefly describes their structure and habitat, and Dumont (1981) records 86 species from the Red Sea, sometimes in densities of $5000 \, \mathrm{m}^{-2}$. Hopley (1982) remarks that Bryozoa are not major framework constructors, but states that despite their apparently delicate appearance they are important in reinforcing the reef fabric by adhering strongly to limestone by mucopolysaccharide adhesive. They appear to be able to bind together fairly stable coral fragments and coarse sediments, but are not able to bind very unstable substrates. It is still difficult to quantify the present importance of bryozoans in this respect, although in geological times they were a much more important component of tropical reefs. They are still important community components in cryptic areas and in deep parts of Red Sea reefs where illumination is low (Fricke and Schuhmacher 1983).

Sponges were until recently thought to be at best neutral in the reef building process, (and are mentioned below under bioerosion too) but some are now known to be important to reef construction. Their contribution in the Caribbean has been studied, and appears to fall into two parts: firstly deep living, very dense forms of sclerosponge contribute directly to the deep reef framework construction, but secondly and more importantly in shallow, illuminated areas, several species provide considerable binding action. They achieve this by stabilizing sediments and rubble long enough for chemical and microbial action to bind them together. In this respect they have a similar role to bryozoans and sessile foraminiferans, but to an even greater extent and over a wider range of particle sizes. No reports of this are available for reefs in the Arabian region, though the near-universal occurrence of thin sheets of sponge on Red Sea reefs suggests that they probably have a similar result there.

Although ubiquitous on Arabian reefs, sponges are commonly excluded from ecological studies of the region, partly because of taxonomic difficulties. Wilkinson (1980a, 1987) studied ecological and physiological properties of reef sponges, including those near Port Sudan, and found that they are trophically important for two reasons. Firstly they trap and consume a particularly abundant, minute size class of food particle and may be one of the principal groups preventing the loss of this material from the tightly-cycled reef system. Secondly many species contain symbiotic Cyanophyta, and add moderately to the primary production and nitrogen fixation. The latter feature is found in sponges from illuminated, shallow depths on the reef, while sponges in deep or cryptic habitats do not exhibit such symbiosis. Head (1987b) sketched a reef profile showing the principal location of burrowing sponges, photosynthetic sponges, boring and etching sponges, encrusting cryptic species and the deeper calcareous forms. It is mainly the encrusting forms and possibly the burrowing forms which consolidate sediments.

(2) Red algae

Filamentous and leafy red algae are discussed later under grazing and algal lawns, as well as in Chapter 6 on algae and seasonality, while the importance of calcareous red algae in resisting waves in the spur and groove zone is emphasized in Chapter 3. In addition, a further functional group of mildly calcareous and encrusting red algae are abundant throughout the photic zone, and these also stabilize sediments. Walker (1987) attributes the greater part of these algae in the Red Sea to the genera *Porolithon* and *Melobesia*. These genera also form the rudimentary spurs in the few locations in the Red Sea where wave energy is adequate for their development. Where wave energy is not sufficient for this, these algae merely coat considerable proportions of the substrate. Sheppard (1983a) plotted the proportion of such algae with depth on exposed, mid shelf and sheltered patch reefs in the central Red Sea. This study showed (Figure 5.1) that whether or not these algae developed spurs, their cover of substrate ranged from about 15 to 40%, with higher values nearer the surface. No quantitative work on coralline algal cover appears to exist for the coral reefs and coral communities of Musandam and the Gulf, although in both locations the group is locally abundant. Observations show that in the southern part of Oman, cover may be appreciable in some areas (Sheppard 1986a), where crusts of coralline and fleshy encrusting algae cover non-reef limestone and other rocky surfaces between scattered corals as well as providing similar cover of substrate on true reef structures.

The factors influencing the development of such algae appear to closely resemble those for corals, except that several coralline forms are much more resistant to intense insolation on reef flats. Because of this, they are commonly abundant on reef flats of the Arabian region which experience some of the most severe extremes of insolation, heating and desiccation found anywhere in the world. It has been reported that in open reef structures, mixes of coral and algal rubble with live coralline algae provide strongly cemented infill between the structures built by corals (Hopley 1982).

B. Grazing and bioerosion

Establishment and maintenance of algal lawns (turf algae) by the defensive actions of some herbivorous fish, is described in the previous chapter. The effects of this behaviour, however, extend further than their trophic roles.

(1) Grazing effects of fish and echinoderms

Scraping herbivores, of which parrotfishes are the most important among fishes, are major agents of bioerosion on reefs (Gygi 1975, Frydl 1979). With each mouthful they ingest some carbonate material as they scrape algae from the reef. Some also scrape at living corals to some degree. By far the majority by weight of material within a full gut is inorganic. Ingested material is ground up in the

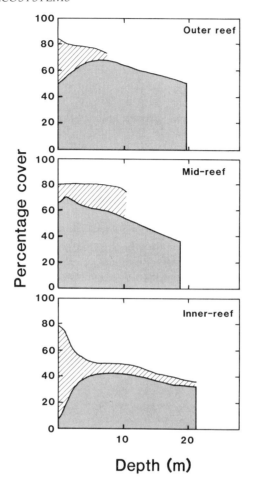

Figure 5.1 *Distribution of encrusting red algae in different reef conditions, at Yanbu, central Red Sea. Hatched areas = encrusting red algal cover, shaded areas = coral cover. Extent of hatched and shaded areas indicates the depth range surveyed.*

pharyngeal mill, a bony apparatus in the throat. This physical breaking open of algal cells is thought to help with digestion since parrotfishes lack the enzyme necessary to break down cell walls (Ogden and Lobel 1978). Its importance in the present context is that it results in carbonate material being finely powdered prior to excretion. It has been estimated that parrotfish grazing accounted for 0.6% of total bioerosion on a fringing reef in the Caribbean, prior to mass mortality of the echinoid *Diadema antillarum* (Frydl and Stearn 1978). Comparable figures are not available for Indo-Pacific reefs but it is thought that the importance of bioerosion by parrotfishes will be considerably greater (Hutchings 1986).

Fishes may affect rates of bioerosion indirectly too. Probably the most important bioeroders are echinoids which ingest large quantities of inorganic material whilst grazing algae (Frydl and Stearn 1978, McClanahan 1988). Fishes are important predators of echinoids, in particular triggerfishes (Balistidae), pufferfishes (Tetraodontidae) and emperors (Lethrinidae). It has been widely suggested that removal of such predators by fishing has resulted in high population densities of echinoids (Messiha-Hanna and Ormond 1982, Hay 1984, McClanahan 1988). Subsequent overgrazing by urchins has been claimed to have led to net erosion, for example on the reefs around Hurghada in the Red Sea (Messiha-Hanna and Ormond 1982).

In the central Red Sea, several small urchins graze on the coral rubble and reef substrate, including *Echinostrephus molaris*, *Echinometra mathaei*, *Eucidaris metularia*, *Diadema setosum*, *Echinothrix diadema*, *Heterocentrotus mammillatus* and *Parasalenia poehli*. The first two of these excavate deep burrows in rock, while the others live in exposed locations, though some emerge at night only. As pointed out by Head (1987b) probably all are responsible for some degree of bioerosion during feeding. The relative importance of fish and urchin bioerosion differs between Indo-Pacific and Caribbean reefs. Prior to the mass die-off of the urchin *Diadema antillarum* in the Caribbean during the early 1980s, urchins accounted for more than an order of magnitude more bioerosion than did fishes in this region (Frydl and Stearn 1978). By contrast, Indo-Pacific reefs generally have lower population densities of echinoids, and higher densities of herbivorous fishes which scrape the substratum for food (Sammarco 1985). Although Red Sea and Gulf reefs follow the Indo-Pacific pattern on the whole, echinoid densities may be very high in some places. For example, Dart (1972) found that on Sudanese reefs, *Echinometra mathaei* reached densities of 60 per 100 m^2, and he argued that higher levels of grazing by this and other species noted above led to a reduction in the cover of algae in the shallow reef zone usually associated with algal lawn, and that this led to improved recruitment of corals. Further north in the Gulf of Aqaba, Benayahu and Loya (1977a) discovered that echinoids are the main invertebrate grazers of the turf algae. A significant negative correlation existed between annual turf algal cover and density of the urchin *Diadema setosum*, whose thorough rasping activity regulates the algal growth. However, studies elsewhere have shown that the rasping activity of echinoderms also reduces survivorship of newly settled corals, so that the overall picture is not straightforward.

There are also indirect effects of territory defence by herbivorous fishes on rates of bioerosion and calcification. Vine (1974) noted how in the Red Sea the substratum within territories of surgeonfish *Acanthurus sohal* contained much loose material compared with adjacent unprotected areas. He proposed that this was due to the lack of calcareous algae within territories. Such algae play a major role in reef calcification, cementing dead skeletons of corals together into solid structure of the reef. They are unable to do this effectively within herbivore territories due to competition from fast-growing filamentous species.

Within the Red Sea and Gulf, territories of *A. sohal* may also act as refuges for the sea urchin *Echinometra mathaei*. For example, in the northern Red Sea, *E. mathaei* densities are higher within reef-edge colonies of *A. sohal* than elsewhere on the reef. Within *A. sohal* colonies, urchins have eroded complex networks of passages open at the top to the light. By contrast to surrounding areas of dense algal turf, these passages are heavily grazed and lined with calcareous and blue-green algae. Unlike damselfishes in the Caribbean, which exclude the most important urchin grazer, surgeonfishes and damselfishes of the Arabian region seem incapable of excluding urchins (Williams 1979, Eakin 1988).

The result of this complex network of interactions is that the reef within surgeonfish colonies is typically more structurally complex than areas of adjacent reef outside territories. A lack of accreting coralline algae, coupled with boring by urchins, has led to colony areas having numerous cracks, fissures and caves below them. This is not the case for all areas of the region though. In some parts of the central Red Sea, colonies have few echinoids and are relatively featureless, lacking the structural complexity characteristic of those in the north (Roberts 1986). Causes of differing densities of sea urchins among regions are as yet unknown.

The crown-of-thorns starfish *Acanthaster planci* has a notorious and well justified reputation for consuming coral polyps in many areas of the Indo-Pacific. Following "plague" outbreaks, coral mortality may approach 100% over whole reefs and over many kilometres of fringing reefs. Long-lasting community changes and erosion inevitably follow such perturbations. The literature on this is now extensive and much is summarized in Moran (1988), though knowledge is concentrated on the Great Barrier Reef and several Pacific island groups. *Acanthaster* is currently also a problem in some parts of the Maldive Archipelago. Much of the early work was carried out in the Red Sea (Ormond *et al.* 1973, Ormond and Campbell 1974, Campbell 1987). In initial investigations in the Red Sea, it was determined that numbers were generally between 4 and 44 individuals along each kilometre of reef face, though in areas of apparent outbreak, densities reached 1000 per kilometre, or 1 per metre. Clearly such densities have important effects, though these levels are still considerably lower than some extreme plague levels found on the Great Barrier Reef. Throughout the 1980s, however, observations by all of us in numerous widespread localities in all seas around Arabia found *Acanthaster* densities to be very low, and the species extremely scarce or absent in most reef localities. As Campbell (1987) remarked, this species has not caused the concern of late in the Red Sea as it has in areas further east.

The species is not listed for the Gulf at all (Price 1982a), but is found in Oman. Just as the presence of this species in Sudan helped to initiate a new phase of reef research in the Red Sea, so too concern about its numbers stimulated the first survey of coral reefs in Oman, reported in Glynn (1983). There it occurs in aggregations, though not commonly in "plague" abundance. Along the Capital Area of Oman, it was found also in moderate densities feeding on soft corals in rocky habitats subjected to cold water and devoid of coral reefs (Sheppard 1986a).

(2) Effects on benthic community structure

Activities of feeding fishes are not only important in terms of bioerosion. Impact of herbivorous fishes on algal communities has already been described. However, grazing fishes also affect corals and other sessile organisms. Such effects result mainly from partitioning of the reef environment between territorial and non-territorial species. Differences in the way members of the two groups feed (scraping versus cropping) account for their influence.

Outside territories, the reef surface is intensively scraped by parrotfishes. Their strong beak-like jaws are capable of removing not only algae, but newly-settled coral spat and recruits of other sessile organisms (Birkeland 1977). Vulnerability to removal differs among species and depends on time since settlement. Birkeland (1977) argued for the importance of rapid early biomass accumulation in corals as an adaptation to this form of mortality. Sammarco and Carleton (1981) found that settlement success of corals was five times higher within damselfish territories than in areas exposed to scraping herbivores on the Great Barrier Reef. Corals settling within the territories of surgeonfishes and damselfishes enter a completely different kind of environment. Here the biggest source of mortality is overgrowth by algae. However, there may also be direct effects from territory holders themselves. Wellington (1982) undertook a series of experiments to determine the control of reef zonation on the Pacific coast of Panama. He found that shallow areas were dominated (80–85% cover) by branching pocilloporid corals, whilst deeper parts of the reef had richer assemblages, but were dominated by the massive coral *Pavona gigantea* (approx. 18% cover). Shallow parts of the reef were home to the territorial damselfish *Stegastes acapulcoensis*. They defend algal turfs from other herbivores and open up areas of free space for algae by killing corals. *Pavona* suffer higher mortality than *Pocillopora* because their massive, open surfaces make it easier for fish to remove tissue. By contrast, on the deeper reef slope, juvenile *Pocillopora* suffer higher mortalities than *Pavona* from grazing parrotfishes, and corallivorous pufferfishes. Damselfish territories were also characterized by higher coral diversities at this Pacific site than undefended parts of the reef.

Similar processes operate on Arabian reefs. Particularly notable is the damselfish *Stegastes nigricans*, one of the most pugnacious in the world. They live in tight-knit colonies on the reef flat and lagoonal habitats from which they exclude virtually all other fish species. Territories typically contain branching corals of the genera *Acropora*, *Pocillopora* and *Stylophora* at higher cover levels than areas of adjacent reef. Few other corals occur within these territories.

Other damselfish and surgeonfish have similar effects, especially in the shallow reef margin where they sometimes occupy more than 80% of space (Roberts 1985). In the Red Sea their territories generally have a lower overall coral cover and diversity than areas deeper on the fore-reef (Ormond *et al.* 1984a,b, Roberts and Ormond 1987).

(3) Grazing by other groups

In the intertidal region, bioerosion is marked. Overhanging limestone cliffs are the normal condition throughout the northern and central Red Sea. There is commonly an etched region within the upper intertidal and splash zone which is excavated about 1 m back from the cliff face, and which extends from the high tide level up to 2 m high. In these sites, the large chiton *Acanthopleura haddoni* and limpet *Cellana rota* are abundant, and each feeds by rasping. Although both can almost exclusively be found in depressions of their own making, there are no estimates of the importance of their erosion, either in absolute terms or in relation to erosion by the waves. In the corresponding region in much larger limestone cliffs of Musandam, erosion from mussels *Lithophaga cumingiana* has been estimated (Vita-Frinzi and Cornelius 1973) and is believed to average $0.25\,\mathrm{cm\,y}^{-1}$ which certainly exceeds the rate of erosion from wave action in these very sheltered khawrs. The result is marked undercutting of the cliffs with common falls of rock. Similar rocky cliffs are found along several other parts of the Oman coast, but in these cases the aspect is towards the high energy waves of the Arabian Sea, and so effects of mussels are less important than abrasion by waves which bear considerable loads of suspended sand (Sheppard 1986a).

Sponges, mentioned above for their role in cementation, also include species noted for bioerosion. As noted by Head (1987b), the genus *Cliona* bore by chemical secretion, and in the process create large quantities of fine sediment. *Cliona* is a particularly abundant component of dead coral skeletons and areas of coarse coral rubble on Red Sea reefs. Forms which etch and erode are most abundant in reef crest and reef flat zones and shallow, well-illuminated areas of coral reefs.

A review by Borowitzka (1981) examines the importance of grazing, and although he concentrates mainly on fish grazing and community effects resulting from it, he also indicates that grazing by chitons and limpets complicates the picture further. These molluscs are territorial animals, and it appears that their removal of epiphytes encourages the survival of underlying coralline algae which generally thrive in shallow reef areas. This is observed by the occurrence of such animals in the middle of a patch of crustose coralline algae devoid of turf algae. Although the animals also rasp coralline algae, it is much more easily regenerated.

C. Soft corals

Alcyonaria are quantitatively extremely important in the Red Sea, where they provide substrate cover approximately equal to that of corals, and in many areas well exceeding it. They are locally important in parts of the Arabian Sea where upwelling is not severe, but are greatly restricted where it is. The Gulf is curiously almost completely lacking in soft corals. While the stressed environmental conditions of the latter might be an obvious reason, more detailed examination of habitat preferences for soft corals in the Red Sea makes this a less plausible reason.

Almost all of the work on soft corals in the region is taxonomic and physiological, concerning reproductive physiology and strategy (Benayahu and Loya 1983, 1984a,b, 1985, 1986). The latter also cite numerous taxonomic works for areas in the Red Sea, mainly by the first author and J. Verseveldt. Alcyonaria may be locally common on reef flats and in very shallow water, especially members of the genera *Sinularia* and *Lobophytum*. The band of greatest diversity in the northern Red Sea is from below the surf zone (about 2 m) to about 8 m deep, where numerous species of the genera *Sarcophyton*, *Alcyonium*, *Parerythropodium*, *Cladiella* and *Nephthea* are abundant (Benayahu 1985).

The most common genus of all, however, is *Xenia*, and its relative *Heteroxenia*.

The commonest of these, *Xenia macrospiculata*, form extensive carpets, especially in sheltered and even sedimented locations, and are conspicuous by their mass of rhythmically pulsating feathery tentacles. This species is highly fecund, has an early onset of reproduction, is opportunistic in settlement and reproduces rapidly asexually. It is also able to grow over stony corals, move actively over the substrate (Benayahu and Loya 1981) and because it is toxic, suffers little from predation. This results in the species being one of the most abundant coelenterates in the northern and central Red Sea. In back reef areas such as those dominated by enormous colonies of the scleractinian coral *Porites*, dead basal parts are often completely covered by *Xenia* spp. Similarly, most patch and fringing reefs of the northern and central Red Sea have areas dominated by coral rubble where the steep upper slope becomes more gentle at 5–15 m deep; this rubble is also heavily colonized by *Xenia*, both on exposed reef slopes as well as lagoonal areas. *X. macrospiculata* has been considered to define a zone in a study of both corals and soft corals in Sanganeb atoll in the central Red Sea (Mergner and Schuhmacher 1985), though in similar studies by Sheppard and Sheppard (1991) it is considered to be too widespread and ubiquitous a species to usefully define a zone. It occurs almost literally everywhere where there is hard substrate or rubble within the photic zone, in patchy and often high abundance. Both approaches, however, illustrate the abundance of this coelenterate.

On Red Sea reefs, patchy distribution is one of the most remarkable features of alcyonarians according to Benayahu (1985), though such a distribution applies to many scleractinian corals also. Patchy spatial distribution of soft corals results from a short pelagic phase and gregarious behaviour, and this in turn leads to several distinct soft coral assemblages being identified. Large patch sizes may also result from aggressive behaviour and interspecific dominance, discussed later.

Deeper on reefs, the very attractive *Dendronephthya* and related *Stereonephthya* develop large, coloured and translucent colonies. Small examples are common even on sheltered reef crests in the central Red Sea, but in water below 20 m they become more abundant, continuing down to at least 45 m deep. These genera are not zooxanthellate, unlike those noted above which are abundant on shallower parts of reefs.

In the southern Red Sea, soft corals gradually disappear along the mainland coast (Figure 5.2). From high cover values in the north, high diversity and cover still occur at about 20°N on both Arabian and African sides (Schuhmacher and Mergner 1985), namely at Sanganeb, Port Sudan and Jeddah, where *Xenia* is particularly abundant in lagoonal areas, providing over 75% cover over several hectares. South of Jeddah, soft corals then disappear very rapidly. They cover about 40% of reef slopes at 2–4 m deep as far as Al Lith (20°N), but immediately south of the latter, some reefs have a cover of <5% but most have none at all. Where soft corals do occur in the south, they are invariably *Xenia* sp.

In the Gulf of Aden and southern Arabian Sea, no published information is available on soft corals. It might reasonably be presumed that some tolerant species exist there, and some are found off Oman. There, the fleshy *Sarcophyton* becomes locally abundant, at least as far north as the Strait of Hormuz where it occasionally provides cover of 80% over large areas (Anderlini 1985, Sheppard 1986a), though only very patchy cover is found in areas subjected to upwelling.

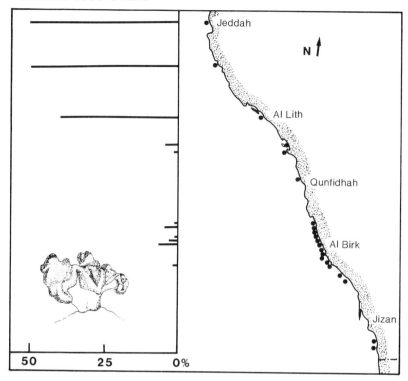

Figure 5.2 *The decline of soft coral abundance in the southern Red Sea. Locations of sampling sites on the coast are marked. Values are mean cover on reefs. From Sheppard (1985a).*

Dendronephthya is widespread but rarely abundant. In the Gulf, it has been commonly asserted that no soft corals occur, but on the west coast of Musandam, strictly speaking inside the Gulf, as well as further west, some including *Dendronephthya* do occur. They are inconsequential, however, and do not penetrate as far as the central part of the Gulf.

The factors controlling the distribution of soft corals remain problematic. Firstly, low temperatures in winter may be important. The southern Gulf of Suez, as with all parts of the northern Red Sea, has abundant soft corals, including *Xenia*, and conditions in the southern Gulf of Suez experience a fairly wide temperature range. As such, the conditions are not markedly different from those in the Gulf where soft corals are essentially absent. It does appear from very limited data, however, that further north in the Gulf of Suez soft corals do decline and disappear. As remarked in several other chapters, these similarities of environment have led to many parallels in biota, for example with corals and seagrasses. The similarity therefore may extend to soft corals as well, at least as far as the northern part of the Gulf of Suez is concerned. It may be concluded from this, therefore, that soft corals generally are not tolerant of low winter temperatures. Exceptions must be made for the two genera *Dendronephthya* and *Sarcophyton* which occur, sometimes abundantly, in areas affected by cold upwelling water in the Arabian Sea. This would be expected of *Dendronephthya*, since it is non-zooxanthellate.

Low temperatures, however, cannot explain their absence from the southern Red Sea. Therefore sedimentation might be suggested as being important as both of the above areas and the southern Red Sea are fairly heavily sedimented. But this is not a convincing explanation given that soft corals of several genera are very abundant in sedimented lagoons and in sharms and mersas throughout the northern and central Red Sea. Indeed, *Xenia*, *Sinularia*, *Sarcophyton* and *Lobophytum* are common in areas such as "holes" in the reef flat where coral diversity is greatly reduced due to high sedimentation and lack of water movement. On the face of it, there is no common cause for the absence of soft corals from such widely varying geographical localities. It may be argued that it is unwise to group together all soft corals, but at present insufficient is known about the ecological constraints on individual genera to do differently.

Where conditions are favourable for soft corals and stony corals, competition for space presumably occurs between them, and as noted above, turf algae are also an equally important space competitor. Studies on competition between these components are reviewed in Sheppard (1982a). Early Red Sea work on soft coral competition is by Benayahu and Loya (1981), though there has been a greater volume from Australia (Coll *et al.* 1982, Sammarco *et al.* 1983, 1985, La Barre *et al.* 1983, 1986) which demonstrates that alcyonarians exhibit growth modification, movement away from contact, stunting, scarring and bleaching, probably by allelopathy. These characteristics, caused by allelopathy and toxin secretion, appear to be effective in allowing soft corals to compete successfully against stony corals and hence in determining space partitioning on reefs in these areas.

Such work could be extremely important to several questions of reef maintenance. Where soft corals such as *Xenia* are overwhelmingly dominant to the point of exclusion of stony corals, as is the case in many embayments in the central and northern Red Sea, it is difficult to presume that reef growth is continuing as adequately as where soft and stony corals are more equally distributed. Present knowledge can only assume that sedimented and low energy areas favour the soft corals, but although they are known to possess several physiological attributes which would tend to improve their success (toxicity, movement, rapid vegetative reproduction), there is considerable room for further work on how these are exploited.

Summary

Arabian reefs contain high diversities of cryptic invertebrates, which are treated in this chapter by their known function in the ecosystem.

Cementation of carbonate sands is a crucial function, as important to reefs as the initial coral growth. Various chemical and biological mechanisms solidify particles, many modified by macrobiota which smother and stabilize sediment so that cementation may occur. Foraminifera are important in this respect, with 86 Red Sea species, in recorded densities of up to $5000\,m^{-2}$. *Homotrema rubrum* forms widespread, reddish coloured sheets on up to 25% of all concealed substrate and basal parts of corals, and forams also provide much sediment themselves. In addition, they provide a living veneer of tissue over dead coral skeleton which is otherwise prone to bioerosion. Bryozoans similarly bind coarse sediments. It is difficult to quantify their importance, though it is undoubtedly less than in geological times. Sponges, which were until recently thought to be neutral at best in the reef building process, are now known to be locally important, especially to deep reef construction. In the Red Sea they are also trophically important. Firstly they trap the abundant class of minute food particles and may be one of the principal groups preventing loss of this material from the tightly-cycled reef system. Secondly many contain symbiotic Cyanophyta, and add to primary production and nitrogen fixation.

Spur formation by calcareous red algae is very important constructionally, but in addition a group of mildly calcareous forms are abundant throughout the photic zone, which also stabilize sediments. Factors influencing development of such algae appear to closely resemble those for corals. Several coralline forms are highly resistant to intense insolation on reef flats where they may be particularly abundant.

Bioerosion is an incidental function of many groups, mainly as a consequence of feeding methods. Parrot fish are conspicuous eroders, but probably the most important bioeroders are echinoids, whose numbers themselves interact with numbers of their predators. Removal of predators has resulted in high population densities of echinoids, whose subsequent overgrazing may lead to erosion. Several small species graze in densities of up to 60 per $100 \, \text{m}^2$ on coral rubble and reef substrate. Not only does the complex network of interactions affect algal cover, but the reef under different combinations of grazer and predator may become structurally more complex. Some central Red Sea reefs lack the structural complexity found in the north because of this. Community effects by parrot fish include control of coral mortality by grazing of algal overgrowth, and influence of the morphology of the dominant coral species. Intertidally too, bioerosion averages 0.25 cm per year, more than the rate from wave action, in many places.

Alcyonaria are extremely important in the Red Sea, and parts of the Arabian Sea where upwelling is not severe. They are almost completely absent in the Gulf. The genus *Xenia* is very abundant, forming extensive carpets in sheltered and sedimented locations. It is highly fecund and competitive, and displaces reef building corals in such areas. Where soft corals dominate, it is difficult to presume that reef growth is as vigorous as where soft and stony corals are more equally distributed. However, in the southern Red Sea where soft corals gradually disappear, so too does vigorous reef growth. In the Gulf, soft corals occur only at the extreme eastern end. Low winter temperatures may be the important control here, but this cannot explain their absence from the southern Red Sea. Factors determining space partitioning on reefs in these areas are largely unknown, but could be extremely important to several questions of reef maintenance.

CHAPTER 6

Seaweeds and Seasonality in Arabian Seas

Contents

A. **Substrates and major distribution controls** 122
B. **Distributions of free-living seaweeds of Arabia** 127
   (1) Turf algae and algal lawns 127
   (2) The *Sargassum* community in the Red Sea and Gulf 131
   (3) Non-reef green algal flora 134
   (4) The seasonal kelp communities of southern Oman 135
C. **Microscopic forms and productivity** 137
Summary 140

Several areas with hard substrate around Arabia are not dominated by corals but by macroalgae instead. This may occur in shallow coral reef areas, when the algae tend to be filamentous greens and small browns which grow as "algal lawn". While such lawns are widespread on coral reefs of the Indo-Pacific, on reefs of the northern and central Red Sea they are almost universal. Elsewhere, large browns over 1 m tall may dominate. Areas where this occurs include the Gulf where severe natural stresses exist from salinity and temperature, for example in the Gulf of Salwah which lies between Qatar and Bahrain, while other macroalgal dominated areas are found along the Oman coast of the Arabian Sea. Similar conditions are likely to be found along parts of the Yemen coast, though this remains unconfirmed. Brown algae also dominate extensive parts of shallow hard substrate in the southern Red Sea.

Algal communities in most of these areas show a strong seasonality. Many appear to be annual, at least in terms of their fronds though many have substantial stipes and holdfasts which endure for several years. Their seasonality is correlated with water temperature which, for the Red Sea and Gulf, is coldest in winter but which, in the Arabian Sea, is coldest during the summer upwelling. Any discussion of algal communities and periods of vigorous and seasonal growth must include the question of competition, and in these seas the main competitor for hard substrate is corals which, of course, are not annual but endure for periods of many years and even centuries. Characteristics of coral communities of areas where algal seasonality is important are discussed in Chapter 12 where there is a synthesis of the effects of severe natural stresses on sublittoral communities, especially corals. The following focuses on algal communities.

A. Substrates and major distribution controls

Much, but by no means all, of the hard substrate dominated by macroalgae is non-reefal. Non-reefal hard substrate is defined here as that which is at present not accreting or growing. Obviously it includes all non-limestone surface rock, and all examples of these where algae have been studied to any extent are from central and southern Oman. Examples of these rocky substrates include ophiolite which provides important shoreline and sublittoral substrate; ophiolite is a broad suite of igneous rocks which originate from oceanic crust which has overridden continental crust due to tectonic movements. It is relatively rare on the surface (Clarke *et al.* 1986) and the Oman example is one of the largest coastal deposits known. Other non-limestone rocks in the sublittoral occur in southern Oman where the Dhofhar highlands meet the sea. These are mainly Tertiary rocks: shale, limestone and gypsum with chert and marly beds, and igneous rock. The latter occur particularly on the coastal terraces and in the Kuria Muria Islands (Clarke *et al.* 1986).

Non-reefal hard substrata may include ancient limestone formations as well. Indeed, on the ophiolite are limestone masses, termed "Oman exotics" (Clarke *et al.* 1986) which are believed to be parts of oceanic reefs brought to land when the oceanic crust overrode the continental crust. Other ancient limestone substrates in the sublittoral of Arabia include Tertiary limestone and dolomite strata of Musandam and parts of the Capital Area of Oman. In Musandam, the limestone forms strata which are tilted up to 10° from the horizontal, and these stratifications are as clearly visible underwater as they are above; corals which thrive on the steep slopes are attached to the Miocene limestone and there appears to have been no Recent limestone accretion.

Other limestone and sedimentary sublittoral rocks exist as platforms or raised domes which do not appear to be accreting. These are numerous in the Gulf near Bahrain, Saudi Arabia and Qatar. Many of the hard substrate structures close to shore along the Saudi Arabian Gulf coast, like those near Bahrain, are raised domes of limestone, some of which are uplifted by underlying salt deposits. As discussed in Chapter 1, similarly uplifted domes occur in the southern Red Sea. These have all been termed reefs in the past but most of the Gulf examples which lie nearest shore, as well as many southern Red Sea examples amongst the Farasans, are almost certainly not accreting. In the case of those structures within about 20 km of the shoreline of both areas, it is clear that there is not a sudden jump from reef-constructing coral communities to those which are dominated by macroalgae and which are not accreting. In the western Gulf, a group of several of these limestone structures were described in Tetra Tech (1982), whose important material is reproduced in McCain *et al.* (1984). Coral cover on these domes correlated with mean depth of water at their base; in other words, coral cover was greatest in sites where the vertical extent of hard substrate was greatest. Additionally, coral diversity increased with distance of the site from the shoreline.

In the Saudi Arabian series as with those off Bahrain, there is usually a gradation from a coral dominated to an algal dominated condition as stresses increase. From the point of view of the corals and reef growth, a severe constraint is the increasing presence of macroalgae. Where algal growth increases, either in seasonal pulses or as sustained communities, factors involving competition with corals become more and more important, and contribute to the change of dominant community from the limestone producing corals and calcareous red algae to fleshy green and brown algae. These aspects are discussed later.

Whether competition between corals and algae is direct or whether seasonal nutrient enrichment simply favours algae and disfavours corals has been studied in the Gulf, but without firm conclusion (Coles and Fadlallah 1991). Whatever the direct reason, there is a consequent demise of both corals and reef growth. Environmental conditions around Arabia include numerous cases where reef growth stops, and elsewhere in this book the annually averaged or prevailing environmental conditions where this happens are termed "marginal". Of course, what becomes marginal for corals becomes central from the point of view of some of the main communities of attached algae in the Arabian region. The point at which the shift takes place is of considerable interest ecologically, but unfortunately has not been particularly well studied mainly, it seems, because it is a point which is central neither to coral biology nor to phycology; it is at an uncomfortable extreme for both.

Although the presence of hard substrate is an obvious, principal requirement for algal attachment and growth, many tropical seaweeds, including some which appear to be important to productivity in Arabian seas, also grow well on banks of rubble and even on sand where disturbance is low. This applies to all three of the main groups considered here namely the Rhodophyta, Chlorophyta and Phaeophyta, given the general group names of Reds, Greens and Browns.

Aspects of control of algal distribution by temperature and upwelling are discussed later in this book. Of relevance here is control by light since this strongly affects vertical distribution, which interacts with control by grazing, and this is briefly discussed here. As is well known, tropical sea water differentially absorbs different wavelengths of light. Red light is absorbed within a very few metres, and blue light extends to depths of over 60 m in almost all parts of the Arabian region, and to over 200 m in most parts of the Red Sea. Dissolved salts and thus the varied salinities found around the Arabian peninsula do not themselves affect light absorption, but organic material does so strongly. Red Sea water is typically tropical in that it has low dissolved organic content except at its southern end, which distinguishes it to an appreciable degree from water in the Arabian Sea. Most information on the characteristics of illumination in water comes from work in temperate seas rather than tropical, and in temperate seas especially, even in clear conditions, the deepest penetrating wavelengths may be greenish hues due to the fact that dissolved organic substances ("Gelbstoff") strongly absorb blue light. Some of this material derives from terrestrial sources where there is run-off, but in addition a considerable contribution comes from phytoplankton itself. This is also very important in the Arabian Sea at least in the summer period as noted in Chapter 10, and briefly later in this chapter. Figure 6.1 shows profiles of light absorption which have been measured against depth for various parts of the Arabian region, with comparison to different Jerlov water types. Light in the marine environment is discussed very comprehensively in Drew (1983) who shows numerous plots of several additional parameters of

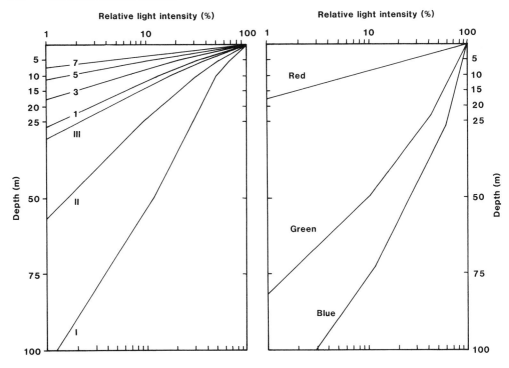

Figure 6.1 *Relative light intensity with depth. Left-hand graph shows relative intensity at different depths in different Jerlov seawater types. Northern and central Red Sea, including water over coral reefs, is typically Oceanic Type I or II. Gulf water offshore is typically Type 1–3, while inshore waters are typically less clear, around Type 5 or even 7. Right-hand graph shows relative transmission of different parts of the visible spectrum in northern Red Sea water (same axes as left-hand figure). Data are from Drew (1983) and Walker (1987).*

transmission of different wavelengths, photosynthetically active radiation (PAR) and light energy with depth under different conditions and water types.

Of possible importance to considerations of depth distributions of algae is the fact that each deeply penetrating blue quantum contains more than double the energy that a red quantum does, though there are indications that it is the individual quanta absorbed rather than total energy on which photosynthesis actually depends (Drew 1983). Green algae require more of the shorter wavelengths, and are thus restricted to shallower conditions. The depths at which algal lawns are found would seem to be clear evidence of this, but grazing is a very important complicating factor in the pattern. Vine (1974) used settlement plates in Port Sudan to determine the effects of grazing on algal growth, and from his data it can be seen that fastest growth rates of filamentous algae (on plates protected from grazers) occur at shallowest depths, and that growth decreased rapidly with increasing depth, until there was virtually no growth at 20 m (Figure 6.2). It is seen that the decline of ungrazed algal growth with depth matches almost exactly the decline in illumination. In contrast, the grazed algal distribution on his settlement plates resembled that seen on reefs naturally and

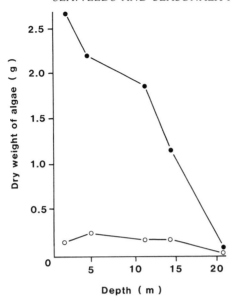

Figure 6.2 *Growth of filamentous algae on protected (closed circles) and unprotected (open circles) plates placed at different depths off Port Sudan and left* in situ *for 30 days. Values shown are g/225 cm². From Vine (1974).*

has no relation at all with illumination, whether measured as PAR, quanta or total radiation. Thus illumination controls the potential depth distribution of this very important algal component, but other factors modify it greatly.

Brown algae, which are also important components of the small turf algae, are also generally restricted to shallower water, but as pointed out by Walker (1987) this is also because they tend to have a higher proportion of non-photosynthesizing tissues and need greater intensity of illumination. The large browns *Sargassum* and *Turbinaria*, for example, thrive only in very shallow water. Red algae utilize the longer and more deeply penetrating blue and green wavelengths. It is tempting to suppose that the greater energy of blue light is an important consideration in determining the deepest penetration of red algae (see Drew 1983), but whether important or not, these red plants also tend to have energy conserving growth patterns with very low biomass, thin fronds and small stipes, which allows the leafy reds to grow to the base of the reef slope and beyond. The fact that they can do so at all is because of their accessory pigments. These absorb radiant energy from blue light and transfer it by a very efficient process of electron resonant transfer to special forms of chlorophyll a, which is the only pigment capable of mediating the conversion of radiant energy to chemical energy. The accessory pigments are numerous, but important to deeper groups of plants are phycobilins which are found in reds and in Cyanophyta, and carotenoids which are found in most algae. Figure 6.3 illustrates their active absorption regions, together with that of chlorophyll, as well as the existing spectrum at different depths in clear water of the Red Sea. Drew (1983) notes that if the photosynthetic pigments are removed from the photosynthetic apparatus,

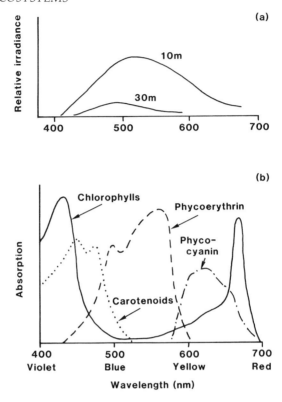

Figure 6.3 *The value of accessory pigments. (a) Relative irradiance of the visible light spectrum at different depths in the Red Sea; (b) absorption spectra of accessory pigments (same axes as in a). Data from Drew (1983), Walker (1987).*

they will re-emit the quanta they absorb from blue light as red (by chlorophylls) and orange (by phycobilins). It is not uncommon to see such fluorescence on red algae and even on corals at depths where ambient light is dim in the Red Sea, and indeed elsewhere in the Indian Ocean.

An exception to this pattern of red algae is found with the calcareous red algae, which tend to be restricted to very shallow areas, as discussed in Chapter 3.

Whether or not habitat is suitable for algae also depends on water movement. For many species, it appears that vigorous water movement is desirable or mandatory. It is possible that the low levels of dissolved nutrients in the sea water of the Red Sea and Gulf mean that only with adequate water flow past the fronds will the algae manage to absorb sufficient for their growth and maintenance. However, dense algal growth may also be found on unconsolidated substrate in many locations, and clearly the stability of such substrate depends on low water movement. Because of these factors, and others such as abundance of grazers, it is extremely difficult to predict the amount of algae that may be found in any particular area. Apparently similar areas differ widely, possibly because of differences in water movement, proximity of a bay with sediments and associated nitrogen fixing blue-greens, and other factors.

B. Distributions of free-living seaweeds of Arabia

According to Walker (1987), who has reviewed information on algae in the Red Sea, the overall productivity of green algae is considerably less than that of the browns. However, while this certainly applies to several special cases described in the following sections, the importance of the algal lawns in shallow waters, which are largely made up from green algal species and Cyanophyta, should be emphasized. Not only do they occupy an important and extremely widespread shallow water zone, but their high occupancy of substrate, commonly reaching over 80%, indicates a high productivity and high trophic importance. Generalizations therefore may be unwise, and most of the measurements reviewed in the following sections refer to specific locations and often to specific times of year too. Variability with seaweeds is extremely high, though generalizations are made where possible.

(1) Turf algae and algal lawn

Three groups of particularly abundant or conspicuous algae are discussed in separate sections later; the following deals with the remaining, usually small species. These are assemblages variously called turf algae, macroalgae and crustose algae (Steneck 1988) which, when seen covering extensive areas in the shallow reef slope, are commonly described as algal lawns. These assemblages are composed of a diverse group of leafy and filamentous species which mostly attain heights of 1 cm or less, and include greens, reds and browns in various proportions. Steneck (1988) discusses the importance of this diverse group, noting that research has shown them to provide significant sources of food for herbivores, and that high levels of biomass interact in many other reef processes, including a reduction of larval recruitment and growth of sessile invertebrates. They may, however, support enormous densities of small herbivores such as polychaetes. The following section describes abundances and production of turf algae where this is known. In most cases, the data come from studies of interactions of this group of algae with grazers, especially fish and urchins, and only a few though now increasing number of studies are devoted to the algae themselves.

The ground is now to some extent converted to a slippery sea-weed steppe, and the naked feet . . . slip among the luxuriant vegetation . . .   Klunzinger 1878

A) DIVERSITY AND DISTRIBUTION

Papenfuss (1968) recorded nearly 500 species of algae from the Red Sea, in a list derived from numerous earlier collectors. Basson *et al.* (1977) and Basson (1979) record only about a tenth of this number of fleshy algae from the Gulf, and there has been little systematic collecting subsequently to indicate what the true numbers are. That they are undoubtedly much higher is suggested by Jones (1986a) who records 13 greens, 21 browns and 21 reds from seashores alone in

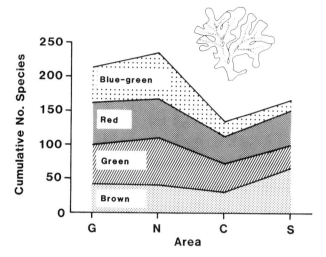

Figure 6.4 *General distribution pattern of algae in the Red Sea. Data from Walker (1987). G, gulfs of Aqaba and Suez; N, north; C, central; S, south Red Sea.*

Kuwait. In the Red Sea, the collections have been concentrated around the Sinai in the north and the Dahlak Archipelago in the south, which allowed Walker (1987) to remark on their latitudinal distribution. Biogeographical patterns are examined in a later chapter, but an important point here is that there is a latitudinal division across the central area, from about Suakin to Jeddah. From tables in Walker (1987) the pattern can be shown (Figure 6.4). All three groups are least diverse in the central area, greens are common mainly in the north, while browns have by far the greatest diversity in the south. How much this pattern is derived from incomplete and poor sampling in the central area is unclear, but it is possible that the greater range of temperature, and cooler winter temperatures in the northern half, restrict some tropical species and result in a north–south differentiation. This is dramatically observed, for example, with the *Sargassum* band along reef crests, as discussed later. Further, the demarcation for algae also follows that for coral communities and for fishes (see Chapters 3, 4) suggesting the action of strong environmental controls rather than a simple artefact of sample size.

Most species comprising the algal turf in northern and central areas of the Red Sea are macroscopic, non-calcareous forms of greens with browns and reds. One of the commonest species is the brown *Sphacelaria tribuloides*, which itself serves as a substrate for other epiphytic and turf algae including numerous species of greens. The reds provide the highest diversity in northern areas at least (see Benayahu and Loya 1977b).

In Oman, there is an increase in diversity of large brown algae towards Dhofhar, and these rich algal communities affected by upwelling are discussed in detail later. Intertidally in the Dhofhar region, the algal-colonized zone extends upwards from the high tide level by 1–3 m due to the severe wave splash experienced in the region (Barratt *et al.* 1986), though this zone has only a poor algal diversity. Four main leafy species manage to grow well in the intertidal region while wave energy is high and while the upwelling continues, but these die back rapidly from desiccation and high temperatures in early November, from which time small filamentous and encrusting algae replace the leafy species.

B) PRODUCTIVITY AND BIOMASS

Lewis (1977) reviewed algal productivity on reefs generally, though without many data on reefs from Arabian Seas where few direct measurements have been made. Generally, the standing stock of algae is small compared to the rate of production, due largely to intense grazing pressure. In general, algal lawns in the northern Red Sea and Gulf of Aqaba have standing stocks of approximately 25 g (dry wt) m^{-2}, a value similar to that of the epiphytic algae in seagrass beds (Walker 1987). The latter seems surprising until it is remembered that the area available for colonization in a seagrass bed is up to 12 m^2 for each 1 m^2 of seabed (Hulings 1979). The algal lawns have, therefore, a very high rate of production with a high turnover. Production rates of algal lawns in the Gulf of Aqaba are about 1–3 g (dry wt) m^{-2}d^{-1} (Walker 1982, in Walker 1987) which suggests that standing stock is only about 1–2 weeks growth. (Algae in areas not protected by territorial fish have even more startling standing stock to production ratios of perhaps 3–4 days growth. Standing stock is lower in such "unprotected" areas.) It is still fairly difficult to compare such values with other tropical areas, not so much because the persistent, non-standardized use of different units add confusion (e.g. wet wt, dry wt, g C, per day, per hour or per year, net and gross values) but because there is such a great range of values in the literature from different areas. Lewis (1977) for example tabulates gross production from several localities and observes a range of 319 to 7300 g C m^{-2}y^{-1} for a Hawaiian coral reef and a Caribbean lagoon respectively, with most values falling between 2000 and 5000 g C m^{-2}y^{-1}. It may be concluded that northern Red Sea values of productivity of algal turf fall within the global range of tropical reefs, but probably well towards the lower part of the range.

South from the Gulf of Aqaba, standing crops of turf algae on reef flats and reef slopes have not been measured in as much detail. While some data suggest that it might decline on reef flats, Roberts (1986) shows that on reef slopes standing crop was ten times higher on southern Red Sea sites. Well developed, shallow algal lawns in the central region have high cover too. Dart (1972) found that in the central region at Port Sudan, there was a maximum standing crop of 20 g m^{-2} at 3 m deep, which is in the zone where algae may cover over 90% of the substrate, and this decreased to 1 g m^{-2} at 10 m deep.

In the absence of production measurements, estimates of algal cover may be useful, though the amount of visible cover relates as much to the presence or absence of grazers as it does to the rate of production. In the central Red Sea at Yanbu (Sheppard, 1983a and unpublished) turf algae generally provided around 4% cover below 5 m deep, a similar value to that found in many sites further north in Sinai. In the algal lawn zone about 1–3 m deep in contrast, this component was 50–80% of the substrate of many reefs. Benayahu and Loya (1977a,b, 1981) recorded monthly averages of living cover of turf algae at several locations in Sinai, and found similar values on reef flats of 5–40% cover in most sites, rising to 70% and sometimes more in a few cases. On reef slopes algal cover values were 5–20% between the surface to 3 m, and "negligible" below 4 m in at least one site in the Gulf of Aqaba, and values of less than 1–2% were found elsewhere in the Red Sea by Mergner and Schuhmacher (1985). In view of the frequent high cover values, it is surprising that this component of surface cover has commonly been ignored completely in several ecological studies. This may indicate a general tendency to overlook small algae in studies on, for example, coelenterate cover, but the wide range of measured values shows that there is no doubt a considerable patchiness too.

These values refer to apparently clear and unpolluted sites, without evident eutrophication or significant disturbance from pollution. It is widely reported also that eutrophication leads to very high cover by algal turf as well, but extensive belts of algal lawn exist without such disturbance.

C) SEASONALITY

Seasonal changes in turf algae were measured in the Gulf of Aqaba by Benayahu and Loya (1977a). This is one of the most northerly reef sites of the region and so seasonal climatic fluctuations are fairly marked, though not as severe as in northern parts of the Gulf. They observed that turf algae communities existed throughout the year on reef flats and on the upper parts of reef slopes. Those in the latter region did not exhibit seasonal patterns, but provided a cover of about 10–20% throughout the year. Reef flats exhibited marked seasonal changes, however, with highest cover in winter and spring, from February to May for most sites, and with a significant increase towards the end of this period suggesting an intensive settlement of turf algae in April and May. This period of high cover is fairly short lived, and all reef flats showed lowest average living cover in June, July or August. The decrease of algal cover towards and during the summer is followed by widespread detachment of thalli from the substrate, and new young thalli start to appear again at the beginning of winter. Cover values varied considerably from site to site. Typically, summer low cover values were >10% of the substrate, while the spring high cover values varied from about 40% to 70% in two cases.

One of the most important larger algal species on numerous Red Sea reef flats is the brown *Turbinaria elatensis*. During the summer this is seen only infrequently in the north, and Benayahu and Loya (1977a) reported that it is completely absent on reef flats of the Gulf of Aqaba, though some holdfasts might be found. It is increasingly persistent towards the south and central Red Sea where seasonal differences of abundance evidently decline. In the far north, Benayahu and Loya (1977a) showed that the echinoid *Diadema setosum* is the main invertebrate grazer, and a significant negative correlation existed in density of *D. setosum* and annual turf algal cover. The echinoid *Echinometra mathaei* was found to be more abundant, but affected algal cover less.

In the Gulf of Aqaba, the seasonal cycle of algal development on reef flats usually starts with a drop in winter temperature to about 20–21 °C, which occurs in November and December, and algal cover continues to increase until about May when water temperature reaches 24–25 °C. Summer water temperatures occasionally exceed 30 °C and appear to exceed the tolerance of many turf algal species. From Chapter 2, it is seen that surface water temperatures are not as low in the central Red Sea. Temperatures as low as 20 °C are almost never reached, and values of 24–25 °C are general February values (equivalent to May values in the north). Summer temperatures in the central and southern regions are correspondingly higher, and because of these differences, reef-flat algae generally provide much lower cover than in the northern area. Seasonality is observed, but mainly because of a near absence of algae in the hotter months. This does not apply to reef slopes where water temperatures are both cooler and more constant, nor does it apply to the reef crest where a *Sargassum* zone exists, as described in the next section.

(2) The Sargassum community in the Red Sea and Gulf

A feature which is characteristic of "marginal" reef conditions is increasing dominance of some of the Fucales, large brown algae which include the genera *Sargassum*, *Cystoseira* and *Hormophysa*. The term marginal reef is commonly taken to mean higher latitude reef, and examples of well studied localities include the southern coasts of both eastern and western Australia (see Sheppard and Salm 1988 for summary), the Indian Ocean coast of southern Africa, and Japan. Around Arabia, the "condition" of higher latitude is not essential. Instead, both along the Red Sea and Arabian Sea coasts of Oman, increasingly marginal conditions (from several viewpoints) occur in a southerly direction, i.e. towards the equator. Also, similar conditions occur in the Gulf, where marginal conditions are generated more by extremes of environmental conditions, rather than by heightened conditions accompanying greater latitude. In all the above mentioned areas, there is a dramatic increase in the Fucales.

A very conspicuous trend which occurs on the crest of the fringing reefs in the Red Sea is the southward increase in large brown algae, mainly *Sargassum* (Figure 6.5). The greatest increase is seen in the south where there is a corresponding decline in the extent of reefs and coral diversity too (Sheppard 1985a). Growths of *Sargassum* of the extent seen in the south are not known from the northern Red Sea. The first notable occurrence of patches of this seaweed occur just north of Yanbu on the Saudi Arabian coast, where it grows in sheltered sharms, along with equal abundances of the brown alga *Turbinaria*. Just south of Yanbu, thick growths of *Sargassum* begin to appear fairly abruptly on the seaward edges of the fringing reef flats but these are not extensive until the region of Jeddah, where *Sargassum* reaches lengths of up to 0.5 m tall and forms a belt 3–5 m wide (Ormond *et al.* 1986b). South of Qunfidah this algal community is very well developed, and grows on either coral rock or on calcareous red algae along the reef crest. The plants are typically 1–2 m in length and form dense belts 5–15 m wide along the crests of reefs or on top of offshore patches of calcareous algal reef. The growths are almost universal in the southern Red Sea, at least on the fringing reefs and on the numerous patches of non-accreting limestone. In particular, where old and emergent reef flats dip into the sea, there will invariably be dense *Sargassum*. The southern patch reefs made from calcareous red algae

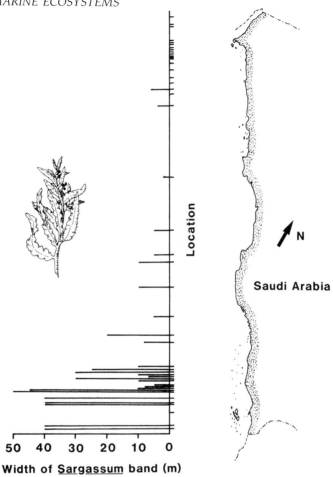

Figure 6.5 *Southward increase in width of* Sargassum *belt on reef crests. On y-axis, dashes to right side of axis show sample sites, indicating all sites including those where no* Sargassum *occurred. Data from Roberts (1986).*

noted in Chapter 3 also support dense growths of brown algae.

The *Sargassum* grows in thick stands, with numerous plants in each square metre, up to 2 m tall. The plants possess air bladders, so that the fronds reach the surface. The band in which this grows is generally 5–25 m wide but may reach 50 m wide in the far south, and it lines the extreme edge of the reef flat, continuing down the reef slope to not more than about 2 m deep. In the central and southern Red Sea it is dense enough to cause a noteworthy obstruction to swimming, and is very clearly visible on satellite images of the region; false colour chlorophyll enhancement shows a coloured line approximately delineating the edge of the reef flat. In the south, the patches of *Sargassum* may completely cover the tops of small patch reefs or submerged platforms.

The substrate on which the *Sargassum* grows varies with latitude (Sheppard 1985a). In the central region, it grows on reef crests. Where this is so, other reef crest fauna and flora are greatly reduced, and adjacent sites without the *Sargassum* are fully colonized by corals and calcareous red algae. *Sargassum* may attach both onto the calcareous red algae, and onto bare reef rock. Further south, calcareous red algae grow more abundantly, and *Sargassum* attaches firmly to this or to the dipping reef flats where they exist. *Sargassum* reaches its most luxuriant condition on the red algal reefs which are known only in the far southern Red Sea. Observations in the central Red Sea suggest that it is a constant component year round, although there is a degree of seasonality and frond shedding.

The Red Sea pattern has a parallel in the Gulf (Sheppard 1985d) at least so far as the *Sargassum* (together with *Hormophysa*) increases with abundance into areas where coral growth is restricted. In the Gulf of Salwah, the algal growth increases in a southerly direction in a manner similar to the Red Sea pattern.

But in both of these cases, no work has been done on the undoubtedly high productivity of the algae, or on the associated fauna, and no work has yet been possible to demonstrate any cause and effect in the reciprocal abundance of corals and algae. Indications of causes may come from work performed elsewhere, especially in the Houtman Abrolhos Islands off Western Australia (Crossland 1983, Hatcher 1985, Hatcher and Rimmer 1985, Johannes *et al.* 1983). In Houtman Abrolhos, seasonal growth of dense algal stands is high enough to overgrow the corals and kill them, probably by shading, suggesting that competition between the two main groups is important. These authors also questioned the traditional model which cited low temperature as being the main constraint on corals, as although temperatures in this location do drop seasonally below 18 °C, the area also exhibits high levels of dissolved nutrients which favour algal growth and inhibit coral growth.

Coles (1988) examined the question of temperature versus competition on the Gulf coast of Saudi Arabia where seasonal blooms occur which become dense enough to provide heavy shading of corals for several weeks at a time. *Sargassum* occurred in conjunction with several other algal species, but by itself covered up to 85% of the shallow reefs in some areas. In his study area, Coles (1988) found water temperature ranges of 13.5–36 °C for inshore reefs and 17–34 °C for offshore reefs, salinities of 39.5–41 ppt and 39–46 ppt for inshore and offshore respectively, and low levels of measured phosphate, nitrate, ammonia and silicate. Coles remarked that although substantial and seasonal algal growth occurred which led to considerable shading, some observations made were not consistent with the competition model of Johannes *et al.* (1983). There was, for example, no elevation of nutrients above levels normal for reefs, indicating that algal growths in the Gulf do not require stimulus from raised nutrient levels. Temperature stimulus was considered also, and it was found that maximum growth of *Sargassum* and *Colpomenia* (not a Fucales but a Dictyosiphonales) occurred on nearshore shallow reefs during winter months, which is opposite of the Australian condition where maximum *Sargassum* growth occurred in the annual maximum water temperatures. Coles (1988) had to conclude that controls of macroalgal blooms are complex and unclear. Certainly, different algal species respond in different ways, but the relative importance of competition and physical controls "will be established only by field and laboratory experiments on the long term effects of seasonal macroalgal growth on reef growth and survival."

Extrapolation of experimental work done in the Gulf to the Red Sea would be difficult, due to the different physical conditions. The southern Red Sea does not have such strong seasonal temperature differences as the Gulf, and the general decline in corals and even greater decline in soft corals (which are not present at all in the Gulf, however) may have totally different causes. On this question, cause and effect is still very unclear, even if the actual patterns and distributions are now fairly well known.

(3) Non-reef green algal flora

In Chapter 2, the Arabian Sea upwelling was described, and shown to be most marked in southern Oman and southern Yemen, and had effects which reduced in intensity northwards. The coast between about Masirah Island and the easternmost part of the Arabian mainland, Ras al Hadd, appears to be where there is a marked reduction, but cold, nutrient-rich upwelling continues at least as far as Muscat in the Gulf of Oman. Temperature data given in Green and Keech (1986) clearly indicate that upwelling effects lasting for a few weeks reach Muscat, and this effect also provides a simple explanation for reduced reef growth on the Capital Area of Oman.

East of Muscat, the shoreline remains rocky. Partly it is non-limestone and sedimentary, and partly it appears to consist of Tertiary limestones. It is dominated sublittorally, in winter at least, by dense growths of filamentous algae (Sheppard 1986a). The shoreline is of spectacular, vertical cliffs up to 100 m high, accessible only by sea except where interspersed by wadis cut through the rock and by sandy embayments. Hard substrate continues sublittorally to at least 8 m deep beneath the cliffs. In the north, this area supports a 95% cover of filamentous green algae, with conspicuous tufts of bushy red algae. Attached fauna includes <2% of both hard and soft corals.

Further southeast, encrusting red algae increases to provide 10% of the substrate in less than 3 m deep, but otherwise the same mix of green and red algae is dominant. This pattern continues southeast, with a gradually increasing abundance of soft corals, which eventually provide 80% cover as the easternmost tip is approached. Both soft and hard corals of this area are aposymbiotic and bleached. Sand patches in this area are thickly covered with films of blue-greens.

Clearly the algal-dominated substrates are highly productive, though no measurements of productivity are available from northern and central Oman. Densities of the grazing sea urchin *Diadema* reflect the high production, however: measured *Diadema* density was 3 per m^2 in all sites where the green algae dominated, tailing off to 1 per 25 m^2 where the algae was replaced by the soft corals (Sheppard 1986a).

The reciprocal nature of green algae/soft coral abundance may have a clear parallel with the *Sargassum*/hard coral abundance already noted in the Gulf and southern Red Sea. Again, the Oman environment is a complex one of water temperatures which are close to, or beyond, the tolerances of the coelenterate fauna which thrives only a few kilometres further northwest, and of competition between the floristic and faunal groups. Other areas of Oman support dense green algal lawns on non-reefal substrate. Especially notable perhaps is the far south, beyond the location where large browns and kelps dominate, to at least as far as the border with Yemen. In such areas, greens commonly provide over 75% cover of the rock, often in conjunction with browns, and are heavily grazed by both fish and invertebrates. No quantitative data are available regarding these undoubtedly important areas.

(4) The seasonal kelp communities of southern Oman

The rocky sublittoral of coastal promontories and of the Kuria Muria Islands of southern Oman have been described by Barratt *et al.* (1984) as being unique ecosystems. They are certainly unusual, and they probably extend well into Yemen, and in some aspects and some species, to Socotra Island and possibly also to the north African coast. Essentially, the sublittoral in those areas which experience the strongest degree of upwelling are dominated by macroalgae.

Some obvious parallels may be drawn with the southern Red Sea and Gulf *Sargassum* communities, especially since *Sargassum* is a ubiquitous component of the Oman marine flora too. In the Red Sea and Gulf, increased density of large brown algae, and their greater abundance and growth, appears to be a result of localized good water exchange and/or increased nutrients, though good data clearly demonstrating this are lacking. Also, both the Oman and Red Sea brown algal communities support understories of diverse, flourishing red algae, giving more than a merely superficially similar appearance. But the similarity may go little further than this, because the Arabian Sea brown algal communities are predominantly seasonal, stimulated by the summer upwelling when incoming water is both colder and nutrient rich, and because it involves the only kelp plant known in the region.

Broadly, there are two distinct algal communities (Barratt *et al.* 1984, 1988). Firstly there is one which is found mostly in the shallow sublittoral, dominated by the species *Sargassopsis zanardinii*, which is endemic to the Arabian coastline. This species contains air bladders, and coexists with *Sargassum*. *Sargassopsis* has a perennial holdfast, with new blades developing each spring. Most vegetative growth occurs in the summer monsoon, and in August growth rates of the single lamina of the plant reach 2 cm d^{-1} in plants which are 0.5 m tall. Growth of the lamina stops at about 0.8 m tall. A central reproductive spike develops in August and becomes the largest and most dominant part visually from about September to December, when it can reach over 1 m long. It contains air bladders, resulting in a floating canopy. After December, the whole erect part breaks off to form a floating canopy. New plants are initiated from zygotes as well as from regeneration of the remaining holdfasts.

Occurring deeper in the sublittoral zone is the kelp plant *Ecklonia radiata*, a genus found otherwise only in the southern hemisphere around Australia, New Zealand and southern Africa where its physiology has been studied (e.g. Novaczek 1984, Probyn and McQuaid 1985). The latter demonstrated the dependence of the kelp on high nutrient inputs, and Hatcher *et al.* (1987) point out that this genus has higher optimal and maximal temperatures than do most kelp genera, a characteristic presumably essential for its existence in the Arabian Sea. With regard to its location deeper in the sublittoral, Wood (1987) has shown that in Western Australia, sporophytes of *Ecklonia radiata* are strongly susceptible to damage from ultraviolet radiation. This excludes the plant from shallow parts of the sublittoral in that region, and the same undoubtedly applies in the Arabian Sea where latitude and radiation are comparable.

As with *Sargassopsis*, the growth of *Ecklonia* is also strongly seasonal. Although holdfasts of this species are perennial in southern hemisphere communities, in Oman most of the plants behave as an annual due to very severe dieback, though Barratt *et al.* (1988) have shown that some survive through the pre-monsoon period without growing. Frond growth commences with production of a single blade sporophyte which reaches 0.5 m long. Secondary blades develop laterally from this, and growth is roughly balanced by continuous erosion of the frond tips. Tertiary blades then develop, producing large plants, which can reach 1 m tall. These produce zoospores, which fall to the substrate and grow into creeping, filamentous gametophytes. The latter produces the large sporophyte.

Biomass and productivity of the algae are high. Most reported measurements are for *Sargassopsis*, for which over 60 g (dry wt) $m^{-2} d^{-1}$ was recorded from a sheltered site (Barratt *et al.* 1988). Maximum growth occurs during the monsoon when sea temperature is at its lowest and nutrients are at their highest levels. Production for *Ecklonia* is lower, at a maximum of just over 4 g (dry wt) $m^{-2} d^{-1}$. Of considerable interest are the total seasonal or annual production rates of the two plant dominated communities, calculated by the above authors. For *Sargassopsis*, an average value of 1.0 ± 0.38 kg (dry wt) m^{-2} was obtained for the season's production, though values were less at 0.69 ± 0.22 kg m^{-2} in more sheltered sites. Results were also calculated for different depths in the total, mixed community of a fairly exposed site. Expressed as kg (fresh wt) m^{-2}, at 3 m deep mean biomass (of all algae) was 4.9 ± 1.0, at 9 m it was 5.7 ± 1.6 while at 12 m deep it was 4.3 ± 1.2. Similar values to the last indicated also that the biomass sustained similar values to at least 25 m deep. *Sargassopsis* was the main component at 3 m, while at 9 m, the coralline red alga *Amphiroa* was a significant contributor too. At 12 m, the *Ecklonia* was dominant, contributing 3.2 kg of the total 4.3 kg (fresh wt) m^{-2}. In general, highest values were found in exposed and moderately exposed locations. The values are high during the growing season, and this is attributed in part to the very seasonal pattern of growth which occurs in Oman.

The flora beneath the algal communities includes at least 90 other species, with coralline red algae prominent in most depths. In shallow water bays on the Mirbat peninsula, a varied mix of smaller brown and green algae provide dense cover. Included in many areas are the green algal lawns of a kind apparently similar to those which are widespread in the Arabian region on reefs. Beneath the canopies of *Sargassum* and *Sargassopsis*, foliaceous reds and filamentous greens provide an abundant understorey of algae, along with, and possibly competing with, coelenterates, especially corals.

Of the diverse fauna, two invertebrate grazers are particularly important in community energetics: the abalone *Haliotis* sp. and the urchin *Echinometra mathaei*. Numbers recorded in transects by Barratt *et al.* (1988) were impressive, with densities of 4–6 individuals m^{-2} for the echinoderm, and up to 10 individuals m^{-2} for the mollusc. Other important grazers were the rabbitfish *Siganus oramin*, the Arabian pinfish *Diplodus noct* and the bream *Rhabdosargus sarba*. The first of these feeds mainly on *Sargassopsis*, which provides 66% of its diet, with *Ecklonia* providing another 12%. This grazer also appears to time its breeding to permit the young to find optimum shelter in the macroalgae. Another notable group of grazing fish are several species of parrotfish which, however, feed mainly on the algal lawns rather than on the macroalgae. Importantly, the green turtle *Chelonia mydas* is believed by Barratt *et al.* (1988) to be the most significant grazer in quantitative terms. While it was not seen grazing, it is known to be fairly abundant in the area (see Chapter 10) while seagrass, commonly a very important component of its diet, is not.

Loss of live plant material to grazing was estimated to be 22% for *Ecklonia* and 10% for *Sargassopsis* to which an additional 14% loss from damage occurred. In addition to this food input, these plants produce much beach detritus which support very large numbers of isopods, amphipods, oligochaetes and nematodes. The tentative summary of the fate of frond material (Barratt *et al.* 1988) indicates the importance of these plants to the region's total energetics. *Sargassopsis* loses 30% of its material in a year's cycle – 15% by the green turtle, 7.5% by the rabbitfish, 6% by urchins and 1.5% by molluscs. An additional 20% may be lost by necrosis and leakage of dissolved organic material while the plants float freely, and when stranded a further 15% is lost to crustaceans and 29% to oligochaetes. In the case of *Ecklonia*, 10% is estimated to be grazed while the plant is attached, 2% by the green turtle, 6% by rabbitfish and 1% by both echinoderms and molluscs. With this alga, about 25% erodes away from the distal parts of the fronds, and of this debris, urchins and amphipods each consume about 5%, with a further 1% taken by molluscs. In short, the ecosystem is strongly dependent on the primary productivity of these two brown algal species, as is that of temperate kelp forest sites.

C. Microscopic forms and productivity

The subject of phytoplankton is covered mainly in Chapter 10, where several productivity estimates are given, but some points concerning the seasonality of this are relevant here where general primary productivity is concerned. In Chapter 10, it is seen that measurements of productivity are few and are spread widely both geographically and seasonally, and may give a misleading impression. Certainly one or two sets of data presented there are not easily reconcilable, and seasonal patterns are still difficult to deduce in most parts of this region. In drawing together the very limited data on seasonality of pelagic primary productivity, it is shown that some parts of the Arabian region exhibit large changes which are more in line with temperate marine ecosystems than with tropical ones. Very little of the information derives from benthic algae, with the exceptions covered in earlier sections, so the conclusion for this is mainly inferred from the planktonic pattern.

The annual pattern of planktonic algae in the northern Red Sea follows only very roughly that described earlier for benthic, reef-flat algae as determined by Benayahu and Loya (1977a). Benthic algae showed their highest cover in winter and spring. They showed greatest recruitment at the end of this period in April–May, and had least cover in the summer. As noted, however, factors such as grazing are just as important in determining substrate cover by benthic algae as are factors determined by temperature and dissolved nutrients. In the northern Red Sea, the pattern of phytoplankton biomass is likewise low in summer and high in autumn in terms of cells m^{-3}, with values of 180 and 14,500 cells m^{-3} respectively (see Weikert 1987). In terms of productivity, measured as gCm^{-2}, the pattern is one of low values in both summer and winter, but it is three times higher in autumn. Further differences are shown in data of Ponomareva (1968) and Dowidar (1983). The former determined that summer was the most productive season in the Red Sea, while the latter determined that the peak was in winter with a secondary peak in summer too, while Weikert (1987) concluded that the production peak is in winter. At present, the consensus is that peak production and biomass is in winter, fertilized by increased turnover in the water column, especially near coastlines and, in the north, around the entrances to the Gulfs of Suez and Aqaba where the main flow overturns, and fertilized in the south from nutrients and short lived plankton which are swept in from the Gulf of Aden. Weikert (1987) estimated that about half of the annually averaged pelagic production of the Gulf of Aqaba ($160\,gCm^{-2}$) occurs in the three winter months of December to February.

These data cannot be interpreted easily at present, other than commenting that coastal effects may be complicating the pattern observed by different authors, and that productivity and biomass estimates do not show a better correlation in the planktonic system than they do in the benthic system. In the Gulf, Basson *et al.* (1977) showed that winter plankton biomass is about double that of the summer values (the overall numbers being many times greater than those of the northern and central Red Sea). In both areas, it appears also that zooplankton abundance lags behind that of phytoplankton by a few weeks.

In the southern Red Sea, overall values of production and biomass differ considerably from those in the north, though there may be a similar seasonal pattern. In the southern part, there are much greater densities of phytoplankton (see Chapter 10 for details) but this is caused to large extent by the inward flow from the Gulf of Aden. This is especially important in the winter when the complex contra-flow system in the Bab el Mandeb results in a strong influx of surface water (see Chapter 2). Much of the incoming plankton survives only a few weeks, but it contributes the greatest component of the measured values, and its decomposition provides more nutrients. Because values for the northern and central parts of the Red Sea are similar and small, Weikert (1987) assumed that southern Red Sea production and biomass was similar to northern production and biomass, and estimated that the difference between the latter and measured biomass gives the amount due to influx from the Gulf of Aden. According to Weikert, the input of mesoplankton is 96×10^{13} organisms, or 6×10^4 tonnes dry weight per month at peak times, and an equally important influx of zooplankton also occurs (Kimor 1973). This input is more than enough to camouflage any internal seasonal changes which take place within the southern Red Sea.

Another factor complicating the interpretation of data on seasonal planktonic productivity and biomass is blooms of the "red tide" alga *Oscillatoria erythraeum*. Blooms are common in both the Gulf and Red Sea mainly in summer and autumn, when productivity of other planktonic groups may be lowest and when nutrients are lowest (Weikert 1987, Shaikh *et al.* 1986). The ability of this species to fix nitrogen makes it partly immune from the seasonal factors which control pelagic productivity. Their abundance, which may produce concentrations of 9×10^5 cells l^{-1}, and their patchiness, both hide any seasonal pattern. It is tempting to speculate that some of the very high plankton density values noted here and elsewhere are the result of such blooms.

In the Arabian Sea, the benthic productivity cycle has been described. Planktonic cycles remain largely unmeasured, since emphasis has been mainly on the high values in the summer upwelling period, with little quantitative data for the low winter levels. As with benthic aspects in this area, however, it is clear that the upwelling leads to a marked seasonality, whose peaks are temporally reciprocal to those in the Gulf and Red Sea. It is interesting to note that most of the flux of plankton from the Gulf of Aden into the Red Sea, which so greatly increases the planktonic biomass of the latter, occurs when there is no upwelling in the Gulf of Aden. Therefore, even in its winter months of lowest productivity, the Arabian Sea has a similar, or more probably greater, planktonic productivity than does the Red Sea in the latter's most productive months. In the Arabian Sea in winter, zooplankton biomass has been recorded as being between 6 and $25 \, \mathrm{mg \, m^{-3}}$ dry weight, which is rather higher than available Red Sea values of 30–$40 \, \mathrm{mg \, m^{-3}}$ *wet* weight (Weikert 1987). In the Southwest Monsoon when upwelling is strongest, pelagic productivity in the Arabian Sea is considerably greater still.

Zooplankton seasonality has been little studied. The observation was made above that zooplankton peaks followed phytoplankton peaks by a few weeks, so that spring and early summer sees greatest abundance. However, invertebrate larvae may contribute well over 90% of the total zooplankton found in inshore waters of the Red Sea (Shlesinger and Loya 1985) and probably of coastal waters of the Gulf as well, so that seasonality of zooplankton in general is determined by the reproductive pattern of benthic invertebrates. As noted in an earlier chapter, for many benthic species this peaks in the spring, though many species continue to spawn for extended periods from spring to autumn.

The Red Sea has a poor pelagic productivity in general (Weikert 1987) if the Gulf of Aden influx is excluded. In this sense it is typically tropical, and is typical of coral reef dominated ecosystems elsewhere. In contrast, pelagic productivity of the Gulf and Arabian Sea is considerably higher. How much higher depends on whose data are used, but values range from double, to one or two orders of magnitude greater (see Chapter 10). In this respect, pelagic productivity values approach those of temperate marine habitats. In the case of the Arabian Sea this is not surprising, since temperatures during the upwelling fall to temperate sea values. In the Gulf it may be attributable to the shallow nature of the water and to the fact that the sediments throughout are within the illuminated zone. The interesting point is that in the Arabian Sea, which is the body of water nearest to the equator, the seasonality is reversed.

Summary

Areas dominated by macroalgae are extensive on both reefs and non-accreting substrates. Brown algae in most depths are of small species, while large forms occur on reef crests and in the Arabian Sea where upwelling is important. Greens and reds are ubiquitous, the latter including members which grow deeper in the Red Sea than anywhere else due to their utilization of blue light and energy conserving growth patterns. Calcareous red algae are mainly restricted to very shallow areas. For many species vigorous water movement is essential, though dense algal growth also occurs on unconsolidated substrate, which it helps to stabilize.

All groups contribute to algal lawns. The latter commonly provide over 80% cover in a narrow zone where they have a corresponding trophic importance. As well as being a food source, their high cover affects coral recruitment and growth while supporting enormous densities of small herbivores. On reefs, the principal control by light on algal distribution is strongly complicated by grazing. Whereas in an absence of grazing algal lawn growth declines with depth in parallel with illumination, in natural conditions algal cover and illumination correlate poorly. Algal lawns on some reefs have standing stocks of approximately 25 g (dry wt) m^{-2}, and production rates of about 1–3 g (dry wt) $m^{-2}d^{-1}$, so standing stock is equivalent to only about 1–2 weeks growth. This productivity falls well towards the lower part of the range of reefs globally. Standing crop at 3 m might be 20 g m^{-2}, but it decreases to 1 g m^{-2} at 10 m deep. Another measure, cover, similarly drops from 50–90% of substrate in shallow water to 4% cover below 5 m deep in many Red Sea sites. These values, however, exhibit seasonal changes, ranging in one set of observations from a summer low value of >10% to a high spring value of 40–70%. Seasonal changes are not determined solely by temperature but by complex interactions with grazers.

In Oman, growth of kelp and other large brown algae is also strongly seasonal. Production is 4–60 g (dry wt) $m^{-2}d^{-1}$. The vegetation beneath the algal canopies includes at least 90 other species and many areas show algal lawns similar to those on reefs. Grazing and erosion of plants is high, with a significant part of the production entering the energy flow via detritus.

In the Gulf there is usually a gradation from coral to algal domination on limestone platforms as stress increases, causing a demise of both corals and reef growth. In these conditions, usually termed "marginal", algal dominance arises from shading, greater tolerance to temperatures below 18 °C and to high levels of dissolved nutrients.

Seasonality of pelagic primary productivity is similar to that of benthic productivity, though more marked. Biomass is low in summer and high in autumn with one comparable set of values being 180 and 14 500 cells m^{-3} respectively. Peak production and biomass in winter is fertilized by increased turnover in the water column, and by nutrients swept in from the Gulf of Aden in that area. About half the pelagic production of the Gulf of Aqaba (160 g C m^{-2}) occurs in three winter months. In the Gulf too, winter phytoplankton biomass is double that of the summer values, with overall values being much greater than for the Red Sea. In the Arabian Sea, summer upwelling leads to productivity peaks, but even in the winter months, values are higher than those of the Red Sea at its peak. Zooplankton biomass shows a broadly similar pattern.

CHAPTER 7

Seagrasses and Other Dynamic Substrates

Contents

A. **Introduction** 141
B. **Seagrass species along environmental gradients** 143
    (1) Species composition 143
    (2) Distribution of seagrass beds 144
C. **Ecology and dynamics of seagrass beds** 145
    (1) Community structure 145
    (2) Development and growth 146
    (3) Standing crop and productivity 148
    (4) Epiphytes 150
    (5) Consumption of seagrass and energy budgets 150
    (6) Faunal assemblages 151
D. **Role of seagrasses in the Arabian marine environment** 153
    (1) Importance and role in fisheries production 153
    (2) Energy and nutrient interactions with other ecosystems 155
E. **Other soft subtidal ecosystems** 156
    (1) Physical features and community composition 156
    (2) Faunal abundance and biomass 158
    (3) Soft substratum coral sand faunas 159
Summary 159

A. Introduction

Seagrasses are attached plants found on soft substrata, with leaves above-ground and interconnected stems, or rhizomes, and roots below-ground (Figure 7.1). Seagrass beds resemble underwater meadows, often of rather uniform appearance. However, closer study reveals the structure and ecological workings of this aquatic vegetation to be heterogeneous and highly complex.

Found in virtually all seas except in polar areas, seagrasses are among the commonest of shallow, coastal ecosystems. Seagrasses are the only group of flowering plants able to withstand permanent submergence in the sea. They belong to two families (Potamogetonaceae and Hydrocharitaceae) of a single order of the Monocotyledons. Worldwide, about 50 species within 12 genera are recognized (den Hartog 1970, 1979, Phillips and Menez 1988). Despite their name, seagrasses are not true grasses, but more closely related to pond weeds. Like mangroves and saltmarsh halophytes, they comprise an assemblage of species

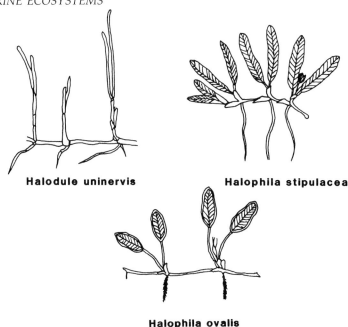

Halodule uninervis

Halophila stipulacea

Halophila ovalis

Figure 7.1 *Diagram of* Halodule uninervis, Halophila stipulacea *and* Halophila ovalis, *showing the main external morphological features of seagrass plants (from Jones et al. 1987).*

that are not closely related. The term "seagrass" therefore has only ecological, not phylogenetic, validity (see Barnes and Hughes 1982).

Even definition of a "seagrass ecosystem" is problematic (see den Hartog 1980, IUCN 1987). Seagrasses occur with other habitat types (e.g. rock/macroalgae, sand and coral), sometimes within an area of a few square metres. Such ecosystems are clearly difficult to classify convincingly, and can only be described as mixed, or named according to the particular dominant habitat (see Basson *et al.* 1977, IUCN 1987). Seagrass ecosystems are taken here to include those dominated, qualitatively or quantitatively, by monospecific or mixed stands of seagrass plants. Recent conceptual models characterize seagrass ecosystems according to structural, functional and evolutionary attributes (see Phillips and Menez 1988); hence as complex dynamical systems.

The depth range of seagrasses varies according to species, with some individual species occupying a wide range (Crossland *et al.* 1987). Light is believed to be the most important factor determining depth stratification (Backman and Barilotti 1976, Hulings 1979), although pressure may also be important (Beer and Waisel 1982). While seagrasses may extend exceptionally to 90 m depth (Barnes and Hughes, 1982), the majority are restricted to shallower, better illuminated waters. Some hardy species (e.g. *Halodule uninervis* and *Halophila stipulacea*) in the Arabian region can even withstand brief emersion and exposure on the lower intertidal zone. Sediment grain size is an important factor controlling the composition of grass beds, as in the case of benthos in general (Jones *et al.* 1987).

Seagrass beds are much less spectacular visually than coral reefs. However, their ecological significance may well be comparable, although still not well understood (Phillips and McRoy 1980). Their importance as primary producers in coastal environments, for instance in sustaining fisheries, was proposed as far back as the turn of the century (Petersen 1913, 1918). However, Petersen incorrectly believed the role of seagrass to be primarily one of providing a direct food source for fish, crustaceans and smaller prey animals. For this and other reasons, interest in seagrasses thereafter declined (Phillips and McRoy 1980, Jones *et al.* 1987). Only in recent years has there been a resurgence of seagrass research. It is now recognized that seagrasses are one of the richest and most productive ecosystems, rivalling even cultivated tropical agriculture in productivity. Like other marine macrophytes and coral reefs, productivity is very high in comparison with their relatively limited areal extent. Such ecosystems, sometimes termed "critical marine habitats" (Ray 1976), may therefore be regarded as "coastal food factories" (Barnes and Hughes 1982). Often their productivity is enhanced by encrusting algal epiphytes.

The following are important ecological attributes of seagrasses: stabilization of the seabed against wave action and other erosional forces; promotion of sedimentation and accumulation of organic and inorganic matter; providing a direct source of food for a few herbivorous animals such as some urchins and fish, green turtles and dugongs and, more importantly, as a major input to detrital food chains, which provide an indirect source of food for many more marine organisms; providing a nursery or refuge area (and often also feeding area) for resident and migratory fauna (adult and juvenile), including fish, crustaceans, molluscs and other invertebrates, many of which are economically important (e.g. penaeid shrimp, pearl oysters); providing the site for important biogeochemical processes (e.g. the sulphate cycle).

This chapter mainly considers the ecology of coastal ecosystems of the Arabian region in which seagrasses are a dominant or conspicuous element, but also includes a brief account of soft subtidal ecosystems devoid of seagrasses, other macrophytes or reef corals.

B. Seagrass species along environmental gradients

(1) *Species composition*

Eleven species of seagrass have been recorded from the Arabian region. These belong to seven genera, the total number known for the tropical Indo-West Pacific region. Details of species composition within each area of the Arabian region are summarized in Table 7.1. Morphological features of three common species are shown in Figure 7.1 (see also den Hartog 1970, Basson *et al.* 1977, Jones *et al.* 1987, Phillips and Menez 1988).

The Red Sea, including its two northern Gulfs, is relatively diverse (11 seagrass species), whereas only four species have been reported from SE Arabia and the Gulf. Other biotic groups in the Gulf display similar impoverishment (see Chapters 3, 4, 6, 11 and 12; Basson *et al.* 1977, Price 1982a,c, Coles and McCain

Table 7.1 Seagrass species within different areas of the Arabian region

Seagrass species	Gulf of Suez	Gulf of Aqaba	Red Sea	SE Arabia	The Gulf
Halodule uninervis	+	+	+	+	+
Cymodocea rotundata	+	?	+		
Cymodocea serrulata		+	+	+	
Syringodium isoetifolium	+	+	+	+	+
Thalassodendron ciliatum	+	+	+		
Enhalus acoroides			+		
Thalassia hemprichii	+	+	+		
Halophila decipiens	+				
Halophila ovalis	+	+	+	+	+
Halophila minor (= H. ovata)			+		
Halophila stipulacea	+	+	+		+
Total number: 11	8	7 (?8)	10	4	4

1990). Most seagrasses in the Arabian region are widely distributed in the Indo-Pacific, while *H. decipiens* is also found in the tropical Atlantic (den Hartog 1970, Phillips and Menez 1988). Another species, *Posidonia oceanica*, has been reported from Yemen (Banaimoon 1988), but this record is incorrect (D.I. Walker, pers. comm.). The occurrence of individual seagrass species is probably controlled by low temperature, high salinity and complex interactions of other environmental factors. However, small-scale spatial variations in species occurrence are apparent which cannot yet be explained fully (Price *et al.* 1988).

(2) Distribution of seagrass beds

The distribution of seagrass beds has been mapped along the eastern Red Sea. A progressive southerly increase ($p<0.05$) in abundance occurs, despite the reverse trend shown by some individual species (Price *et al.* 1988). The overall increase in development to the south is attributed principally to the wider and shallower shelf, as well as the greater prevalence of unconsolidated sediments. These factors, together with the generally less extreme temperatures and salinities, may have enabled seagrass beds to develop to their fullest extent in the south in the same manner as mangroves and the reverse pattern shown by corals.

Seagrass distribution along the coast of SE Arabia is poorly known. Grass beds appear to be uncommon along much of the Oman coast (Barratt *et al.* 1984, Jones 1985, Salm and Dobbin 1986, Salm *et al.* 1988). However, limited areas of dense seagrass have been recorded in central Oman (Clarke *et al.* 1986), southern Oman (Barratt *et al.* 1984), and Gulf of Aden (Hirth *et al.* 1973).

Seagrasses in the Gulf show a complex distribution pattern, reflecting the heterogeneous nature of the seabed and fluctuating oceanographic conditions. In Saudi Gulf waters, well-developed stands occur within a number of shallow

(<10 m) coastal embayments (Basson *et al.* 1977, IUCN/UNEP 1985a, IUCN 1987). However, the total area within these waters may constitute only a small proportion (1%) of the subtidal zone (IUCN 1987, Price *et al.* 1987a). There is a significant correlation ($r_S = 0.44$, $p<0.01$) between total seagrass cover and latitude (Price 1990). This is much stronger than that between seagrass cover and salinity ($r_S = -0.05$, $p>0.05$), even though salinity and latitude together show strong negative correlation. Around Bahrain, where water depths are mostly <10 m, seagrasses are more common. Their distribution has been mapped in detail based on satellite imagery (Vousden 1988), a technique also used on the Saudi Gulf coast (IUCN 1987). Well-developed stands in Bahrain occur principally along the southeast coast. Seagrasses are also extensive in southern parts of the Gulf, off the coasts of Qatar and UAE, but less so on the coast of Iran (Jones 1985), and Kuwait (IUCN/UNEP 1985a).

C. Ecology and dynamics of seagrass beds

(1) Community structure

On the eastern Red Sea coast seagrass assemblages have been identified from cluster analysis using species cover data (Price *et al.* 1988). At a broad level, this revealed three groupings, separating by latitude, suggesting biogeographic trends. To further define community structure, cluster analysis was performed using abundance data from 212 sites in the southern Red Sea and this indicated six distinctive assemblages. Three assemblages were dominated by a single seagrass species (*Thalassia hemprichii*, *Halophila ovalis* and *Halodule uninervis*).

Along the Sinai coast Lipkin (1977) differentiated communities of seagrasses on species composition. In a later study (Lipkin 1979) seagrasses were divided into three communities according to biomass characteristics. The first was composed of monospecific *Thalassodendron ciliatum*, reaching (remarkably) over 100 kg dry wt m^{-2}. The second group was dominated by *Thalassia hemprichii*, *Cymodocea rotundata*, *Syringodium isoetifolium* and *Halophila stipulacea*, as well as all subtidal and intertidal communities dominated by *Halodule uninervis* (biomass several hundred g to a few kg dry wt m^{-2}). The third group comprised all communities dominated by *Halophila ovalis*, some intertidal communities of *Halodule uninervis* and seedling populations of *Halophila stipulacea*; biomass of these communities reached only a few dozen g dry wt m^{-2}. Further information on community structure of seagrasses in the Gulf of Aqaba is given in the literature (e.g. Lipkin 1977, Hulings 1979, Hulings and Kirkman 1982, Wahbeh 1980).

In the Gulf, most seagrass communities are dominated by *Halodule uninervis*, although mixed stands also occur (Basson *et al.* 1977, Price and Coles in press).

(2) Development and growth

i) Grass beds as a dynamic system

Seagrass beds are highly dynamic. They reflect the balance of seagrass colonization, growth and subsequent trapping of sediments; and recession, through erosion caused by water movement and sediment loss, and through biological factors, such as senescence, consumption and competition. The dynamics of seagrass beds in the Gulf have been described qualitatively (Basson *et al.* 1977). In sheltered locations, flat, homogeneous meadows are formed. In environments of stronger wave or tidal action, erosional forces create several different topographical features. Off open beaches, sand belts parallel to the shore, and rip-current channels aligned perpendicular to them, are features typically associated with subtidal seagrass beds. In bays and other sheltered environments, where tidal streams rather than wave action are the main form of energy, "blow-outs" and tidal channels are produced (Basson *et al.* 1977).

The canopy structure of the larger seagrasses in turn influences the abiotic environment. In particular, the presence of vegetation dampens water movements, promoting sedimentation. The magnitude of this effect is a function of both seagrass cover and species composition. The more massive, broad-leaved species, such as *Enhalus acoroides* and *Thalassodendron ciliatum*, clearly have a greater dampening effect than smaller, narrow bladed forms such as *Halodule uninervis* (Jones *et al.* 1987).

ii) Controlling factors

Seagrass vegetation and species composition is controlled by several interactive factors. In the Red Sea (Lipkin 1977, Jones *et al.* 1987, Crossland *et al.* 1987) and the Gulf (Basson *et al.* 1977), its presence is probably governed by light, a function of depth and water turbidity, substratum type, and water movements.

Light and depth: In the Gulf, which is turbid in comparison with the Red Sea, seagrasses are often restricted to depths of 8–10 m or less, but exceptionally occur at 17 m depth (Basson *et al.* 1977). In the better illuminated waters of the Red Sea and Gulf of Aqaba seagrasses extend deeper. *Halophila stipulacea* and *Halophila ovalis* are known to occur in the Gulf of Aqaba at depths of at least 70 m (Lipkin 1979) and 28 m (Hulings and Kirkman 1982) respectively. Perhaps surprising is the occurrence of *Halophila decipiens* at 30 m depth in the Gulf of Suez (Jacobs and Dicks 1985), since this body of water is relatively turbid and and shares many biophysical features with the Gulf (Price 1982a). The majority of species in the Red Sea region, however, are restricted to depths of 10 m or less (Jones *et al.* 1987). Depth ranges of individual species are discussed further in the literature (Lipkin 1977, 1979, Wahbeh 1980, Crossland *et al.* 1987, Jones *et al.* 1987). While some species (e.g. *Halodule uninervis* and *Halophila stipulacea*) extend to intertidal areas, the majority are subtidal. Intense solar radiation and high temperatures during summer, particularly in the Gulf, are thought to be a major factor controlling the upper depth limit.

The morphology of certain seagrass species varies with depth, at least in the

Red Sea. In *H. stipulacea*, intertidal populations tend to have shorter and narrower leaves, whereas those of deeper, poorer lit waters are generally greener with longer and broader leaves (Lipkin 1977, 1979, Hulings 1979, Jones *et al.* 1987) at least down to 20–25 m. This suggests that larger leaves at depth might be an adaptation to reduced light. However, at depths greater than about 25 m leaf length and leaf width appear to decrease, a pattern that cannot be easily explained (Hulings 1979, Hulings and Kirkman 1982). Observed differences probably result from interactions of several environmental factors (e.g. light, habitat) and/or inherent characteristics (see Hulings 1979, Lipkin 1979). In *T. ciliatum*, height in shallow water (c. 60 cm), compared to deep water (up to 60–100 cm), is attributed to decreased wave action at depth, allowing greater plant growth (Jones *et al.* 1987). Other likely effects of depth include the changes in leaf:rhizome and root biomass ratios in *H. stipulacea* (Lipkin 1979, Jones *et al.* 1987). In intertidal populations the ratio is approximately 1:2, whereas at depths of 10–30 m the ratio is about 2:1. This reflects the need for proportionally greater root development and anchorage in the turbulent shallows, and greater light capture at depth (see Jones *et al.* 1987).

Substratum: Seagrasses of the Red Sea, as elsewhere, have colonized a range of unconsolidated sediments (Lipkin 1979, Price *et al.* 1988). In the Gulfs of Suez and Aqaba, fine to coarse sands are favoured, although certain species (e.g. *Halophila stipulacea*) have also colonized environments with fine-grained sediments (Hulings and Kirkman 1982), and even lagoons containing fine mud (Lipkin 1979). On the eastern Red Sea coast, *Thalassodendron ciliatum* and *Thalassia hemprichii* mostly inhabit coarser sediments, while *Enhalus acoroides* is characteristic of very soft mud substrata (Price *et al.* 1988). The remaining species are associated with sediments of intermediate grain size (Price *et al.* 1988). In the Gulf, seagrasses are generally associated with relatively fine-grained sediment types. Highest biomass is associated with finer sediments (see section 3 below), although sediment preference of individual species is not known in detail.

Water movements: Some effects of water movement on development have been described above. In the Gulf of Aqaba, *Halophila stipulacea* grew better under protected and shaded conditions, whereas growth in *Halodule uninervis* was better under exposed conditions and high illumination (Wahbeh 1980).

Other factors: These include temperature and salinity, both of which show greater extremes in the Gulf than in the Red Sea (Chapter 2). These factors may be largely responsible for the restricted flora (four species) in the Gulf, although seagrass biomass is not significantly correlated with salinity (see section 3 below). In the Gulf, in particular, the development of seagrass beds is influenced strongly by seasonal weather and environmental patterns (Basson *et al.* 1977, Vousden 1988). During winter, cover may drop from 90–100% to only 30% or less (Vousden 1988).

iii) Growth and reproduction

Studies of *Halophila stipulacea* in the Gulf of Aqaba indicate leaf biomass turnover rates to be 1.1–1.4% d^{-1}, based on clipping experiments, and 0.9% d^{-1} based on growth of leaves (Wahbeh 1980). The lower values compared with studies elsewhere were attributed partly to the effects of depth: 3–20 m in the Gulf of Aqaba, compared with less than 2 m in the other studies (Wahbeh 1980).

Reproduction takes place asexually through vegetative growth, and less commonly sexually by flowering, pollen production and fertilization, followed by seed production and dispersal. The frequent occurrence of monospecific stands and limited incidence of truly mixed populations in the Red Sea probably reflects this phenomenon (Lipkin 1977, Jones *et al.* 1987), although this is not a universal pattern (Price *et al.* 1988). In the northern Red Sea the majority of seagrasses flower between June and September (Jones *et al.* 1987).

(3) Standing crop and productivity

A wide range of values (1.6–3100 g m^{-2} dry wt) of standing crop is reported for seagrasses in different parts of the world (Phillips and McRoy 1980). McRoy and McMillan (1977) give figures of 20–8100 g m^{-2} dry wt for tropical seagrass beds, and values for some Red Sea species are even greater (see section C1 above). The high variablity is a reflection of differences in size or biomass among species (Jones *et al.* 1987) and a wide range of environmental factors, including season, depth and temperature. Unlike algal macrophytes (e.g. *Sargassum*, *Hormophysa*), seagrasses have a vascular system, enabling nutrients to be taken in from sediments through the roots, and also from the water column through the leaves, and then transported internally. Cycling of nutrients between the host seagrass plant and associated epiphytes may enhance productivity of the seagrass ecosystem as a whole (see section 4 below).

Studies on seagrass standing crop in the Arabian region have been undertaken mostly in the Red Sea, particularly in and around its northern Gulfs. Data are summarized in Table 7.2. Apart from *Thalassodendron ciliatum* and *Thalassia hemprichii*, which have relatively high biomasses, other species studied in the Red Sea have biomass values well below 1 kg dry wt m^{-2}.

In the western Gulf preliminary biomass estimates have been determined for *H. uninervis* dominated communities (Table 7.2). These are comparable to figures for the Red Sea and other parts of the world, where environmental conditions are generally less extreme than in the Gulf (Price and Coles, in press). Seagrass biomass in the Gulf was found to be significantly negatively correlated with depth and sediment grain size, but not with season, salinity or nutrient concentrations NH_4^+, NO_{3+2}^-, PO_4^{3-}, SiO_4^{4-} and measured heavy metals (Cu, V) (Price and Coles, in press). These authors found the third highest mean biomass was associated with high salinity (53.4 ppt). In contrast, seagrass biomass in other regions has been found to be negatively correlated with salinity (e.g. Walker 1985, Walker and McComb 1990).

Standing crop measurements provide a useful index of food available, directly

Table 7.2 Estimates of mean standing crop for seagrasses in the Red Sea and the Gulf.

Dominant species	Region	Standing crop (kg dry wt m^{-2})	Author
Thalassodendron ciliatum	Gulf of Aqaba	>100 (max.)	Lipkin (1979)
Thalassia hemprichii	Gulf of Aqaba	2.5	Lipkin (1979)
	Red Sea (Jeddah)	c. 0.5[a]	Aleem (1979)
Syringodium isoetifolium	Gulf of Aqaba	1.2	Lipkin (1979)
Cymodocea rotundata	Gulf of Aqaba	0.5	Lipkin (1979)
Halophila stipulacea	Gulf of Aqaba	0.33	Lipkin (1979)
	Gulf of Aqaba	0.26 (max.)	Wahbeh (1980)
Halodule uninervis	Gulf of Aqaba	0.23	Lipkin (1979)
	Gulf of Aqaba	0.40 (max.)	Wahbeh (1980)
	The Gulf	0.05–0.24[a]	Price and Coles (in press)
Halophila ovalis	Gulf of Aqaba	0.04	Lipkin (1979)
	Gulf of Aqaba	0.01 (max.)	Wahbeh (1980)

[a]Derived from wet weight determinations.

or indirectly, to consumer organisms. Of equal or greater ecological significance, however, is primary productivity. Crossland *et al.* (1987) estimate gross productivity of Red Sea seagrasses to be in the upper range of global estimates ($2.0–5.5\,kg\,C\,m^{-2}\,y^{-1}$), especially in well-developed *Thalassia* beds. Net productivity values typically range from 120 to $320\,g\,C\,m^{-2}\,y^{-1}$ (Mann 1982), but may reach $1000\,g\,C\,m^{-2}\,y^{-1}$ or more in dense beds of turtle grass, *Thalassia testudinum* (McRoy and McMillan 1977, Mann 1982, Jones *et al.* 1987). Net productivity values determined in the Gulf of Aqaba, using oxygen release methods, appear to be relatively high. Figures are as follows (from Wahbeh 1980, Jones *et al.* 1987):

Halodule uninervis	$1326\,g\,C\,m^{-2}\,y^{-1}$
Halophila stipulacea	$617\,g\,C\,m^{-2}\,y^{-1}$
Halophila ovalis	$11\,g\,C\,m^{-2}\,y^{-1}$

Wahbeh (1980) also showed that of the gross primary productivity in *H. stipulacea* ($880\,g\,C\,m^{-2}\,y^{-1}$), 29.9% was lost in respiration, 28.5% was used for production of leaves, and 13.6% for the production of rhizomes. The remainder (28%) was thought to be divided among storage, production of reproductive structures and excretion as dissolved organic matter (DOM). Further details of the fate of productivity within the seagrass plant, and to the surrounding environment, are discussed by Wahbeh (1980) and Jones *et al.* (1987).

In the Gulf, Basson *et al.* (1977) derived tentative productivity figures from standing crop estimates of seagrass leaves. The estimated value for the area investigated (Tarut Bay) was $256\,g$ dry wt $m^{-2}y^{-1}$ (c. $100\,g\,Cm^{-2}y^{-1}$), and excludes any productivity contribution of roots and rhizomes, which can be substantial. These authors also estimated the energy equivalent of grass beds to be about $1.4 \times 10^{11}\,kCal\,y^{-1}$, and equated this to the energy from about 95,000 barrels of oil. These calculations included many assumptions and were largely hypothetical. Problems in deriving productivity figures from standing crop estimates are discussed by Lipkin (1979). Nevertheless, the figures provide at least a crude basis for quantifying the magnitude of local seagrass resources (see also Chapters 13 and 15).

(4) Epiphytes

Seagrass blades are generally coated by a film of surface algae, known as periphyton. Often dominant are Cyanophyta, in particular species of *Oscillatoria*, *Phormidium* and *Lyngbya* (Walker 1987). Diatoms are also important. Physiological interactions in the form of nutrient transfer between the host plant and attached epiphytes suggest the association to be symbiotic, rather than merely mechanical attachment. Utilization of phosphate from sediments by epiphytes via the seagrass host has been shown to be feasible (Harlin 1973). Similarly, it has been demonstrated that nitrogen can be fixed by epiphytic Cyanophyta and transferred to the host seagrass (Goering and Parker 1972). This could contribute to the high productivity of seagrass ecosystems in the nutrient-poor waters of the Red Sea (Walker 1984). Photosynthetic epiphytes may account for up to one-fifth of the primary productivity of a seagrass bed. While algal and other associates can enhance overall primary productivity of the seagrass ecosystem, a study on *Zostera* beds suggested that a coating of epiphytes actually depressed photo-synthesis of the host seagrass (Penhale 1977).

A detailed study of seagrass epiphytes in the Gulf of Aqaba (Walker 1984) revealed 94 epiphyte taxa, a number comparable to results from studies in other regions (e.g. Humm 1964, Ballentine and Humm 1975). In the Gulf of Aqaba, only nine species were restricted to seagrasses, while the great majority are also found to colonize other habitats. Maximum species richness occurred during spring. Biomass of epiphytes ranged from 5 to $50 \, \mathrm{g \, m}^{-2}$ dry wt, reaching a maximum in summer when light intensities are at a peak. This is also the season of peak productivity of both seagrass and epiphytic algae (Walker 1984) in the Gulf of Aqaba. A composite picture of the role of epiphytes in seagrass productivity is not yet possible, due to the limited information available for many parts of Arabian region.

Encrustation on seagrass blades also includes fauna such as sessile, often colonial invertebrates such as hydroids, bryozoans, sponges, barnacles and tunicates. These in turn attract other fauna (e.g. gastropods, polychaetes, crustaceans), which form the basis of important food chains within the seagrass ecosystem (see sections 5 and 6 below).

(5) Consumption of seagrass and energy budgets

Seagrass beds in the Arabian region constitute an enormous store of biomass available, directly or indirectly, to consumer organisms. Direct herbivory, i.e. consumption of live plant material, is relatively uncommon, due perhaps to the presence of distasteful phenolic compounds (Jones *et al.* 1987), and the apparent indigestibility of cellulose to most organisms. For most consumers, the action of bacteria and other microorganisms is needed to render seagrass material digestible and available to consumers via detrital food chains (Barnes and Hughes 1982, Mann 1982). Mechanical breakdown also assists in the production of detritus.

In the Arabian region, among the larger grazers of seagrasses are dugongs (*Dugong dugon*) and green turtles (*Chelonia mydas*), the former feeding principally on *Halodule uninervis* and to a lesser extent on *Syringodium isoetifolium* in the Red Sea (Gohar 1957, Lipkin 1975). *C. mydas* is also known to feed on algae. Of the smaller grazers, the urchin *Tripneustes gratilla* is important in the Red Sea, and some grazing is reported by surgeonfish (*Zebrasoma xanthurum* and *Ctenochaetus striatus*) and rabbitfish (*Siganus rivulatus*) (Wahbeh and Ormond 1980). In a quantitative study (Wahbeh 1980), *T. gratilla* was present in a bed of *Halophila stipulacea* at a density of $1\,m^{-2}$. Urchin consumption was equivalent to a third of the total seagrass growth ($124\,g\,C\,m^{-2}\,y^{-1}$). In contrast, consumption by fish amounted to less than 5% of total plant growth. The remainder (c. 60%) was assumed eventually to enter the detrital food web (see Jones *et al.* 1987). While herbivorous fish, such as *Z. xanthurum* and siganids, are known to occur in the Gulf (Basson *et al.* 1977, Smith *et al.* 1987), their importance as a direct consumer of seagrass has not been determined.

(6) Faunal assemblages

i) Red Sea region

General accounts of seagrass fauna are available for the Red Sea (Fishelson 1971, Mastaller 1978, Wahbeh 1981, Price *et al.* 1988). Within beds of *Halophila stipulacea* in the Gulf of Aqaba, Wahbeh (1981) recorded 49 species of invertebrates, of which about 70% were molluscs. These were either epifauna living attached to the seagrass plant (e.g. *Cerithium*, *Phasianella* and *Smargdia*), or infauna (e.g. *Mitra* and *Mitrella*). Many of the gastropods recorded are predators and scavengers and included *Cerithium rostratum*, *Pyrene variana* and *Mitrella blanda*, although some (e.g. *Strombus*) are herbivorous. Bivalves, which are mainly deposit and filter feeders, include *Linga semperiani* and *Parvilucina fieldingi* (Wahbeh 1981). Molluscs are also prevalent in seagrass beds of the eastern Red Sea (Price *et al.* 1988), a region in which 91 species have been identified during recent surveys (McDowell, unpublished).

Apart from molluscs and direct grazers of seagrass, other important groups associated with Red Sea seagrass beds include polychaetes, crustaceans, echinoderms and fish. Of significance is that seagrass beds harbour juveniles of various commercial fish (e.g. lethrinids and lutjanids) and crustaceans (e.g. *Penaeus semisulcatus*) (see Bemert and Ormond 1981, Vine 1986, Jones *et al.* 1987, Price *et al.* 1988), and therefore similar to those elsewhere (IUCN/UNEP 1985a) in functioning as nursery areas. Availability of food, shelter and protection from predators within the seagrass lattice contribute to the nursery function. Young fish and crustaceans may later spend their adult lives in other ecosystems. The resident fish fauna is relatively limited, at least in the eastern Red Sea, although other species migrate into seagrass beds by day or night to feed (Crossland *et al.* 1987).

ii) The Gulf

In the western Gulf, a total of 530 floral and faunal species associated with seagrasses was recorded by Basson *et al.* (1977), with 31% (164 spp.) being molluscs. The total number is less than reported for subtidal sand (638 spp.) or subtidal mud (610 spp.). Greater overall species richness in sand was also observed by McCain (1984b), although faunal abundance was greater in seagrass ($450-36,200 \, \text{m}^{-2}$) compared to sand ($840-9670 \, \text{m}^{-2}$). Seagrass biota contained fewer species, partly because many of the polychaetes in the samples were not identified (McCain 1984b). This study revealed, surprisingly, a significant positive correlation between the abundance of organisms in seagrass and salinity. Seagrass beds in the Gulf also have an important nursery function, for commercial penaeid shrimps, pearl oysters and other organisms (Basson *et al.* 1977, Price 1982b, IUCN/UNEP 1985a, Vousden and Price 1985, IUCN 1987, Vousden 1988). Adults of many species (e.g. *Penaeus semisulcatus*) also inhabit grassbeds as well as other habitats.

A recent study in the Gulf revealed a total of 834 species associated with seagrass and sand/silt substrata (Coles and McCain 1990). Mean numbers of individuals in seagrass beds in the northern stations (up to c. $52,000 \, \text{m}^{-2}$) were comparable to those found in the Manifa–Safaniya area (up to c. $36,000 \, \text{m}^{-2}$) during an earlier study (McCain 1984b). However, Coles and McCain (1990) report significantly greater numbers of species and individuals in seagrass than in sand/silt substrata. This was attributed partly to the finer grained sediments of the seagrass areas. Overall mean biomass (ash free dry wt) of benthos within seagrasses ($1.03 \, \text{g} \, 0.1 \, \text{m}^{-2}$) was also significantly greater than within sand/silt ($0.36 \, \text{g} \, 0.1 \, \text{m}^{-2}$). Hydrocarbon and other pollution-related parameters indicated relatively low concentrations in sediments. The only significant correlation indicated increased abundance of seagrass organisms in areas of higher sediment petroleum, resulting in lower species diversities. This pattern was believed to reflect primarily the increased occurrence of a polychaete, *Perielectroma zebra*, which may be an indicator of increased petroleum concentrations (Coles and McCain 1990).

iii) Regional comparisons

The study of Coles and McCain (1990) in the Gulf and that of Wahbeh (1981) in the Gulf of Aqaba are broadly comparable, since similar sampling methods were used in each. In the Gulf of Aqaba densities of seagrass organisms reached only c. $2000 \, \text{m}^{-2}$ and of the total 49 species, about 60–70% comprised gastropods, 10% bivalves and about 5% polychaetes. By contrast, abundances in the Gulf study reached nearly $52,000 \, \text{m}^{-2}$ in seagrass, and the order of dominance was bivalves (52%), polychaetes (26%), gastropods (6%) and crustaceans (5%). Coles and McCain (1990) suggest that benthic fauna (within seagrasses and sand/silt) in the Gulf are principally suspension feeders which utilize more abundant organic particulates than occur in the clearer waters of the Gulf of Aqaba. These authors suggest this as a possible reason for the differences. Few quantitative studies of seagrass have been undertaken elsewhere in the Arabian region.

D. Role of seagrasses in the Arabian marine environment

(1) Importance and role in fisheries production

Species occurrence data for different intertidal and subtidal ecosystems of the Gulf are presented by Basson *et al.* (1977). Despite the uncertain or incomplete identity of many taxa, preliminary comparisons can be made between different ecosystems on the basis of species composition.

Analysis of these data indicates that at least 48 of the 530 species recorded in seagrasses (i.e. c. 9%) were recorded only in seagrass beds and not in other ecosystems. Approximately 50% of the seagrass "specialists" are molluscs. Seagrasses might therefore be an essential ecosystem for these species. The relationships between seagrasses and other ecosystems can be explored further by means of cluster analysis, using species presence: absence data (Price *et al.* 1987a). Using Ward's (1963) method, at a dissimilarity level of 0.40 three groupings are recognized in the resulting dendrogram (Figure 7.2), one of which (Group III) comprises all subtidal sedimentary ecosystems, i.e. seagrasses, subtidal mud and subtidal sand. Within this grouping, seagrass biota share greater affinities with mud than with sand ecosystems. Group I is composed of all intertidal and subtidal hard ecosystems, except coral reefs, while Group II consists of only coral reefs.

The above suggests that the seagrass ecosystem in the Gulf shows some distinctiveness, with a relatively small proportion of seagrass "specialists", but a larger proportion of "generalists". However, the observed patterns may well be modified by analysis of further data from recent studies which include additional species (McCain 1984b, Coles and McCain 1990). Whether seagrasses are critical at some phase of the life cycle to species inhabiting several different ecosystems, i.e. the seagrass "generalists" (see Price *et al.* 1987a) is undetermined. For some animals (e.g. *Dugong dugon* and *Chelonia mydas*) dependence on seagrasses is undoubtedly great, since this vegetation forms their staple diet. In the case of many other species, the importance of seagrass seems less clear. For instance, pearl oysters (*Pinctada* spp.), as adults, live attached to hard substrata. Along the Saudi Arabian Gulf coast, their spat appear to settle preferentially on grass beds (Basson *et al.* 1977), so seagrass would seem important to juvenile phases of their life cycle. However, in nearby Bahrain waters, many young pearl oysters are found attached to *Sargassum* weed too (Vine 1986).

Similarly, the degree of dependence of juvenile shrimp (*Penaeus semisulcatus*) on seagrass beds is still unclear. The earliest benthic juvenile stages have been found in thickets of macroalgae (*Hormophysa* and *Sargassum*) in different parts of the Gulf (Basson *et al.* 1977). In Saudi Arabian waters, older juveniles are found only among seagrasses. In Kuwait, where macroalgae are prevalent, seagrasses are less abundant, yet Kuwait waters support a major fishery for penaeid shrimp.

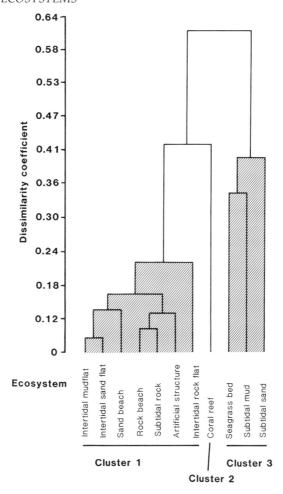

Figure 7.2 *Dendrogram resulting from cluster analysis using species presence/absence data for seagrass beds and other ecosystems. Three groupings are recognized at a dissimilarity value of 0.40 using Ward's (1963) method (modified from Price* et al. *1987a).*

This suggests Kuwaiti populations may complete their life cycles using mainly macroalgae, in which case seagrass may not be critical. Alternatively, migration of postlarvae or other stages might occur between Kuwait and nearby northern Saudi Arabian waters, where extensive grass beds do occur (Basson *et al.* 1977, IUCN 1987, Coles and McCain 1990). In this case, the seagrass ecosystem would seem critical here too. Studies elsewhere have shown that larval migration of shrimp can occur even against general water movements (Penn 1975). Adult *P. semisulcatus* principally inhabit subtidal/sand mud, where they are fished, although some individuals remain in grass beds throughout their lives (see Basson *et al.* 1977).

Possible consequences of seagrass degradation, for instance on the Gulf's fisheries, have been described qualitatively (e.g. IUCN 1987, Madany *et al.* 1987,

Sheppard and Price 1991; see also Chapter 14 B2a and b), although there appear to be no published quantitative accounts. The overall role of seagrass ecosystems in the coastal ecology of the region therefore cannot yet be fully evaluated. A study is currently in progress to assess temporal and spatial changes in seagrasses and other coastal ecosystems in Bahrain, using time series satellite imagery (Wilson unpublished). A non-linear dynamical model is being developed to determine the extent to which changes in shrimp and fish catches reflect the effects of habitat recession (e.g. productivity decline), compared with removal of individuals by fishing.

(2) Energy and nutrient interactions with other ecosystems

Since seagrass beds constitute a dynamic ecosystem, physical and biological interactions with adjacent ecoystems might be expected. Many of the interactions discussed by UNEP (1985) are probably applicable to Arabian coastal ecosystems. In general, physical interactions are considered to be more significant than biological ones (UNEP 1985).

Hulings (1979) reports that in parts of the Gulf of Aqaba *Halophila stipulacea* stabilizes otherwise unstable sandy substrata, increasing the surface area for algal colonization up to $12\,m^2$ for each $1\,m^2$ of substrate at a depth of 1–2 m. Leaf turnover at about 35 days (Wahbeh 1980) results in a continuously produced substrate for epiphytes and other colonizers (Walker 1987). In addition to altering the properties of underlying sediments, seagrasses actually produce sediments through their calcareous epiphytic algae (Walker and Woelkerling 1988). Water movements may later transport these away from the site of production.

Migratory and other animals provide a major link between different ecosystems, which often results in nutrient and energy transfer, through feeding–defecation cycles. The migration of shrimp (e.g. *Penaeus semisulcatus*) in the Gulf from macroalgae to seagrass, and thence to other ecosystems has been mentioned above. Migratory feeding on a diel basis is common between coral reefs and seagrass beds for several taxa including snappers (Lutjanidae) and Green turtles (*Chelonia mydas*), barracuda (*Sphyraena* spp.) and others (Crossland et al. 1987). Herbivorous fishes and urchins often create "halos" around reefs (bare sand and heavily grazed seagrass areas) as a result of nocturnal foraging. The net effect is transfer of nutrients to adjacent ecosystems such as coral reefs. Nutrients and organic matter may also be transferred to the intertidal zone. On the Gulf coast of Saudi Arabia, seagrass litter was observed at 59% of 53 shore sites visited (IUCN 1987). Hence, seagrass detritus may become widely distributed to both intertidal and subtidal communities. Detritus originating from seagrasses and other ecosystems may also be transported offshore and remineralized in the open sea (IUCN 1987). However, this pathway is probably not significant, at least in the Gulf, because nutrient levels do not appear to increase dramatically in deep water offshore, and sediments are low in organic carbon (<2%) (IUCN 1987).

Marine ecosystems seldom occur in isolation, but rather as an array of ecosystems juxtaposed in various combinations. According to Crossland et al.

(1987) in the Red Sea the scale of integration may extend for hundreds of metres (e.g. narrow fringing reefs in Gulf of Aqaba) or tens of kilometres (e.g. Wedj Bank and Farasan Bank in Saudi Red Sea). The resulting integrated systems may be classified according to physical factors (e.g. wave exposure, tidal flushing and water exchange). Interactions with benthic sediment types provide the environment in which ecosystems have developed. Under the scheme of Crossland *et al.* (1987), seagrasses in the eastern Red Sea occur in semi-open systems.

E. Other soft subtidal ecosystems

Vast tracts of the offshore Arabian marine environment are composed predominantly of sand or mud, devoid of attached macrophytes, or corals. Subtidal sandy ecosystems are prevalent in high energy environments (e.g. near coral reefs), whereas mud is more common in low energy environments associated with slow water movements (e.g. shallow embayments). Except in the Gulf, studies have focused more on community composition than on process-oriented ecology.

Although unspectacular, these ecosystems may harbour rich biological assemblages in terms of both species abundance and diversity. The occurrence of rich sand and mud faunas suggests that primary productivity (e.g. from benthic diatoms and Cyanophyta) is high in some areas. In addition to sessile epiphytic and infaunal benthos, sand and mud ecosystems are a major habitat for motile fauna (e.g. demersal fish and crustaceans) during some or all phases of their life cycle. Among the important factors controlling the nature and distribution of communities are light penetration (governed largely by depth), sediment grain size, sediment stability/mobility (controlled by water movement) and salinity (particularly in the Gulf). Despite their small size (0.04–0.01 mm), the meiofauna can exhibit high turnover of biomass (high production/biomass) and are taxonomically diverse (Mann 1982).

(1) Physical features and community composition

In the Gulf, subtidal sandy ecosystems extend down to at least 30 m, whereas mud occurs in depths of 6 m and more and is also the principal benthic ecosystem at depths greater than 30 m (Basson *et al.* 1977). Details of the processes controlling these and other sedimentary environments in the Gulf have been described in Chapter 1. Unlike SE Arabia and the Red Sea, deep sea clays and oozes have not formed in the Gulf which is a shallow sea.

Among the early biological studies in the Gulf were the Danish Scientific Investigations in Iran (e.g. Mortensen and Gislen 1941), which yielded useful information on subtidal sand and muddy benthos. Subsequent research provided further details of community composition. As noted above, species richness is reported to be high (>600 species) in both subtidal sand and subtidal mud, surpassing that of all other ecosystems, even coral reefs (543 species) (Basson *et al.* 1977). Polychaetes followed by gastropods were the dominant fauna groups. However, other research has revealed different patterns of species richness and dominance (Coles and McCain 1990). McCain (1984b) reports a significant positive correlation ($p<0.05$) between salinity and the number of major taxonomic groups within subtidal sand. At least four or five biological communities inhabiting subtidal sand and muddy ecosystems of the Gulf have been described (Basson *et al.* 1977). Two of these are found in mud, and have been named after the dominant fauna. One is the *"Murex/Cardium"* community (dominated by two molluscs: *M. papyraecum* and *M. küsterianus*), and the other is the *"Brissopsis/Amphioplus"* community (dominated by two echinoderms: *B. persica* and *A. seminudus*).

In the Red Sea, Fishelson (1971) and Jones *et al.* (1987) report a *Hippa* (mole crab)/*Mactra* (bivalve) community in shallow water (<3 m) on coarse sands and gravels mixed with mud, which includes several macroalgae (e.g. *Caulerpa serrulata*, *Cystoseira myrica* and *Padina pavonia*). Another community, dominated by molluscs, is the *Operculina/Turritella* community. This occurs on gravels and shell overlain by silty mud in deeper waters (30–260 m) (Fishelson 1971). At intermediate depths sediments are often colonized by seagrasses. Published data suggest that species richness of shallow subtidal benthos may be lower in the Red Sea (Fishelson 1971, Jones *et al.* 1987) than in the Gulf (Basson *et al.* 1977, McCain 1984b, Coles and McCain 1990).

The Red Sea and southern Arabian deep sea sediments, or oozes, are characterized by carbonate components, originating in particular from Foraminifera, Pteropoda, Heteropoda and planktonic Protozoa (Thiel 1987). Calcareous algae are not indicated by Thiel (1987) as a significant contributor to deep sea oozes, but can be an important component of coastal sediments (Walker 1987). The organic content in sediments is generally higher nearer the coast (<1%) and decreases towards central regions (Thiel 1987). Fauna associated with deep sea sediments in the Red Sea extends to depths of at least 2000 m. Most major groups are represented (Thiel 1987), and 30% of the species may be endemic (Turkay 1986). Most species occurring in the deep sea are invaders from shallow water rather than true benthic deep sea species. Possible reasons for this pattern are discussed by (Thiel 1987). The deep sea epifauna appears to be sparse, and limited to organisms such as anemones (*Halcurias sudanensis*), sponges, hydroids and possibly ascidians (Thiel 1987). Dominant mobile fauna are fish and shrimps, while other fauna (e.g. crustaceans, gastropods and echinoderms) are rare (Thiel 1987).

As a result of sampling in nearshore and offshore waters of SE Arabia and the Red Sea during the John Murray Expedition, considerable information became available for certain faunal groups, such as echinoderms, and several other expeditions have also visited the southern Arabian region (Campbell and Morrison, 1988). Ansell (1984) reports a restriction in the subtidal benthos in Oman, possibly from irregular upwelling of cool oxygen-depleted water.

(2) Faunal abundance and biomass

Comparison of benthic abundance in different soft substrate ecosystems of the Gulf has already been mentioned. Broadscale comparisons between benthic biomass (sand and mud) in different regions of the Gulf and adjacent areas are summarized in Table 7.3 (see also Neyman *et al.* 1973). The figures suggest that benthic biomass is reduced in mud, and with distance from the open ocean (Jones 1985). Reduced benthic biomass (and diversity) towards the north of the Gulf is also apparent from other studies (Aoki 1974, Enomoto 1971). Highest biomass is reached in the northern Arabian Sea (up to $>500\,\text{g m}^{-2}$).

Table 7.3 Benthic biomass (g m^{-2}) in the Arabian region (from Jones 1985, except where otherwise indicated).

	Red Sea	Northern Gulf	Western Gulf	Central Gulf	Southern Gulf	Gulf of Oman Central region	Arabian Sea north coast
Coarse sand benthos		26.5		67	130[a]		
Sand/silt	5–50[b]		36[c]				35[d]
Unspecified benthos	>100[d]						(max. >500)
Muddy benthos		8		13	18	65	

[a]After Thorson (1957), [b]Jones *et al.* (1987), [c]Coles and McCain (1990) and [d]Neyman *et al.* (1973).

Based on sampling at four sites near Jeddah, faunal biomass of shallow subtidal benthos ranged from 5 to $50\,\text{g m}^{-2}$ near Jeddah, while for the Red Sea as a whole a value exceeding $100\,\text{g m}^{-2}$ is reported (Table 7.3). These faunal biomasses are broadly comparable with those in similar habitats in other parts of Arabia (Table 7.3) and elsewhere in the world (Jones *et al.* 1987).

Quantitative sampling of deep sea infauna of the Red Sea has revealed the macrofauna ($>1\,$mm) to be of low diversity, density and biomass. Figures for biomass range from 0.05 to $0.5\,\text{g m}^{-2}$ (Thiel 1987), which are lower than for subtidal benthos in shallower areas (Table 7.3). The abundance of meiofauna, although higher than for macrofauna, is also reported to be low compared with comparable environments in other parts of the world. Figures for meiofaunal biomass range from 15 to $180\,\text{mg}\,10\,\text{cm}^{-2}$ (i.e. 1.5–$18\,\text{g m}^{-2}$) (Thiel 1987). The low standing stock of deep sea benthos in the Red Sea may be due to several factors. Low standing stock of plankton in deeper waters, together with the relatively high temperatures (probably resulting in high respiration rates and energy loss), may result in little energy remaining to become fixed as biomass. It is suggested that the benthos and pelagic environments together form a unified oceanic environment, which in the Red Sea is characterized by low standing stocks, high metabolic rates and low productivity (Thiel 1987).

(3) Soft substratum coral sand faunas

Soft substratum faunas around coral reefs represent essentially a modified sand fauna, showing few affinities with the biota inhabiting the solid framework of a coral reef. Sediments generally comprise coarse sand with shell and coral fragments. In the Gulf, echinoderms (e.g. *Pentaceraster mammillatus*, *Euretaster cribrosus* and *Temnotrema siamense*), a planktivorous sipunculan worm (*Aspidosiphon* sp.) and two species of solitary corals (*Heterocyathus* sp. and *Heteropsammia* sp.) are common associates. Also common around coral reefs and other sandy environments are "sand volcanoes". These occur in several areas, and are inhabited by callianassid and/or thalassinid shrimp (Basson *et al.* 1977, de Vaugelas and Saint-Laurent 1984, Price *et al.* 1987b). In the Red Sea, Fishelson (1971) describes a rich community in coral rubble and sand dominated by a gastropod (*Gena varia*), other molluscs and an urchin (*Echinometra mathaei*). Interactions with adjacent habitats result from migrations of fish crustaceans, urchins (e.g. *Echinometra*, *Diadema* and *Tripneustes*) between crevices and more open areas.

Sand substrata in the vicinity of coral reef and other lagoons are often colonized by seagrasses (e.g. *Thalassia hemprichii* and *Thalassodendron ciliatum*) and may also be closely associated with other ecosystems, such as mangroves, particularly in the northern Red Sea.

Summary

Eleven seagrass species are known for the Arabian region, of which *Halodule uninervis* and *Halophila ovalis* are the most prevalent. Diversity is greatest in the Red Sea proper (10 species) and lowest in SE Arabia (4 species) and the Gulf (4 species). Seagrass beds also attain greatest development in the Red Sea, particularly towards the south, despite the reverse trend shown by certain species. Probable factors controlling species composition include light (related to depth), substratum, water movements, temperature and salinity.

Studies on seagrass standing crop have been undertaken mostly in the northern Red Sea. Highest biomass is associated with *Thalassodendron ciliatum*, *Thalassia hemprichii* and *Syringodium isoetifolium* (>1 kg dry wt m^{-2}). Biomass of *Halodule uninervis* dominated communities in the Gulf ranged from 0.05–0.24 g dry wt m^{-2}, and are comparable to figures for similar species in the Red Sea and elsewhere. Net productivity estimated for *Halodule uninervis* has been determined in the Gulf of Aqaba (>1300 g C m^{-2} y^{-1}), and compares favourably with figures from other regions. Seagrass productivity can be enhanced by the presence of epiphytic algae.

Seagrasses provide a mostly indirect food source and habitat for both resident fauna and temporary visitors, including commercially important fish and crustaceans (e.g. *Penaeus semisulcatus*). Despite regional variation, available data suggest that both species richness and abundance of fauna are greater in the Gulf than in Red Sea, at least in its northern parts. Benthic fauna (within seagrasses and sand/silt) in the Gulf are principally suspension feeders, which utilize more abundant organic particulates than occur in the clearer waters of the northern Red Sea. This may partly explain the Gulf's greater abundances. The biota of seagrasses show greatest affinities with that of subtidal sand and subtidal mud, at least in the Gulf. It is characterized by a small proportion of seagrass "specialists" and a larger proportion of "generalists". The overall importance of seagrasses and role in the maintenance of populations, including fisheries production, is discussed in detail earlier.

Plant–animal interactions include nutrient and energy fluxes, as well as movements of biota from seagrasses and other ecosystems. They form important links among and between ecosystems, particularly nutrient transfer.

Subtidal sandy and muddy benthos, including deep-sea oozes and soft substrate coral faunas, are described above, together with patterns of faunal biomass. The subtidal benthos, including seagrass ecosystems, exhibits considerable spatial heterogeneity.

CHAPTER 8

Intertidal Areas 1. Mangal Associated Ecosystems

Contents

A. **Introduction**   161
B. **Overall ecological role of mangals**   162
C. **The Arabian mangal**   162
   (1) Mangrove species and distribution   162
   (2) Mangal development and environmental conditions   163
D. **Ecology of the mangal in the Red Sea and Arabian Gulf**   166
   (1) Productivity   166
   (2) Biota and horizontal zonation   167
   (3) Vertical zonation of biota   169
   (4) Comparisons with other mangal biotas   169
   (5) Specificity of biota   172
   (6) Interactions with adjacent ecosystems   172
Summary   173

A. Introduction

Mangroves and other coastal vegetation are a conspicuous and important biological feature fringing the upper parts of many shorelines. Their juxtaposition, both horizontally and vertically, between the terrestrial and marine provinces renders these ecosystems an important ecological transition zone. In the arid climatic zones of the Arabian region, the salt-tolerant shrubs and trees form a dramatic visual and biological contrast to the comparatively barren terrain of hinterland desert.

In tropical environments, mangrove swamps generally predominate in such locations, whereas in more temperate regions these give way to saltmarshes, reed swamps or other freshwater-dependent vegetation. Parts of the Arabian region, in particular the Gulf, are unusual in that mangroves and saltmarshes coexist. The Gulf, except along the Iranian coast, is also unlike many tropical marine environments in having only small areas occupied by mangroves. This is a reflection of both natural conditions and human pressures (see Chapter 14).

This chapter considers the ecology of Arabian mangroves, and their associated biotic assemblages, collectively known as the mangal ecosystem (Mann 1982, IUCN 1983, Por and Dor 1984). While considerable information is now available for the Red Sea, knowledge of mangals is more limited in other parts of the region.

B. Overall ecological role of mangals

Mangals, one of several recognized critical marine habitats (Ray 1976, IUCN 1983, Dugan 1990), assume great ecological significance. Their fallen leaves and sticks, via detrital pathways, often make appreciable contributions to inshore and estuarine productivity. Net primary productivity values typically range from 0.35 to $0.5\,kg\,C\,m^{-2}\,y^{-1}$ (Mann 1982). In a study in Thailand, the annual production of organic matter above ground for a stand of *Rhizophora* was equivalent to $7\,tons\,ha^{-1}\,y^{-1}$ of leaves and $20\,tons\,ha^{-1}\,y^{-1}$ of wood (Christensen 1978). Like other macrophytes and coral reefs, mangals function as "coastal food factories" (Barnes and Hughes 1982), although their occurrence is limited in several parts of the Arabian region.

Mangal ecosystems are also associated with the maintenance of biota, thereby assuming importance as a genetic reservoir. However, it is the concentration of individual species, rather than their diversity, which characterizes the mangal (Por *et al.* 1977, Dugan 1990). The major nursery function of mangrove roots (e.g. for penaeid shrimp and fish) highlights this, and is a feature often exploited by artisanal fishermen and aquaculturists. Mangroves also provide a refuge and breeding area for birds and other marine and terrestrial wildlife. In addition, they provide a range of useful forestry, medicinal and other products, and exert a controlling influence against coastal erosion. More detailed accounts of mangal ecology and conservational importance include IUCN (1983), Por and Dor (1984), Hutchings and Saenger (1987) and Dugan (1990).

Although the mangal is generally treated as a distinctive ecosystem, its biota shows strong affinities with species of other intertidal habitats (see review in Price *et al.* 1987b). Mangals also exhibit strong ecological interactions with adjacent terrestrial, intertidal and subtidal ecosystems (UNEP 1985). Further, mangal ecology is often influenced markedly by human interactions, and the ecosystem is characterized by complex exploitation patterns.

C. The Arabian mangal

Mangals of the region are significant not just biologically, but also in a historical context. Mangals in the Gulf were the first mangroves ever to be reported in the world literature, by Nearchus and Theophrastus over 2000 years ago (Baker and Dicks 1982). As discussed, however, most research has been undertaken in the Red Sea, particularly along the Sinai coast (e.g. Fishelson 1971, Por *et al.* 1977, Dor and Levy 1984, Ish-Salom-Gordon and Dubinsky 1990), but also in other parts of the Red Sea (e.g. Mandura *et al.* 1987, Price *et al.* 1987b).

(1) Mangrove species and distribution

Of the four mangrove species known for the region, *Avicennia marina* is by far the commonest. It is unusually tolerant of harsh environmental conditions, in particular low temperatures and high salinity (Hutchings and Saenger 1987). The

Table 8.1 The total number of mangrove species known for each major area of the Arabian region.

Region	No. of mangrove species	Reference(s)
Gulf of Aqaba	1	Dicks (1986), Por *et al.* (1977)
Gulf of Suez	1[a]	Dicks (1986)
Red Sea	4	IUCN/UNEP (1985a), Ormond *et al.* (1988)
Arabian Gulf	1[b]	IUCN/UNEP (1985b)
SE Arabia	1	IUCN/UNEP (1985a,b)

[a]Absent within Gulf of Suez, but present at mouth of Gulf, including Sinai coast.
[b]Excluding *Rhizophora racemosa* grown experimentally in northern Saudi Arabia (Kogo 1986).

remaining species are *Rhizophora mucronata*, *Bruguiera gymnorhiza* and *Ceriops tagal*. While positive records exist for the latter two species, recent surveys have not been able to confirm that *B. gymnorhiza*, in particular, is still present (see Ormond *et al.* 1988). Mangrove occurrence in the Arabian region is summarized in Table 8.1. Salinity may be a major factor limiting the number of species, although *Avicennia marina* itself may not be salinity limited.

The distribution of mangroves in the region is shown in Figure 8.1, and their estimated total area in different areas is shown in Table 8.2. Mangroves are most prevalent and extensive in the Red Sea, particularly in southern areas.

Along shores of the Gulf and SE Arabia, stands are discontinuous and of patchy occurrence. However, a single stand of 6800 ha is present within the Hara reserve in Iran (IUCN/UNEP 1985a). Mangroves in the Gulf may have been reasonably common 2000 years ago (Ormond *et al.* 1988).

The northernmost stands of mangroves in the Red Sea occur along the Sinai peninsula (lat. 28° N) (Por *et al.* 1977, Dicks 1986). A similar northerly limit of about 27° N for naturally occurring mangroves is set on the Gulf coast of Saudi Arabia (IUCN 1987, Ormond *et al.* 1988). However, mangroves have been grown artificially on the northern Saudi Arabian Gulf coast at Ras al Khafji at about 28° 30' N (Kogo 1986). In both the Red Sea and Gulf these latitudes are close to the most southerly occurrence of occasional ground frost (Ormond *et al.* 1988), suggesting that cold winter temperatures, rather than high salinities (40->50 ppt), limit the northernmost extent of mangroves (Por *et al.* 1977, Chapman 1984). Winter frosts also appear to set the southerly limit for mangroves (*A. marina*) in Australia (Hutchings and Saenger 1987).

(2) *Mangal development and environmental conditions*

In the Arabian region, mangroves occur mainly in sheltered situations behind reef flats of fringing reefs, bays or creeks, in the lee of offshore islands, and on some offshore islands (Ormond *et al.* 1988).

Table 8.2 Approximate areal extent of mangroves in different parts of the Arabian region (figures tentative and taken from several sources including Vousden and Price 1985, Vousden 1988, Salm and Dobbin 1986, 1987, Salm *et al.* 1988, Salm and Jensen 1989, Rabanal and Bueschel 1978, Price *et al.* 1987a, Ormond *et al.* 1988, Banaimoon, pers. comm.).

Location		Area (km^2)
Red Sea		400–500
Saudi Arabia	200	
Other countries	200–300	
Gulf		125–130
Saudi Arabia	4	
Bahrain	1	
Qatar	<5(?)	
UAE	30	
Iran	90	
SE Arabia		50–70
Oman	20	
Yemen	30–50	
Total		575–700

At many sites behind reef flats in the northern Red Sea the sediment in which mangroves grow is only a thin veneer overlaying uplifted rock or fossil reef, where the surface is often dissected and pitted. Within these sediment-filled lacunae only limited space is available for mangrove root growth, a factor probably contributing to the stunting of mangrove bushes in the northern Red Sea. The best developed stands are found in the inner protected lagoons 0.5–1.5 m depth (Dor and Levy 1984). Mangroves growing in the conditions described above are usually termed "hard-bottomed" or "reef" mangals (Por *et al.* 1977, Price *et al.* 1987b, Ormond *et al.* 1988). A diagrammatic section through a Sinai mangal is shown in Figure 8.2.

In contrast are the classical "soft-bottomed", or "peat" mangals, associated with sheltered embayments. Around much of the region such embayments support extensive mangroves. Compared with "hard-bottomed" mangals, there is greater accumulation of finer and deeper sediments, a factor strongly influencing the composition of resident biota (see section D2 below). Soft-bottomed mangals are characteristic of the southern Red Sea (Price *et al.* 1987b). In this region, stands 100–500 m wide occur, with the tallest trees 5–7 m high, contrasting with the scattered mangrove thickets (<2 m high) further north. Along the eastern Red Sea coast, the mean height of *A. marina* shows a significant negative correlation with salinity, although it is unclear if the association is causal (Price *et al.* 1987b). The southerly increase in mangal occurrence and development may also result

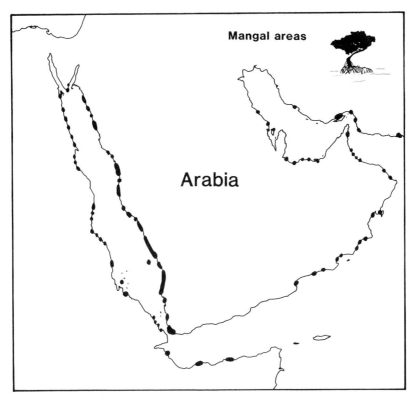

Figure 8.1 *Map showing the approximate distribution of mangal areas in the Arabian region (information for some areas is incomplete; map based on several sources).*

from the wider continental shelf, warmer temperatures, more protected and gently inclining shoreline (at least on the Saudi Arabian coast) and greater freshwater availability (Price *et al.* 1987b). Higher nutrient concentrations in the southern Red Sea (Chapter 2) may also be important. The southerly increase in mangroves (and seagrasses: Chapter 7) also coincides with changes in coral and reef fish communities (Chapters 3 and 4).

On the southern coast of Oman small stands of *A. marina* occur in several *khawrs*, reaching 6 m in height (Salm and Jensen 1989). Densest stands in Oman are found further east towards and within the Gulf of Oman. At Al Qurm a moderate-sized stand has been protected within a reserve. Some dense stands also occur in the region of the Strait of Hormuz. However, there is a pronounced northerly decline in mangal development within the Gulf itself.

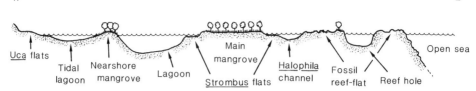

Figure 8.2 *A diagrammatic section through a mangal in Sinai, northern Red Sea (from Por* et al. *1977).*

D. Ecology of the mangal in the Red Sea and Arabian Gulf

(1) Productivity

Compared with the regions such as the Caribbean and SE Asia, knowledge of mangal productivity in the Arabian region is still rudimentary. Experimental data on the productivity of mangrove trees are scarce, as are measurements on associated vegetation in most areas.

Sources of primary productivity within and around the mangal include: mangrove trees, blue-green algae (Cyanophyta), red algae (Rhodophyta) and other algae, photosynthetic bacteria, seagrasses and phytoplankton. On the Sinai coast, estimates of the gross primary productivity of certain communities within a hard-bottomed mangal are shown in Table 8.3; data for mangrove plants were derived from studies in Florida. It is evident that production from benthic macroalgae (e.g. *Caulerpa racemosa, Padina pavonia, Spirydia filamentosa*) is substantially greater than from both microalgae and phytoplankton. *Caulerpa racemosa* showed the highest maximal production rate per organic unit weight (Dor and Levy 1984). Gross primary production rates also showed strong positive correlation with the organic matter content of individual species, and with the chlorophyll-a content in the thallus (Dor and Levy 1984).

Despite the nutrient-poor status of northern Red Sea waters, algal growth is apparently promoted by the good illumination and the availability of hard substrata (see Dor 1975, Dor and Levy 1984). Nevertheless, available data for algae suggest that both gross and net productivity are somewhat lower in the Sinai mangal than in other subtropical and tropical regions (Dor and Levy 1984). Assuming the estimates of *Avicennia* productivity (Table 8.3) to be of the correct order of magnitude, mangrove trees would appear to be the greatest contributors of gross productivity within the mangal, at least in Sinai. However, in terms of net productivity, the picture may be different, in view of the relatively high respiratory rate of mangroves compared with that of algae (Dor and Levy 1984).

Mangal development in the south is known to be greater than further north. Values may approach mid-range of global estimates of gross productivity for well-developed stands ($2.3–5.1\,\mathrm{kg\,C\,m^{-2}\,y^{-1}}$) (Crossland *et al.* 1987). There appear to be no published accounts of mangal productivity from other parts of the Arabian region. In view of their dwarfed condition (1–2 m: Price *et al.* 1987b), gross productivity in the Gulf may be at, or below, the lower values generally cited

Table 8.3 Estimates of the gross primary productivity of various autotrophic communities in the mangal of Shurat Arwashie, Sinai, northern Red Sea (from Dor and Levy 1984).

Primary producer	Total gross productivity in mangal $(kg\ O_2\,d^{-1})$	Relative gross productivity in mangal $(\%)$
Avicennia marina[a]	1690	86
Benthic macroalgae[b]	240	12
Microalgae on soft sediment[b]	15.4	0.008
Phytoplankton[c]	2.4	0.001

[a]Data derived from Lugo *et al.* (1975) for Florida.
[b]Based on changes in dissolved O_2 concentrations in light and dark bottles.
[c]Recalculated by Dor and Levy (1984) as equivalent of O_2 from ^{14}C measurements.

$(0.5\,kg\,C\,m^{-2}\,y^{-1}$: UNEP 1985). As indicated, their overall contribution to Gulf productivity will be minor, because of their limited area.

(2) Biota and horizontal zonation

The most exhaustive studies on mangal biota have been undertaken along the Sinai coast of the Red Sea (Fishelson 1971, Por *et al.* 1977, I. Dor 1984, Dor and Levy 1984), and surveys of the common associates have also been conducted elsewhere in the Red Sea (Fishelson 1971, Hogarth 1986, Price *et al.* 1987b). A summary of the number of species within major groups is given in Table 8.4. Any reduction in total species numbers through recent synonymies may be partly compensated by species additions from studies underway (e.g. McDowell, unpublished). Molluscs and crustaceans account for most species, as in mangals elsewhere. A progressive impoverishment of mangal fauna occurs towards the northern Red Sea, perhaps due to increasing salinities and narrowing shelf area (Price *et al.* 1987b).

In Sinai, a number of habitats within the mangal have been defined, together with details of principal species (Por *et al.* 1977). From this, a characteristic zonation pattern of mangal fauna of the Red Sea has been derived (Jones *et al.* 1987), comprising a landward edge (of mangal), mangrove zone, and seaward edge (of mangal).

The landward edge in Sinai and in southern parts of the Red Sea is dominated by *Uca inversa inversa* and *U. tetraganon*. Extensive Cyanophyta mats suggest that productivity within this zone is high. Dense populations of gastropods (*Pirenella cailliauda* and *P. conica*) occur. Another gastropod, *Cerithidea cingulata*, reaches densities of $579\,m^{-2}$ and shows significant positive correlation with the abundance of Cyanophyta (Price *et al.* 1987b). *C. cingulata* is also common in other

Table 8.4 Summary data for mangal biota in Red Sea, showing approximate number of species in selected groups and examples (from Fishelson 1971, Por *et al.* 1977, I. Dor 1984, Price *et al.* 1987b).

Group	Approx. no. of species	Examples
Mangroves	4	*Avicennia marina, Rhizophora mucronata*
Algae		
(Cyanophyta)	47	*Cyanohydnum sp*(p.), *Lyngbia sp*(p.)
Foraminiferans	10	*Peneroplis planatus, Spirolina arietina*
Sponges	9	*Heteronema erecta, Miemna fortis, Haliclona* sp.
Coelenterates	14	*Cassiopeia andromeda, Zoanthus bertholetti*
Polychaetes	19	*Clymene lobricoides, Perinereis nuntia typica, Nephthys* sp.
Crustaceans	>70	*Uca* (4 spp.), *Dotilla sulcata, Perisesarma guttatum, Metapenaeus* sp.
Molluscs	>45	*Pirinella conica, Cerithium scabridum, Cerithidea cingulata, Saccostrea cucullata, Littorina scabra*
Echinoderms	8	*Tripneustes gratilla, Ophiocoma scolopendrina, Holothuria* sp.
Fish	>30	*Aphanius dispar, Dasyatis uarnak, Lutjanus fulviflamma*

mangrove zones and other coastal ecosystems. A distinctive landward zone is absent along much of the central Red Sea (Jones *et al.* 1987).

Although the southern Red Sea mangrove zone contains a rich fauna, diversity is limited in comparison with well developed mangals in other parts of the world (see section 5 below). In addition to Cyanophyta (section 3 below) may be found red algae of the "Bostrychietum" community (e.g. *Spirydia filamentosa*). Among the crustaceans are fiddler crabs, *Uca lactea albimana* and *U. urvillei*, both of which appear to be more prevalent in the southern Red Sea. Sediment grain size, detrital content, intertidal extent and moisture may control *Uca* distribution. Further details of their distribution and ecology are available elsewhere (Hogarth 1986, IUCN 1987, Jones *et al.* 1987, Price *et al.* 1987b, Ormond *et al.* 1988). The mudskipper, *Periophthalmus koelreuteri*, is confined to the southern Red Sea, probably due to lower salinities or the presence of drainage/tidal channels. Mangroves also provide a nursery or feeding ground and protection for food fish such as snappers (Lutjanidae), mullets (Mugilidae) and porgies (Sparidae). In addition, the habitat provides nesting sites for a variety of shore birds, such as pelicans, reef herons, ospreys, spoonbills and African collared doves.

The seaward edge of the the Red Sea mangal is characterized by fauna of the particular habitat present. Where mud occurs, fauna may include ocypodid crabs such as *Macrophthalmus boscii* and *M. depressus*, the latter occurring at lower levels (Jones *et al.* 1987). Where rocky substrata become exposed, a typical rocky flora and fauna becomes established, in particular the red alga *Digenea simplex*, as well as the large gastropod *Strombus tricornis* and smaller gastropods (e.g. *Clypeomorus clypeomorus, Cerithium coeruleum* and *C. ruepelli*) (Por *et al.* 1977). At lowest shore levels, but particularly within lagoons and embayments, seagrasses are often found (e.g. *Halodule uninervis*) and the biota characteristic of this habitat such as the pin shell, *Pinna muricata, Strombus tricornis* and swimming crabs (e.g. *Scylla serrata* and *Portunus* spp.).

On the Gulf coast only in Bahrain have detailed studies of mangal biota been attempted (Jones 1985, Vousden 1988). Species occurring within the *Avicennia* zone are those typical of other muddy habitats. Three crabs (*Cleistostoma dotilliforme, Metopograpsus messor* and *Eurycarcinus orientalis*) extend throughout the zone, burrow densities of the latter reaching $40\,\mathrm{m}^{-2}$. *Macrophthalmus depressus* is common below the mangal to the low tide mark. The gastropod *Planaxis sulcata* commonly occurs on the leaves and trunks of *Avicennia*, replacing the specialized *Littorina* snail (*L. scabra*) found in the Red Sea (Table 8.4) and elsewhere in the Indian Ocean (see Table 8.5) (Macnae 1968: cited in Vousden 1988). Just to the seaward and landward sides of the *Avicennia* zone, *Cerithidea cingulata* is abundant. The crab, *Ilyoplax frater*, may also be present on the seaward fringe of the mangrove zone. Mudskippers, although not recorded in Bahrain or Saudi Arabian Gulf, are present in other parts of the Gulf including Qatar and Kuwait.

The importance of the mangal of Bahrain to higher vertebrates is suggested by the significant populations of wading birds, and of shoals of shrimp and fish (Vousden 1988). This may be related in part to the higher densities of macrofauna (Vousden 1988). The importance of mangal to fishery resources is also reported for other parts of the Gulf (Rabanal and Beuschel 1978).

(3) Vertical zonation of biota

Mangroves show not only a horizontal zonation (above), but their pneumato-phores and trunks provide a hard surface on which a vertical zonation of typical rocky shore organisms can be seen. These occur as bands of periwinkles/snails (e.g. *Littorina scabra*), oysters (*Saccostrea cucullata*) and at least three species of barnacles. At lower levels sponges and tunicates are often found, together with echinoderms such as *Echinometra mathaei* and *Tripneustes gratilla* around the base of the pneumatophores. Further details of the fauna of mangrove aerial roots in Sinai are given by Por *et al.* (1977).

Details of the vertical zonation of Cyanophyta on *Avicennia* pneumatophores, again on the Sinai coast, are given by I. Dor (1984). Of the 47 taxa recognized, 17 were found on the upper section, 40 on the middle and 19 on the lower section. The midlittoral zone evidently provides optimal conditions. Dor (1984) suggests that despite climatic differences with other regions investigated, there is a similar sequence in the zonation of marine littoral Cyanophyta. Although not spectacular visually, Cyanophyta are known to be highly productive and also ecologically important in the Gulf, partly due to their nitrogen fixing capability (Sheppard and Price 1991).

(4) Comparisons with other mangal biotas

Throughout much of the Arabian region, mangroves occur in relatively high salinities (>40 ppt), which may be higher than the adjacent sea. Using the

Table 8.5 Comparison of the major floral and faunal components within the mangal/muddy shores of different parts of the Arabian region with a typical, muddy shore Indian Ocean mangal (from Jones 1985, Vousden 1988).

	Kuwait	Saudi Arabia/Bahrain	Oman	Red Sea	East Africa
LANDWARD FRINGE High water springs (HWS)	Saltmarsh with Zygophyllum Aeluropus–Juncus/	Saltmarsh and Avicennia marina	Avicenna marina Cardisoma sp.	Avicenna marina Rhizophora mucronata Dotilla sulcata Ocypode saratan Uca inversa inversa Littorina scabra	9 mangrove spp. 5 grapsid crab spp. 2 Uca spp. Littorina scabra Cardisoma carnifera
High water springs (HWS)	Halocnemon zonation Sesarma plicatum	Planaxis sulcata	Uca inversa		
HWS to high water neaps (HWN)	Cleistostoma kuwaitensis Uca sindensis		No published data	Uca tetragonon Uca lactea albimana	3 grapsid crab spp. 3 Uca spp. Helice leachii
HWN to mean sea level (MSL)	Ilyograpsus paludicola Cleistostoma dotilliforme Ilyoplax stevensi Ilyoplax frater Eurycarcinus orientalis	Ilyograpsus paludicola Cleistostoma dotilliforme Ilyoplax frater		I. paludicola Paracleistostoma leachii Macrophthalmus depressus M. telescopicus P. koelreuteri Metopograpsus messor	I. paludicola 3 grapsid crab spp. M. depressus 4 mollusc spp. Eurycarcinus nata Metopograpsus thukuhar

	Periophthalmus koelreuteri, Boleophthalmus boddarti, Tylodiplax indica, Metaplax indica	Eurycarcinus orientalis, Metopograpsus messor, Metaplax indica	No published data	4 Uca spp. 2 grapsid crab spp. 15 mollusc spp.
MSL to low water	Macrophthalmus grandidieri, M. depressus, Scartelaos viridis, Cerithidea cingulata, Macrophthalmus pectinipes	M. depressus, C. cingulata, Pirinella conica	Pirinella caillaudi 7 mollusc spp.	
SUBLITTORAL FRINGE	Portunus pelagicus	P. pelagicus	Scylla serrata	Scylla serrata

classification of Por (1972), mangals in the Red Sea (Por *et al.* 1977) and Gulf may be regarded as metahaline. This contrasts with mixohaline mangals, which are characteristic of brackish or euryhaline environments (Por *et al.* 1977), such as in SE Asia.

Mangals in the Arabian region show much biophysical variability, as described above. Even the mangal within one region, such as the Red Sea, is far from uniform. Comparisons with mangal biota from other parts of the world are therefore problematic. Table 8.5 is a broadscale comparison of biota within the mangal/muddy shores of different parts of the Arabian region with a typical, muddy shore Indian Ocean mangal. Mangal systems of the Arabian region may be less diverse compared with many other parts of the Indian Ocean because of the more severe climatic and environmental conditions (e.g. high salinity), in conjunction with the more limited range of suitable habitats and niches.

(5) Specificity of biota

The mangal within the Red Sea in particular represents a composite ecosystem containing a range of hard and soft substrata, inhabited by fauna and flora typical of each. Probably few, if any, of the biota associated with mangals in the Red Sea or other parts of the Arabian region are mangal-specific. For instance, although three species of *Uca* (*U. inversa inversa*, *U. tetragonon* and *U. urvillei*) were confined to the mangal during a recent survey of the Red Sea (Price *et al.* 1987b), the same species do not appear to be mangal specific elsewhere (see Crane 1975). Also of ecological significance is that *Uca* has never been recorded within mangals or other habitats on the Gulf coast of Saudi Arabia or Bahrain (despite intensive surveys); yet further north on the tidal flats of Kuwait, devoid of mangroves, are found two species of *Uca* (Jones 1986).

The presence of mangroves may principally represent a physical modification (e.g. hard substratum or shade) of an otherwise typical muddy, sandy or rocky shore devoid of such macrophytes. However, as described earlier (section B), there are conservational and other ecological attributes which confer value to mangals and distinguish them from other coastal ecosystems. Similarly, while there is little evidence of mangal-specific species, it is sometimes argued that mangal communities in different parts of the world share many ecological features in common, suggesting some distinctiveness of biota (see Por 1984).

(6) Interactions with adjacent ecosystems

Interactions between mangals and other ecosystems are complex, as exemplified by studies in other parts of the world (Ogden and Gladfelter 1983, Crossland *et al.* 1987). The primary features are physical environmental factors (e.g. topography, currents, substratum type and geomorphology), but also nutrient exchange, transfer of dissolved and particulate organic matter (DOM and POM) and animal migrations, all of which may be influenced by human activities (see Crossland *et al.* 1987). Mangals seldom occur in isolation and usually coexist with other ecosystems (e.g. coral reefs, seagrasses), in the Red Sea usually within semi-open systems. Breakdown of mangroves may be rapid, at least in the northern Red Sea: 80–100% degradation and loss of original *Avicennia* material was reported after 100 days (Reice *et al.* 1984: in Por and Dor 1984). Water movement and other hydrographic factors influence the rate of mangrove decomposition into detrital materials.

The better developed southern Red Sea mangals are likely to exert a stronger influence on adjacent ecosystems and make a greater contribution to coastal productivity. In some areas, transfer or export of nutrients (e.g. as DOM and POM) from mangals might be exptected (Crossland *et al.* 1987). Nutrient transport will be enhanced by the movements of fish, other aquatic fauna, birds and even camels, by feeding–defecation cycles. Migratory and other animals therefore provide a major link between mangals and nearby ecosystems (UNEP 1985, Crossland *et al.* 1987). In areas of increased mangal development, stabilization of shoreline sediments will also be enhanced.

Elsewhere in the Arabian region mangals are generally less well-developed and relatively uncommon, so that their contribution to coastal productivity can be only minimal. It is unlikely that they are any more productive than tidal flats devoid of mangroves, although quantitative details are unavailable. However, in view of conservation attributes (section B) and extensive resource-use conflicts associated with many mangal areas, the ecosystem is an important priority for conservation and management (Chapter 15).

Summary

A eurythermal and euryhaline mangrove species, *Avicennia marina*, is dominant, reflecting the high salinities (40–>50%) and extremes of water temperature (12–>35 °C) associated with the Arabian region. In the Red Sea, three other species are known but are uncommon. The northern latitudinal limit (27–28° N) of naturally occurring mangal ecosystems in both the Red Sea and Gulf is attributed largely to cold winter temperatures. Mangals are tallest (5–7 m) in the southern Red Sea, where the continental shelf is wider and the intertidal slopes more gradual, allowing development of better sedimentary conditions. In parts of Oman *Avicennia* reaches 6 m, whereas in the Gulf it is poorly developed and often stunted (1–2 m), at least along western shores.

Productivity of Arabian mangals is generally low, although few quantitative studies have been attempted. Gross productivity of the poorly-developed stands in the northern Red Sea is probably $<1 \, \text{kg} \, \text{C} \, \text{m}^{-2} \text{y}^{-1}$, and the productivity of stunted mangals of the Gulf may be similar or less, at least on western shores. Benthic macroalgae (e.g. Cyanophyta) associated with the mangal contributed an estimated 12% of gross primary production in a study in Sinai. Gross productivity in well-developed mangals of the southern Red Sea may approach the midrange of global values ($2.3–5.1 \, \text{kg} \, \text{C} \, \text{m}^{-2} \text{y}^{-1}$). Gross productivity of mangals in the Gulf may be at, or below, the lower values determined elsewhere ($0.5 \, \text{kg} \, \text{C} \, \text{m}^{-2} \text{y}^{-1}$).

The mangals of Arabia comprise both the soft-bottomed mangals, and the hard-bottomed type, the latter more prevalent in the northern Red Sea. The mangal is a mosaic habitat, inhabited by species typical of muddy, sandy or rocky shores devoid of mangrove vegetation. Compared with other Indian Ocean mangals, the number of mangrove and associated species in the Arabian region is low, although most of the characteristic faunal zones are still present. Low diversity is attributed by the generally severe climatic and environmental conditions (e.g. high salinity), in conjunction with the more limited range of suitable habitats and niches. Few, if any, associated species are mangal-specific.

Interactions between mangals and adjacent ecosystems are undoubtedly greatest in the southern Red Sea, where mangals are best developed and probably make a significant contribution to coastal productivity. Important mechanisms include transfer of nutrients (e.g. DOM and POM) and energy, aided by movements of fauna. Stabilization of shoreline sediments will also be enhanced by mangals, particularly in the southern Red Sea.

CHAPTER 9

Intertidal Areas 2. Marshes, Sabkha and Beaches

Contents

A. Marsh and coastal plants   176
   (1) Supratidal coastal vegetation   177
   (2) Distribution of saltmarsh communities   178
B. Sabkha, pools and salt flats   182
   (1) Physical conditions, tidal levels, salinity   182
   (2) Prokaryotic biota of the sabkha mats   185
C. Sandy and muddy beaches   188
   (1) Physical   188
   (2) Principal biota   189
D. Rocky shores   190
   (1) Principal biota   190
E. Exotic habitats   192
   (1) Hot and sulphur spring vegetation   192
   (2) Coastal fissures   193
Summary   194

Few parts of the Arabian region have a tidal range of over 1 or 1.5 m, exceptions being the northern ends of the Gulf and Red Sea and the coast of Oman. While this might seem to limit the physical extent and biological significance of the non-mangrove dominated parts of the intertidal region, much of the shoreline is so low lying that even a range of 1 m commonly floods and exposes several hundreds of metres. In many places even a kilometre or two of coast is exposed. Such areas are fairly common in the southern Red Sea and Gulf. In addition, this large intertidal area is particularly interesting for the study of ecosystems or species under severe natural stress, which is one of the themes of this book. The intense insolation, desiccation, and heat, or even freezing temperatures in the Gulf in winter, provide severe natural stress. Added to this, in the central Red Sea the only regular tide is seasonal rather than daily. This gives rise to further, sometimes insurmountable, environmental difficulties for many species, some of which can live in this area during winter months only.

Reef flats which are commonly exposed during extreme low tides are mostly covered in Chapter 3, and not considered in detail here. This applies to *Avicennia marina* habitats too, and while these are the most conspicuous intertidal habitats in much of the Red Sea, especially the south, they are not the most common intertidal habitat in most of the region. Other intertidal habitats are sometimes called "minor habitats" but this reflects only the more limited studies on them.

Some, like sabkha, cover areas far greater than do those colonized by mangroves and marshes. In the case of sabkha too, considerable work has been carried out, at least in some parts of the Arabian region. For other intertidal habitats, however, very little is known, and most existing knowledge is summarized here.

The climatic conditions affecting intertidal habitats are very poorly documented, and quantitative details are often missing, and usually no more than anecdotal. Despite numerous comments about "intense desiccation", "scorching heat" and even "hot enough to fry an egg" (though has anyone ever actually tried?), very few data exist of the important physical parameters from most areas, foremost amongst which are surface temperature, and salinity and alkalinity of interstitial water. From what is known, the conclusion may be reached that physiological stress to plants living in such habitats is not necessarily great, because a few centimetres below the surface, water salinity and quantity may be considerably different from that in the hot, salt crusted surface layer itself. The same may not apply to small fauna living on the surface.

Few quantitative measurements of physical parameters are available for coastal and marsh soils. Temperature has rarely been measured despite the relative ease of doing so; we refer here not to surface temperature of open sea water, but the much greater temperature extremes reached, for example, in tidal pools and ponded water, surface temperatures of sand and intertidal rock, and temperature of the surface of prokaryotic algal mats which develop on some sabkha systems. All these environments clearly have a much greater temperature range than air. Many have dark coloured sediments which absorb radiant heat, and attain temperatures which bear little relation to the readings of shade temperatures from airport weather stations which usually have to suffice as an environmental guide over much of the Arabian region. Likewise, it is not unreasonable to suppose that desiccation by wind is a powerful control on intertidal life here. The strength and duration of winds and the low humidity common in the Arabian region would suggest very severe evaporative effects. An example (Jones 1986) is that in Kuwait's marshes, average rainfall is $115 \, mm \, y^{-1}$ while evapo-transpiration reaches $3400 \, mm \, y^{-1}$, which implies considerable interstitial water movement. Further, it is very rare to find any measurements of the important ions in interstitial water. One significant exception to the latter is in the volume *Hypersaline Ecosystems* (Friedman and Krumbein 1985) which contains most known data on sabkha habitats of the Sinai.

The general tidal regime of Arabian seas is covered in Chapter 2. In the following sections, environmental conditions of sabkha, sandy and muddy beaches, rocky shores and other unusual habitats are restricted to their bearing on the ecosystem concerned.

A. Marsh and coastal plants

Coastal vegetation here is considered as those plants other than mangroves within direct influence of the sea. While sometimes this means rooted in the intertidal or near supratidal region, as with marsh plants, it refers also to xerophytes and halophytes within the splash zone, or at least which live within the band of substrate near to shore which is saline and which usually contains a very high content of marine carbonates.

(1) Supratidal coastal vegetation

The region above extreme high tide is perhaps of marginal relevance to this book in biological terms. However, a strong argument may be made that for management purposes the broader coastal zone should be regarded as a whole. In the present section, however, treatment of the supratidal is very brief, because its total biomass of plant material is so small compared with intertidal and sublittoral vegetation. Also there is generally negligible exchange of organic material and nutrients between the supratidal and marine vegetation. Most supratidal coastal vegetation is simply not a very important ecological component of the marine system in the Arabian region, though it is visually very striking, set as it is against a desert background. Physiologically it is very interesting as it demonstrates some extreme adaptations to environmental stress.

The following is limited to indicating the main species and their approximate zonation from the high tide level. A more detailed account may be found in Evenari *et al.* (1985) who worked in the Gulf of Aqaba. Halwagy (1986) provides some details of northern Gulf coastal plants and their zonation. Much information may also be obtained from the "Arabian Flora" by Collenette (1985), and while her magnificent book is mostly taxonomic, it contains habitat data for most species too.

It is difficult to delimit vegetation along the coastal strip from that further inland in many cases, since many shoreline plant species penetrate, sometimes abundantly, inland, especially up wadis or into ravines which open out at the shoreline. Where such wadis exist, notably where erosion has cut deep (>100 m) gorges into the coastal mountains, many "run-on" habitats exist of fairly luxuriant flora right to the high tide level. Of the large trees present in such areas, *Acacia* and date palms are common on areas of looser sand on sea shores. Both grow vigorously, often in large groves where non-saline ground water collects or flows, and countless good examples occur throughout the region. Palm groves also flourish in flatter coastal areas, where, however, there are usually indications that the site lies above an old alluvial discharge point. On Gulf shores, sand dunes commonly extend inland from the high tide storm level, and here a tall beach grass *Halopyrum mucronatum* may dominate.

Most coastal plants are low shrubs. At the high tide level itself, and well within the spray zone, are succulent shrubs such as *Zygophyllum album* and *Z. decumbens*, and the fleshy *Arthrocnemum macrostachyum* and *A. glaucum*. Similarly, several species of *Suaeda* grow both in loose sand adjacent to fringing coral reefs, as well as landward of mangroves where it grows well in compacted sand. In many of these cases, from the Farasan Islands to the Gulf of Suez, examples of these plants may be found touching the high water mark. Conspicuous also is *Nitraria retusa*, a spreading shrub common in the north of the Red Sea and Gulf of Aqaba where there are extensive stands, and species of *Limonium* whose habitat extends from the high tide sand into stony areas.

Further inland, phytogenic hillocks are a common feature caused by several species of coastal plant, notably *Salvadora*. Many thousands of hectares along the coast are dominated by these features, which extend from the highest point of marine influence, and beyond into the interior of the coastal strip. These features are most common on huge flat expanses such as occur along the Gulf, Gulf of Suez and southern Red Sea where fine sand is dominant. Hillocks develop when the plants trap airborne sand and dust, forcing the aerial part of the plant to grow upwards on an increasing mound which can exceed 3 m high; Evenari *et al.* (1985) refer to hillocks up to 6 m high formed by bushes of *Salvadora* on the Gulf of Aqaba. The species of bush found in the harshest conditions is probably *Halocnemum strobilaceum* which occurs in sabkha, including the well studied pool at Ras Mohammed in Sinai (Friedman *et al.* 1985). This plant (sometimes named *Halocnemon*) is also abundant in coastal marshes which become inundated by sea in Kuwait (Halwagy 1986), and is thus widely distributed ecologically.

With the possible exception of the last when it occurs well inland, many of these plants do not appear to be water stressed for most of each year, although they grow on what appears to be dry, often loose sand. While they obtain a proportion of their moisture from dew, and while many have salt excreting leaves, their deeply penetrating roots commonly extend through the dry surface layer, where the salinity of the limited interstitial moisture is high, and into a layer where the moisture has a remarkably low salinity (Figure 9.1). The roots do not generally penetrate into deeper layers still where salinity rises again to that of sea water or greater. As is well documented by Evenari *et al.* (1985), the intermediate layer of lowest salinity may extend from 10 to 75 cm deep and has a total salinity commonly less than 10 ppt, and sometimes as low as 1 or 2 ppt. Further, water content of the soil in this intermediate depth span may range from 2 to 10% by weight, indicating no physiological shortage of water, even when the surface is dry with a high salt content. Interestingly, similar analyses at sites with no plants showed interstitial water with a similar salinity, but with total water content of less than 1% by weight. This indicates that dryness, not salinity, inhibits germination or growth. For germination especially, the surface layer must have moisture, and even where plants are established, germination may depend on years of higher than average rainfall. Obviously, areas where this coastal vegetation is conspicuous are commonly those where even cursory observation shows mouths of wadis or depressions. Despite these observations, it seems likely that some plants at least are especially dependent on dew. It is difficult to explain otherwise how some may survive a year of drought on a fossil reef.

(2) Distribution of saltmarsh communities

Saltmarsh plant communities are not now widespread around most of Arabia, but where they exist they provide localized areas of extremely high productivity — oases on the desert shoreline. By far the largest area is the Shatt al Arab estuary in the northern Gulf, though it is clear that impressive areas were seen in the Gulf of Suez by the Forskal expedition only 200 years ago. Earlier still, it was suggested that the abundance of reeds in the Gulf of Suez led to the name "Reed

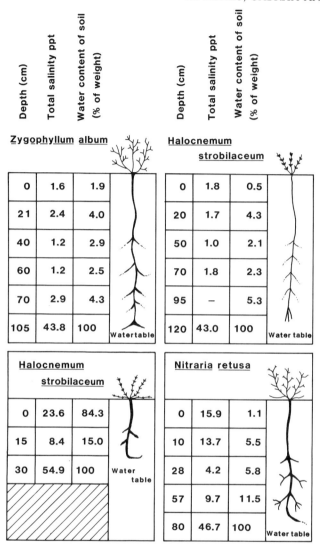

Figure 9.1 *Profiles of several shore plants from Sinai, showing the penetration of roots to conditions of better water supply. From Evenari et al. (1987).*

Sea", which is one contender for the derivation of the name Red Sea. There seems to be no remnant of this now.

Reeds are so common about the Arabic Gulf (= Red Sea) as to have procured the Gulf the name of Jam Suf, *or the sea of reeds, from the ancients . . . It grows with a vigorous vegetation, and in great abundance, in . . . Ghobeybe, where it rises to the height of twenty-four feet . . . nearly opposite to Suez.* Niebuhr 1792

However, although destruction of such sites in the Gulf of Suez and undoubtedly elsewhere has been widespread and in many areas total, in many parts of the region today new marsh communities are appearing where none existed before. These are at sewage outfalls of major cities, which are now supplied with vast quantities of fresh water from desalination plants. Much of this water, greatly enriched with nutrients and organic matter, is discharged into the sea. Typical of these are examples in Bahrain and Jeddah, where enrichment causes greatly increased primary production over many hundreds of hectares. In the latter case especially, effects of this enrichment are exaggerated because the discharge area is low lying and partly retains the effluent before it dissipates into the Red Sea. Most such areas are off-limits to scientists and have not been studied. However, it is clear that coastal vegetation thrives in these areas, and high chlorophyll cover in the Jeddah saltmarshes shows up on remotely sensed images more strongly than that of nearby mangrove stands, and is comparable to the signal from large irrigated parks and estates. Such marsh areas also act as a focus for numerous species of birds, especially migrants in spring and autumn.

As cities grow in the region, and produce ever greater quantities of partly treated or untreated waste water, these sites may become larger and more numerous. Also they may become more important to localized productivity, and it can be speculated, to total nutrient input into wide and spreading areas of the nutrient poor Red Sea. For example, approximate calculations from data of Weikert (1987) suggest that the surface 100 m of the entire Red Sea contains a total of only about 0.66 million tonnes of dissolved nitrate. The demographic pattern along the Red Sea coast is that of a few, high density urban areas and the input from several million coastal residents could add a significant nutrient load to adjacent parts of the Red Sea.

Orme (1982) reports that *Phragmites australis* and *Typha domingensis* dominate reed swamp vegetation in many wadis on the African side of the southern Red Sea. He also remarks that up to twenty community types of saltmarsh vegetation occur, identified by their dominant species. Apart from the above, most of these community types are those of *Halocnemum*, *Limonium*, *Nitraria* and allied forms, which here are noted more under supratidal vegetation. It is clear, however, that these groups flourish in the littoral region along the Egyptian and Sudanese coasts at least, especially where low salinity water occurs near the surface in depressions and in areas such as wadis, and they appear to form some of the most productive zones in such regions.

These areas are not well studied, although ornithological data exist and general zonation patterns have been reported. While the term saltmarsh is appropriate, they are also termed simply eutrophic and mesotrophic zones. Where conditions are appropriate, notably where a tropical river estuary exists, they grade into true marshes and reed beds, which occur nowhere more extensively than in the Shatt al Arab waterway.

A) MARSH PLANTS IN THE TIGRIS/EUPHRATES DELTA

The large deltaic plain of the Euphrates, Tigris and Karun rivers discharges into the area known as the Shatt al Arab. This covers an area of about 18,500 km^2. It is now possibly broadly balanced in sedimentary terms, since arrival of sediments is compensated by subsidence and erosion from currents (Sanlaville 1982), although a series of maps (Admiralty 1944) shows an advance of over 100 km in the last 2000 years. The estuary contains a strong bi-directional tidal current which causes the coast to be dynamic and continually shifting. It is marked by sandbars, heavily vegetated mud flats, swamps and marshes (Wright 1982). In addition, there is a seasonal change in quantity of water flowing through it from melted snow of northern Iraq, which causes seasonal changes in overall water level (Saad 1978). Its water is almost always very turbid with Secchi disk readings of generally less than 80 cm, due to suspended matter of $0.1–0.7 g l^{-1}$ attributable to stirring of sediments. Water is well oxygenated at all times of the year, although Saad (1978) suggests that this is due principally to aeration rather than to phytoplankton photosynthesis. In times of highest river flow, there is a slight increase in acidity of the water, which reduces phytoplankton density.

The overflow of the water, falling into the plains by the sea, makes lakes and marshes and fens.
                                                                    Strabo, circa 5 BC

Vegetation of the sabkhas and more elevated sand areas around the Shatt al Arab appears to have a flora very similar to that described earlier. More recent observations are limited though they suggest that in the past few decades vegetation has been reduced, partly due to increasing cultivation. More information is available from the southern extent of the Shatt al Arab, where Halwagy (1986) briefly describes 11 halophytic community types in northern Kuwait. The most important of these in Kuwait are similar to those of the Red Sea, and include communities dominated by *Halocnemum* and *Nitraria*. Communities of more limited extent in Kuwait include those of *Juncus rigidus* and *Phragmites australis*, which reach 1.3 and 2.5 m tall respectively.

In Kuwait, zonation of several identified communities runs parallel to the shoreline and is controlled by horizontal distance from shore and by elevation. Halwagy (1986) summarizes the communities by describing a lower, middle and upper marsh zone. The seaward edge of the lower marsh where *Halocnemum* dominates is submerged by the sea about 440 times each year, while its landward extent is inundated only about 10 times. This frequent inundation reduces soil salinity. In the lower and middle marsh zones, surface salinity of the soils may exceed 100–200 ppt, but at slightly deeper depths this falls to 4–15 ppt. In upper marsh communities, the water table is deeper and the soil is coarser, so that the saline water table does not influence soil salinity very much; soil salinity in this region is 10–20 ppt on the surface and 1–4 ppt at depth.

These communities become more extensive further north in the Shatt al Arab itself, and the latter species especially covers thousands of hectares. *Phragmites australis* (= *P. communis*) is a perennial reed, with woody, hollow stems, which thrives in the waterway and in the littoral belt, extending also into inland depressions. In the Shatt al Arab it is joined by extensive stands of another rush *Typha angustata*. The distribution and growth of the former at least in Kuwait is strongly enhanced by sewage enrichment, when it may reach heights of up to 5 m tall (Halwagy 1986), and in Qatar it similarly thrives in areas where sewage is deposited. In the Shatt al Arab, nutrients necessary for its flourishing growth come from the waterway itself. The woody stems float, and are extensively used by the "Marsh Arabs" for rafts from which they fish and even live.

B. Sabkha, pools and salt flats

The term sabkha is here used to denote low lying, sometimes intertidal but usually only seasonally inundated areas. From the geologist's viewpoint, Evans *et al.* (1964) describe sabkha as "unconsolidated carbonate sediment with minor amounts of quartz and other minerals, and the site of deposition of various evaporitic minerals". The latter may be mainly salt, or may be dominated by gypsum in parts. (Some authors have included saltmarshes and lagoonal areas in this term as well, although these habitats are regarded separately here.) The term sabkha in general usage now seems to refer mostly to salt flat areas which have free standing water at least occasionally, which is the habitat discussed here. Any discussion of sabkha ecosystems requires considerable reference to the comprehensive reviews contained in Friedman and Krumbein (1985) on sabkha in Sinai, and only a little work other than this has been performed subsequently elsewhere in Arabia.

The shore and country adjoining it, often for miles into the interior, is a howling wilderness, with an arid or saline and friable soil, here and there covered with white salt flowers, as if with snow or hoar frost; the subsoil water is brackish and bitter, almost undrinkable.   Klunzinger 1878

---

(1) Physical conditions, tidal levels, salinity

Sabkha forms very flat plains in the coastal area, commonly with periodically filled pools, crusts of white salts and crusts of what are often, and only partly correctly, called "algal mats". The salts are most commonly concentrated sea salts, notably sodium chloride, which is harvested in many locations. These are recognized by their white, often sparkling appearance, but in some areas, salts are of gypsum, forming extensive mats of crystals up to 1 cm across. In both cases, ponded water may be strongly reddish brown, in some cases from halides but more commonly from the dinoflagellate *Dunaliella*. Also in both cases, beneath a crust of organic mat is a thick black reducing layer with a clear sulphide odour. These areas may be fairly small, as little as 100 m across in many parts of the region, to over 10 km across.

Probably the largest sabkha system so far described is from the UAE (Evans *et al.* 1964, 1969) which includes contiguous sabkha areas up to 32 km wide and 320 km long. This region overlies Quaternary limestone which outcrops in places, but is mostly buried 0.3–3 m by overlying sediments. A part of this system at Khor al Bazam which is 130 km long and 8–16 km wide is mapped and described in considerable detail by Kendal and Skipwith (1968, 1969a,b), who note that the system is slowly prograding to seaward. The latter authors also focused on blue-green algal mats and on their role in dissolving and re-precipitating aragonite sediments. From their descriptions and from others of Sinai (Friedman and Krumbein 1985), it is clear that small differences in degree of elevation, which usually correlate roughly with distance from sea, result in major differences in the nature of the biota and the form taken by the blue-green algal mats.

Other large areas of sabkha include the Bar al Hickmann peninsula in Oman, and along supratidal regions of shores of the Gulf of Suez. Large areas of sabkha are also known from much of the Saudi Red Sea coast, especially in the south. Quantitatively therefore, they are a very extensive component of the Arabian coastal region.

Geologists have studied sabkha much more than have biologists, and have generally required more precise definitions than the above, especially if the evaporite area is to be of use as a palaeographic marker (Purser 1985). The origins of sediments and underlying basement rocks, as well as the overall hydrology of the area, are important. For biologists, it is less important whether the sabkha makes physical connection with the highest tidal area, or whether the sabkha is in a depression separated from the sea by slightly higher drifts or dunes and only subterraneanly connected. In many examples in the Gulf, the seaward edge of the sabkha grades into seagrass beds and complex regions of shallow patch reefs, mud banks and lagoons. Lagoons adjacent to sabkhas studied by Bramkamp and Powers (1955) extended to several metres depth, but still retained a salinity of 200 ppt at depth, and had sediments of depositional lime muds and shelly material bound by "algae". In the Red Sea they may grade into mangroves as well. Some of the best studied in the Gulf of Aqaba are separated partly from the sea by land by greater elevation, and others grade on to beach with fringing coral reef.

Most of the sabkha is usually near the high intertidal, and whether directly connected with the sea or not, is usually immersed at least seasonally. Flooding may be extensive at high water level, and free standing water may disappear for at least 6 months each year. However, several factors tend to keep the sabkha surface moist. Firstly, capillary action from the underlying water table, either marine or from a fresh water incursion, continuously draws water to the surface. Secondly, many of the salts which develop on the crust of the sabkha are strongly hygroscopic and attract a film of moisture even in relative humidities of less than 10%. In addition, many of the photosynthetic and chemosynthetic micro-organisms which have adapted to living on the sabkha surface, have developed compounds and mucilaginous jackets which help them to retain sufficient water for their own requirements. In some Sinai systems further removed from the sea, a horizontal water flow (Figure 9.2) has been suggested.

A) POOLS

Pools are a special feature of the sabkha. With these, as with several aspects of sabkha, terminology and precise definition has overtaken the amount of scientific information about them. Polyhaline-euhaline anchialine pools are one group,

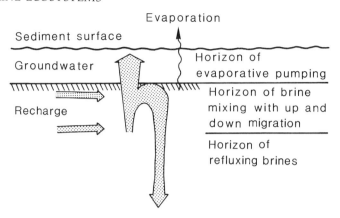

Figure 9.2 *Water movement in the sabkha surface in parts of Sinai. From Gavish* et al. *in Friedman and Krumbein (1985).*

defined (Holthius 1973) as having no surface connection with the sea, although they do have a subterranean one, such that water in the pools fluctuates with the tide in the open sea. A connection as strong as this means that the hydrology and chemistry of the pool is strongly influenced by that of the sea. These pools have a relatively high diversity of benthic species, most conspicuously several decapods which are exclusive to these locations and to subterranean crevices, but very few pelagic species because of problems with stratification (Por 1985).

Pools with the next level of salinity (up to 70 ppt) are described as metahaline anchialine pools. These have a reduced diversity of typical marine biota, including the large medusa *Cassiopeia*, serpulids and grazing molluscs. In the damp littoral of these pools, Cyanophyta are present, but they do not yet develop mats, possibly because of effective grazing.

Pools with salinity of over 70 ppt are termed hypersaline anchialine pools. Even with salinities of this order, pools may be extensive: one near Jeddah is 30 km^2 but is a maximum of 3 m deep and has salinities varying from 51 to 113 ppt (Meshal 1987). These may develop from progressive enclosure of a lagoon by a fringing reef and subsequent partial infilling (as is the case in the above). Where a connection with the open sea remains, as it does in the Jeddah pool for example, the marine biota is still recognizably marine, although this is greatly reduced and increasingly specialized due to osmotic stress, ionic composition changes and eventually because of seasonal drying out of the edges if not all of the pool.

With increasing enclosure, the remaining biota is eventually entirely microbial, and resembles that of pools which did not originate from encircled lagoons but from seepage of water into depressions. This biota is comprised of very euryhaline diatoms, Cyanophyta and photosynthetic bacteria, below which are chemosynthetic bacteria which form characteristic mats of the sabkha region. A later section amplifies this biota. Where conditions are not too extreme, fauna remains, consisting of a few flatworms, nematodes and even one or two species of harpacticoid copepods. It is suggested (Por 1975, 1985) that only animals with resting or anabiotic stages can exist permanently in such pools. Along the gently

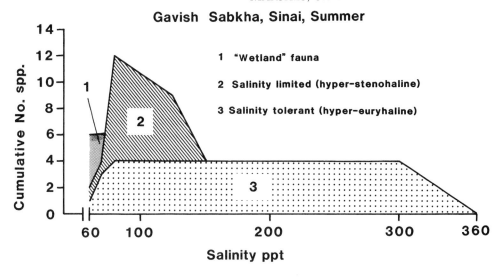

Figure 9.3 *Biota in pools of Sinai with increasing salinity. Drawn from data in Friedman and Krumbein (1985).*

sloping shores of the Gulf, some pools in this group may be extensive.

Basson *et al.* (1977) describe conditions and biota in a pool adjacent to the large embayment of Dawhat Zalum where salinities reach 335 ppt in summer. In this as in many pools, the dinoflagellate *Dunaliella* is the base of the food web, reaching densities of several million l^{-1}, making "red tides". While this is the dominant species, other diatoms and various forms of zooplankton also thrive, including copepods, flatworms and nematodes, the latter even found amongst salt crystals on the lake floor. A quantitative ecological study of nearby saline lagoon, Dawhat as Sayh, is described by Jones *et al.* (1978). The lake beaches in the large Gulf embayments are, like the Red Sea equivalents, covered in mats of "blue-green" algae (next section). Figure 9.3 shows one scheme of the progression of biota in pools with increasing salinity.

(2) Prokaryotic biota of the sabkha mats

As with many aspects of the sabkha region, most of the useful data come from studies in Sinai, where details of the flora described in numerous papers are summarized by Ehrlich and Dor (1985). Some descriptive information for the Gulf appears in Basson *et al.* (1977). It is apparent, however, that sabkha throughout the region encompasses a similar range of tidal positions, physical constraints and algal mats of similar appearance to those analysed in detail from the Sinai, and it is likely that the Sinai work is applicable throughout.

The primary photosynthetic constituents of the surface of the sabkha region are Cyanophyta, also termed cyanobacteria. There are smaller components mainly diatoms and green algae, but while these last groups may be fairly abundant in

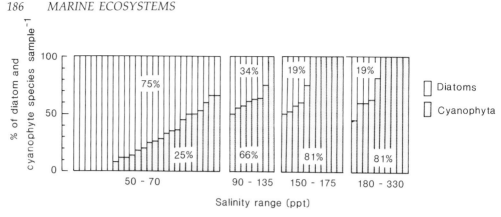

Figure 9.4 *The ratio of blue-green algae (Cyanophyta) to diatoms with increasing salinity in algal mats of the sabkha. Redrawn from Ehrlioch and Dor (1985).*

water with salinity below 100 ppt, they decline in proportion still further as salinity increases above this (Figure 9.4). The most salt-resistant diatoms ever recorded are from Sinai sabkha, which tolerate up to 180 ppt, but Cyanophyta can grow in water up to 300 ppt salinity. Brisou *et al.* (1974) found halophilic bacteria in salinities up to 400 ppt in a hypersaline lake in Somalia. As salinity in water increases, the amount of oxygen which can dissolve falls to less than a quarter that of open sea at the same temperature. Consequently, primary producing biota in high salinity regions often become those which use hydrogen sulphide anaerobically as a hydrogen donor rather than water, resulting in the release of sulphur instead of oxygen.

Thirty-three species of Cyanophyta are recorded from the Arabian sabkhas. Most of these are common in littoral areas around the world (Ehrlich and Dor 1985), In the Sinai, nine to 15 species occur in water with salinity of 80–130 ppt, five to seven species in water of 150–200 ppt, and one in water of 240–320 ppt. (Note that Figure 9.4 refers to abundance of groups, not their diversity.) These species generally form a microbial mat of three different appearances: compact "laminated" mats, soft spongy mats and thin slimy layers. Although the mats are each formed from several species, the appearance which it takes depends on the identity of the dominant prokaryotic organisms. The thin slimy mat form can occur both directly on sediment or on top of another mat. Mats also occur commonly beneath a layer of salt crystals whose deposition has arisen from evaporation of saline water, usually that which seeps in by capillary action. As noted by Krumbein (1985), the layer of Cyanophyta helps to suppress growth of plankton in free-standing water and promotes the size of the salt crystals. This feature is exploited by Bedouin who gather the salt, and who take care not to damage the Cyanophyta layer during harvesting.

A similar number of species (39) of diatoms have been identified, which is much fewer than the diversity found in adjacent open sea water. Their most diverse assemblages on sabkha occur at a salinity of about 60 ppt where 16 species have been found. One species of *Nitzschia* was rare in metahaline water, and its frequency increased in hypersaline samples, reaching a measured occurrence of

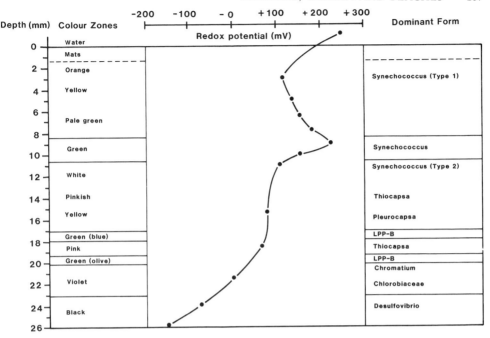

Figure 9.5 *Chemical and physical details through a typical algal mat on a sabkha (composite). Thickness of mat (mm) is shown on left, with colour of the various "sublayers". Redox potential shows anoxic conditions below about 2 cm deep. Main microorganism groups at each depth point are indicated on the right. From Gerdes* et al. *in Friedman and Krumbein (1985).*

50% in samples with salinity of 180–205 ppt. These represent the greatest salinity values in which live diatoms have been found.

Green algae found on sabkha include the common filamentous forms *Enteromorpha* and *Chaetomorpha*, which live in salinity values of up to about 65 ppt. *Rhizoclonium* alone was recorded in salinities between 87 and 132 ppt, while *Dunaliella* is found only in the highest salinities (for this group) of 150–200 ppt.

Mats which develop from various combinations of species in these three groups may occur on permanently damp ground, or in areas which are only dampened by capillary action at high water levels. "Algal mats" are the main photosynthetic group of this habitat, but quantitatively form a thinner layer than do the chemotrophic micro-organisms which flourish in the strongly reducing, sulphurous layer beneath. The total mat may, in damp conditions, include a green-brown layer 1–2 cm thick, lying on the sulphurous black layer. While it is still damp, it may be continuous or in low hummocks, often flocculous, depending on salinity and moisture. As desiccation increases in summer, mats become dry and crisp, later breaking into polygons with upturned edges, which is a characteristic sight on low lying areas throughout the Arabian region. In many cases, a white layer of salt then develops on top of the dried algal mat. While this may represent an extreme amount of salinity, the hygroscopic nature of the salt may actually protect the mat from complete desiccation.

C. Sandy and muddy beaches

(1) Physical

The average slope of the seabed on the Arabian side of the Gulf is about 35 cm km^{-1} (Basson *et al.* 1977), which results in an intertidal region which may be several kilometres wide. This leads to a very long and convoluted coast, despite the appearance of a smoothly curved shoreline on large scale maps of the region. Much of the shoreline contains saline lagoons and mangroves, inland from which are sabkha or broad dunes colonized by grasses, both considered earlier. Details of these are given earlier from lagoons on the UAE coast (Bramkamp and Powers 1955, Evans 1964, Evans *et al.* 1964, Kendal and Skipwith 1969b); this section mentions the remaining soft substrate biota. In the Red Sea, this type of lagoon is rather less abundant except in the south, although Meshal (1987) describes one north of Jeddah and Ormond *et al.* (1984a) describe ponded but slightly more open systems at Shuaiba, to the south of Jeddah. Further north, similar habitat occurs mainly in the Gulf of Suez and in sharms or mersas, but proceeding south, sandy and muddy areas without mangroves become extensive on the mainland and in the Dahlak and Farasan Archipelagoes.

Most sandy shores in the northern Red Sea are narrow beaches adjoining coral reef flats, which themselves are usually narrow. Broader beaches occur where the reef edge swings further out from shore, leaving a lagoon. In extensive parts of the central Red Sea, sand on such beaches has a significant proportion of aeolian sand mixed with marine carbonates. Where reefs decline in the southern Red Sea, beaches become broader and higher, extending deeply into the sublittoral.

This habitat is most important and extensive in the Arabian Sea. Much of the Oman coastline is sandy in the sublittoral, even where cliffs descend to the water's edge, but in the region near the extreme east of the peninsula it reaches spectacular proportions. There the massive sandy desert of the Wahiba Sands has dunes which reach and flow into the sea. The beach along all the Arabian Sea has high energy, and the exposed intertidal and supratidal region extends for hundreds of kilometres along the coast and hundreds of metres inland. Facing the Gulf of Oman, the extensive Battinah coastline is a low lying, sandy shore, similar in nature to that of much of the Gulf. Its substrates are of fine sand, grading to mud in parts.

In the Gulf, most of the beaches are composed largely of carbonate sand, although there is a strong component of aeolian, yellow or red sand. The very gentle slope of the Gulf shoreline results in numerous sand bars, many exposed at low tide, which pond the water but which are mobile and so while they create heated and hypersaline conditions their transient nature does not lead to the formation of sabkha systems. In this area in particular, longshore drift is strong, resulting in numerous, often transient spits. These often overlie recent sandstone or limestone beachrocks which outcrop at intervals.

There are few cobble or rubble intertidal areas around Arabia. The most extensive exception is found in the khawrs of Musandam (Sheppard 1986a), although others are reported from Dhofhar (Barratt *et al.* 1984). These derive from the cliffs surrounding the sheltered inlets and are a very low energy habitat, colonized in particular by anthozoans such as *Palythoa*.

(2) Principal biota

Fishelson (1971, 1973b) provides detailed accounts of intertidal fauna and flora from the northern Red Sea, and in the former, for the Dahlak Archipelago as well. His work is mainly descriptive, and with the exception of the studies by Jones *et al.* (1978), McCain (1984a,b) and Vousden (1988), descriptive rather than quantitative studies have dominated the literature subsequently. The following does not repeat the descriptions in any detail but highlights the main points. Detailed and illustrated accounts of intertidal, soft substrate faunas are found in the above, and for the Gulf in Jones (1985), Jones *et al.* (1978) and Vousden (1988), and for the Arabian Sea in Anderlini (1985).

Throughout the Arabian region, the highest parts of most beaches are marked by conical mounds produced by the ghost crab, *Ocypode saratan.* Other macrofauna include the hermit crab *Coenobita* and numerous amphipods, especially *Talorchestia* which occur beneath detritus. Both Basson *et al.* (1977) and Jones *et al.* (1987) depict characteristic zonation patterns of macrofauna on Gulf beaches. The main large species nearer the low tide level are the burrowing urchin *Echinodiscus* and the crabs *Dotilla* and *Uca.* The latitudinal distribution of the latter is interesting; it is found in the north and south of the Red Sea but rarely or never in the central region where there is no daily tide. In the Gulf, *Ocypode* is joined by the equally common crab *Hippa*, which is known to feed on the mud crab *Macrophthalmus* which it excavates from its burrows. Seasonal differences are considerable, since the shallow lagoons commonly dry out in summer. Annual air temperature ranged from 2 to 50°C over the last 25 years in a lagoon near Dammam, although monthly mean values varied less at 15–35°C (Jones *et al.* 1978). Many fauna in the latter and other similar lagoons are not permanent, therefore, although some species may breed in lagoons in favourable conditions in some years.

In his study around Bahrain, Jones (1985) found that it was difficult to relate the fauna of any beach to sand particle size, and concluded that this probably reflected the mixed nature of sand and rock, together with the poor sorting of sediments in the absence of wave action and large tides. Elements of rocky shore biota appeared intermixed with sandy shore biota and the same mixing is reflected to a considerable degree in the Red Sea. Along Arabian Sea shores, sandy shore fauna is much richer in molluscs than is apparent in the Gulf and Red Sea although this really applies only to macro-molluscs as the interstitial or micro-molluscs have been poorly studied in all parts of the Arabian region.

The density of animals in sandy as well as muddy areas of the intertidal may be extremely high. In more muddy and sheltered regions of the Gulf and Red Sea, there is commonly a covering of algal mat which creates a very high primary productivity to supply dense populations of the gastropods *Pirenella* and *Cerithidea*, the former being able to thrive in very saline conditions of inlets and ponded areas. However, the greatest densities are of interstitial fauna. McCain (1984a) examined 96 cores from the Saudi Gulf coast and determined that each square metre could contain 400,000 animals, each over 0.5 mm in size. Total faunal abundance was significantly correlated to the degree of beach slope, decreasing with increasing slope, probably because of the steepening of a beach that occurs with increasing turbulence. In this study, the number of taxonomic groups increased towards the low tide level as well as with increasing content of carbonate mud in the sand. Altogether, six intertidal faunal communities were identified, and the faunal community became much reduced near sources of oil contamination.

The total number of species in the above study was 147. It was implied that this is a high diversity considering the high temperature extremes experienced by the surface of the sediments. Interestingly, the conclusion reached only a few years earlier (Basson *et al.* 1977) was that the sand biota in this area had a low diversity and that the intertidal sand left a "barren impression"; in the latter case the species count from sand beaches was over 200 species, and the diversity from areas of mixed sand and rock was considerably greater. Diversity is, however, much less than that of rocky shores.

D. Rocky shores

Reef flat communities which dry at extreme low tides are not included in this section, which instead highlights the main biota of higher beach rock and non-reefal hard substrates. These rocky shores are extensive on the Iranian coast, although their biota is extremely poorly known, and elsewhere in the Gulf they are most commonly of the form of beach rock intermixed with sand. The most extensive area of rocky intertidal is found in the Strait of Hormuz, around Musandam, and much of the Oman shoreline is rocky and backed by undercut cliffs, even though these commonly extend only to sand at extreme low water. In the Red Sea, rocky intertidal is more restricted, occurring on undercut raised reefs in the northern half, with expanses of beach rock in the south.

(1) Principal biota

The diversity on rocky shores is considerably greater than that of sandy beaches or mud, although biomass may be less. Fishelson (1971, 1973b) provides detailed descriptions of northern Red Sea shores and Dahlak Archipelago. Lists of species are also provided in conjunction with zonation diagrams, and with descriptions of several fairly clear-cut assemblages of animals. Jones (1986) provides a thorough taxonomic account of the northern Gulf intertidal flora and fauna. This latter is used increasingly in Red Sea areas today as well, although many more species should be expected in the latter than are included in his account of the Gulf.

Basson *et al.* (1977) remark that the rocky intertidal in the Gulf is much less productive than sandy intertidal areas, giving as reasons intense heating at low tides in the summer. This prevents establishment of a vigorous growth of algae which otherwise would provide shelter, thus limiting the rocky fauna to crevice dwellers. Sessile fauna are thus almost all filter and plankton feeders, such as bivalves, tunicates, and sponges. Similarly, these shores support large clusters of attached serpulid worms in calcareous tubes, and the common barnacle *Balanus amphitrite*. Mobile fauna are mainly grazers, feeding on blue-green algal films which are the dominant photosynthetic form of Gulf rocky shores. Important species are top shells *Trochus erythraeus*, turban shells *Turbo coronatus* and the shell-less, air breathing mollusc *Onchidium peronii*. The spider crabs *Grapsus* and *Metopograpsis* are also herbivorous. The commonest predatory species include the molluscs *Thais* and *Drupa*, portunid and xanthid crabs and mantis shrimp *Gonodactylus demanii*. While it is true that rocky fauna is diverse, the extent of this habitat in those parts of the Gulf which are accessible for study is small.

Anderlini (1985), Barratt *et al.* (1986) and Sheppard (1986a) offer descriptive accounts of the shoreline of Oman, and Jones (1985) remarks that the rocky shores of Oman show a general increase in diversity southward. But there has been little taxonomic work done on much of Oman's coastline, a notable exception being a study of echinoderms of the Dhofhar coast (Campbell and Morrison 1988). The Musandam peninsula with its several hundred kilometres of convoluted rocky shoreline of varying exposure undoubtedly harbours a high diversity. While a trend of increasing diversity southward towards the equator might be expected since it is commonplace for diversity to rise with decreasing latitude, in Oman the temperate conditions of the southern part might even suggest a reverse trend as is the case with some fully sublittoral, tropical groups like corals. On the other hand, intertidal regions may be affected less by seasonally lowered temperatures, and in terms of algae at least, which is one of the few well studied groups, there is a marked increase in diversity southward with over 200 species in Dhofhar (Barratt *et al.* 1984). In central Oman, the extensive rocky intertidal has been studied insufficiently for conclusions on diversity.

Throughout the region, rocky intertidal fauna includes large numbers of conspicuous grapsid crabs, rock oysters *Saccostrea cucullata* and the large barnacle *Tetraclita squamosa*, and chiton *Acanthopleura haddoni*. Important to erosion is the mussel *Lithophaga cumingiana* and the piddocks which bore into soft limestones. The latter causes marked undercutting of the cliffs in Musandam (Vita-Frinzi and Cornelius 1973).

Zonation for much of the Arabian Sea shore is broadly similar (Anderlini 1985) although is not always clear because the intertidal region is often narrow and conditions usually turbulent. There is a tendency in the highest regions for there to be large numbers of the gastropods *Littorina* and *Nodolittorina*, and chthamalid barnacles. In mid-tidal regions, rock oysters *Saccostrea* are sometimes very abundant, although they may be mixed with or even replaced by the large barnacle *Tetraclita*. High densities of oysters appear to completely exclude algae in many places. Coelenterates including *Palythoa* and the boring mussel *Lithophaga* occur lowest in the intertidal. Algae are most common on the middle and lower shore, and encrusting reds are the commonest group. Of the distribution of other groups, only that of molluscs is described in any detail to date (Barratt *et al.* 1986).

In the Red Sea, detailed descriptions are given in Jones *et al.* (1987). Much of the available rocky intertidal zone of the north occurs in erosion notches of fossil cliffs. These provide a rather more moist and sheltered habitat than do horizontal expanses of intertidal rock which are more common in the south, and they support a greater range of fauna. Most of the species mentioned above in connection with the intertidal zone of the Gulf and Arabian Sea occur in the Red

Sea too, and in approximately similar patterns. Because this zone is thoroughly described in accessible literature elsewhere, it is not repeated here. It is, in general, a region of very low diversity when compared with the adjacent tops of the coral reefs which, while horizontal and thus more exposed to the sun, are mostly submerged by moving water to at least a few centimetres during the daytime.

Of particular interest is the intertidal zone of the central Red Sea where there is a seasonal but no true daily tide. Here, there is commonly a flourishing barnacle population during winter, which dies off completely when sea level drops over 0.5 m for the duration of the summer. Similarly, numerous algal species attach and grow briefly. Darkened films of blue-green algae are perhaps the only group which survive the prolonged exposure, and these are common only in more exposed locations where there is splash. In this connection, it may be important that the strongest thermal winds occur in summer, causing almost daily splashing for many metres inland, to the extent of forcing up a tidal surge of up to 0.5 m for several days each month as well. This appears to allow the continued existence of the blue-green film and some associated mobile fauna in the seasonally intertidal zone.

E. Exotic habitats

(1) Hot and sulphur spring vegetation

Hot water and/or sulphur springs emerge onto the shoreline in several parts of Arabia. In the south, one such area near Salalah in southern Oman is one of many contenders for a port claimed to be used by the Queen of Sheba. It appears to be connected to an inland lake system, which is another of many contenders for the site where Moses obtained fresh water by striking a rock. On the coast, a sandbar constriction which provides shelter from summer monsoonal waves, allows vigorous growth of mangroves, but of no other notable plant community. In Bahrain (the name itself means fresh water sea), fresh water springs emerge near the coast, and such areas are now heavily cultivated. Springs also emerge offshore where divers used to collect the fresh water in bags.

In Sinai, two small but notable areas are also known. The first is on the Gulf of Suez side of the Ras Mohammed peninsula. Very warm water emerges at an unidentified location on the edge of the reef, possibly not from a single aperture but from seepage over a broad area. This is an entirely sublittoral feature. On the reef flat and slope, there is no living coral for a horizontal distance of over 100 m. Some soft corals survive, but the principal biota is a diverse green algae. Temperatures have not been measured. The seepage is not always present although when it is, it is approximately 45 °C, which is uncomfortably hot for snorkelling.

The second site is further north in the Gulf of Suez near Hammam Fara'oun (hammam meaning bath or spring), at a point where the mountain Jebel Fara'oun descends to the sea. Now becoming a much visited tourist spot, it was studied botanically by Evenari et al. (1985). Water emerging from the mountain is about

70 °C, and runs in several channels into the Gulf of Suez. It has a sulphurous odour, and interstitial ground water contained SO_4 3–5 ppt, or approximately ten times greater than values obtained from most other locations where similar analyses were done (Evenari *et al.* 1985). The main plants in this region are *Zygophyllum*, *Tamarix* and *Arthrocnemum*, but most notably, *Juncus arabicus*, the perennial rush noted earlier as being important in the Shatt el Arab, and whose other known Arabian congeners are found in damp, montaine areas (Collenette 1985). The water is low in salinity, and roots of *Tamarix* and *Juncus* are limited by temperatures of 40–45 °C. There is another well known spring further north (just north of Ras Sudr) named Hammam Musa, "Moses Spring".

(2) Coastal fissures

Throughout the length of the Red Sea, raised Pleistocene limestone is fissured, and where this occurs near enough to the coast, large cracks may be filled with sea water and colonized by an unusual sub-set of marine fauna. In two general areas this has been examined. Best known are the cracks of Ras Mohammed in the Sinai where some open to the surface. A large crack has been briefly described by Por and Tsurnamal (1973), and examined further by one of us (C.S.). According to Holthuis (1973), who describes new species of crustaceans from these cracks, their formation occurred as recently as 1968 as the result of an earthquake.

The surface of the rock in which these flooded cracks are set is about 2–3 m above sea level. The largest is about 40 m long and up to 1.75 m wide, and is located 150 m inland, and at least one other smaller one is known. The water surface is 2–3 m below the entrance lip, and is tidal, with little lag behind the open sea, and has a salinity of 22.6–23.6 g l^{-1} (Por and Tsurnamal 1973). Sedimentation is high and the water easily disturbed and made cloudy. The deepest part of the water so far discovered is 14 m. Walls to at least 5 m deep are dominated with the algae *Codium*, *Valona*, *Botryocladia* and by several red algal species. Attached fauna include bivalves, keyhole limpets, a chiton and a few *Spondylus*. The motile fauna is dominated by two shrimps, a *Periclimenes* and a new genus of blind Hippolytidae called *Calliasmata* (Holthuis 1973), and by a goby.

The existence of these specialized cave crustaceans in these cracks is made more interesting by the fact that some are found also in similar habitats in other distant parts of the Indo-Pacific, suggesting the existence of a limited, specialized, pan-tropical cave fauna. Since these appeared almost immediately following the formation of the cracks, it seems clear that there is an extensive subterranean network of inhabitable cracks and fissures, and that only when these are exposed as after an earthquake can even a small part of them be examined. The tidal movement and salinity shows that they open to the sea, and apart from the specialized fauna, most inhabitants probably colonized via the subterranean access, but the presence of wind-blown seagrasses and other dead drift material raises some possibility of colonization by wind over land as well.

Elsewhere in the Red Sea, the Dahlak Archipelago also contains flooded fissures (Oren 1962, 1964, Por 1968) in areas where uplift is still continuing. One of these fissures is called Devil's Crack, and it also contains a blind atyiid shrimp, known otherwise from Madagascar and the Philippines. This also suggests the presence of an extensive system of subterranean habitat which is under-explored.

Summary

Because much of the Arabian shoreline is low lying, its intertidal zone is commonly hundreds of metres wide. Intense insolation, desiccation, heat, cold, alkalinity and ionic imbalance provide severe natural stresses.

Coastal sand and stony desert supports a wide variety of low shrubs and succulents, with woody trees and palms where ground water rises. Many plants are not water stressed even though transpiration is 30 times rainfall because of interstitial water movement and dew, but mainly because plant roots extend through the dry surface into a layer 10–75 cm deep. Here moisture is greater and only 1–10 ppt saline.

Sabkha is a widespread inter- and supratidal habitat, measuring many kilometres across in places. These are flat plains, with crusts of sodium chloride and gypsum, with important "algal mats" a few centimetres thick, beneath which is a black reducing layer. Mats are complex associations of cyanophytes, bacteria and diatoms, whose composition is determined by small differences in elevation and frequency of immersion. They remain moist from capillary action in the sediments, from the strongly hygroscopic nature of sabkha salts, as well as from physiological adaptations. Pools are a special feature of sabkha. Those with subterranean connection with the sea have a relatively high benthic diversity. In their damp littoral, Cyanophyta are present, but they do not yet develop mats because of effective grazing. With increasing isolation from the sea, diversity falls and the persistent microbial biota then forms typical mat. These are highly productive and fix nitrogen. When desiccated in summer, mats become dry and crisp, breaking into characteristic polygons.

All salt marshes are greatly reduced or eliminated throughout Arabia, though the Shatt al Arab still covers 18,500 km^2. Eleven halophytic community types have been described, each determined by elevation and periodicity of immersion. Inundation reduces soil salinity, though in lower marsh zones surface salinity may still exceed 100–200 ppt, but at slightly deeper depths this falls to 4–15 ppt. In upper marsh communities, soil salinity is 1–20 ppt. Their reed vegetation is strongly enhanced by sewage enrichment, when it may reach up to a height of 5 m. In the Gulf and Red Sea, many new marsh communities are appearing as a result of sewage outfalls whose fresh water derives from desalination plants. Enrichment not only stimulates marsh development but in the case of the Red Sea, also adds significant nutrient. These areas also act as a focus for numerous species of birds, especially migrants.

Most sandy shores in the northern Red Sea are narrow beaches adjoining coral reef flats, which themselves are usually narrow. Broader beaches occur where the reef edge swings further out from shore, leaving a lagoon. Sand beaches are most important and extensive in the Arabian Sea. In the Gulf, each square metre could contain 400,000 animals, each over 0.5 mm in size, and faunal abundance there significantly correlates with slope. Rocky shores typically support a greater diversity of biota, but a much smaller abundance.

Numerous unusual habitats also exist. In the central Red Sea where there is a seasonal but almost no daily tide, a barnacle population flourishes during winter only, and Cyanophyta are the only group which survive the prolonged exposure in summer. Hot, sulphur springs which support reeds may also be found, as may sea filled fissures colonized by an unusual subset of marine fauna, especially crustaceans, which are part of a pan-tropical cave fauna.

CHAPTER 10

The Pelagic System

Contents

A. **The planktonic system**   197
   (1) Pelagic bacteria   197
   (2) Phytoplankton and regional primary productivity   198
   (3) The Red Sea and Gulf phytoplankton   201
   (4) Zooplankton   204
   (5) Vertical migration patterns in the Red Sea   205
   (6) Reef associated and "demersal" plankton   206
B. **Relative importance of planktonic production**   208
C. **Reptiles and mammals**   211
   (1) Sea snakes   211
   (2) Turtles   212
   (3) Dugong   214
   (4) Whales and dolphins   216
Summary   217

The Indian Ocean was chosen for the largest international oceanographic investigation ever made, and from it derived much of the limited information on pelagic systems of the Arabian region. Coordinated by the Scientific Committee on Oceanic Research (SCOR), the International Indian Ocean Expedition (IIOE) lasted between approximately 1959 and 1965. Although it is principally famed for its numerous geological contributions and for studies in physical oceanography, it included some biological work. While studies connected with shallow benthic biota were mostly relatively brief or, by modern standards, scanty, the project became renowned for the wealth of data it obtained in biological oceanography. Included in the areas of study to a greater or lesser degree were parts of the Red Sea, Arabian Sea and the Gulf, and still much of the known information on phytoplankton and zooplankton derives from the work of the expedition, as shown by the papers contained in Zeitzschel (1973).

The Indian Ocean was chosen at that time because it was the least known ocean, and because it has an interesting, reversing monsoonal circulation of its surface waters (Dietrich 1973, Berman 1981). The first condition still applies because work in other tropical oceans has subsequently proceeded much more rapidly, and the second condition is particularly relevant to the Arabian region, especially the Arabian Sea. In retrospect, the latter areas were probably under-investigated, as it was later concluded that the Red Sea and Gulf have an important and unexpected influence on the Indian Ocean. The meridional circulation of the Indian Ocean is initiated by water masses of the Red Sea and

Gulf; density gradients created in these two seas mainly from evaporation, described in Chapter 2, both flow out into the Arabian Sea and help to create an effective convection pattern in the Indian Ocean, especially in winter (Dietrich 1973). High salinity surface water is formed by strong evaporation in the Gulf and Red Sea which results in the driving density gradients.

Most work on plankton and biological oceanography subsequent to the IIOE necessarily has been much less ambitious geographically, sometimes with less scope and attainment. Important results have come from the Gulf of Aqaba and from other small parts of the Red Sea, while work has also derived from the Kuwait and Saudi Arabian Gulf coasts, and some from Oman. However, the total amount of work remains small so that any synthesis must remain tentative. Perhaps the largest set of results derives from work connected with investigations of mining deep sea metalliferous mud in the Red Sea. Longhurst and Pauly (1987) review general tropical plankton studies in their recent review of tropical fisheries, but while their general remarks of trophic patterns and communities apply in large measure to the Arabian region, important differences occur, both in species composition as the latter authors remark, and diversity and abundance.

The sections which follow on plankton are not intended to repeat information found in the earlier works, but are limited to demonstrating the relative importance, or in some cases the lack of importance, of the pelagic ecosystem in the Arabian region, and note some particularly interesting phenomena which occur there. Details of dissolved nutrients and their regional patterns are given briefly in Chapter 2.

A. The planktonic system

(1) Pelagic bacteria

Pelagic bacteria are one of the least studied elements of the pelagic system. Their relative contribution to total pelagic biomass and productivity remains unknown, though their importance may be considerable at least in terms of detrital breakdown and in freeing nutrients and important chemicals from dead phyto- and zooplankton. Only recently has the enormous importance of benthic bacteria to recycling on coral reefs become evident, and in other benthic systems such as sand, densities of up to 200,000 organisms g^{-1} indicate their trophic importance (Basson et al. 1977). Indeed, in benthic systems on reefs and reef-associated sands, production by bacteria is only slightly less than that from some macroalgae, and bacterial populations have a very rapid turnover which ensures their importance to the trophic pattern (Lewis 1982). While pelagic concentrations are much lower than the latter, it is likely that they have a trophic importance in the offshore water column as well.

A very short but important contribution is that of Rheinheimer (1973), relating to the northern Gulf of Oman and Strait of Hormuz. Numbers of saprophytic bacteria ranged from 8 to $470 \, ml^{-1}$, with the highest counts being obtained along the thermocline at 20–30 m deep. Bacteria along the Arabian Sea coast exceeded those along the Iranian coast. Interestingly, bacterial concentration was positively

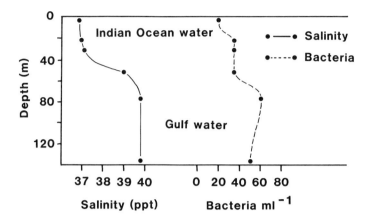

Figure 10.1 *Vertical distribution of saprophytic bacteria and salinity in the southern Gulf. Note increase of both parameters at depth of thermocline. From Rheinheimer (1973).*

correlated with salinity, which was attributed to the fact that Gulf water contained a higher amount of organic material (Figure 10.1). An interesting observation was that as much as 10% of the pelagic bacteria was luminous.

Little other quantitative work appears to have been done on pelagic bacteria of the region, although Karbe (1987) has reported sulphate reducing bacteria from deep Red Sea water, including that near the hot brine pools of the Red Sea. The concentrations ranged from $2.6–7.3 \times 10^4 \, ml^{-1}$, with higher concentrations locally near some of the hot deeps. While Karbe (1987) describes these concentrations as low, they are considerably higher than the values of saprophytic bacteria reported by Rheinheimer (1973) for the Arabian Sea.

(2) Phytoplankton and regional primary productivity

Krey (1973) divided the Indian Ocean into eight plankton-geographical regions. The Arabian Sea was grouped into zone 1 together with Western Australia and Indonesia. This apparently unusual geographical grouping is explained by the presence of upwelling in all three areas, and their phytoplankton is dominated by diatoms, followed in importance by dinoflagellates and some Cyanophyta (blue-greens). Interestingly, the southern part of the Arabian upwelling region off Somalia falls into a different zone, number 3 of Krey (1973). Although it shares the general taxonomic pattern of zone 1, its temporal or seasonal pattern is different. These two upwelling zones are separated by zone 2 which extends from the central Arabian Sea to the Bay of Bengal, whose dominant groups are dinoflagellates and blue-greens, followed by diatoms and coccolithophores. Krey (1973) notes that a dominant species is the Indian Ocean endemic *Oscillatoria erythraeum* (formerly *Trichodesmium*) which frequently appears in strong surface blooms. All are part of the "basic Indo-oceanic complex" in which several dinoflagellates and diatoms are an outstanding feature of the whole of the Indian Ocean (Kimor 1973).

The Arabian Sea is in general the most fertile part of the Indian Ocean, excluding some coastal lagoons, estuaries and shallow water embayments. Table 10.1 shows the general broadscale pattern of productivity. Difficulties arise in comparisons partly because of the different measures of productivity selected by different authors (measurements of mg or g, expressed as m^{-2} or m^{-3}, and production d^{-1} or y^{-1}) and partly also because of problems with experimental conversion factors between, for example, chlorophyll a and carbon. In the present account, measurements of weight and time are unified where sensible or necessary, but volume–area measurements are not; instead original measurements conforming to the same type are used for comparative purposes where possible.

A common measurement is that of chlorophyll a. Concentrations of $>0.5\,mg$ m^{-3} extend from the Arabian shores to India, with coastal regions having values higher than this. This compares with coastal concentrations in other areas of the Indian Ocean of less than half this, while offshore values further distant are one-tenth of Arabian Sea concentrations. Concentrations reach their peak in the summer upwelling period, but may still remain high at about $0.3\,mg\,m^{-3}$, and have the same general pattern in the period December–March as well. Krey (1973) points out that while potential primary production values may be a most useful measure, its derivation from carbon assimilation values involves numerous uncertainties, and direct *in situ* values are few, while simulated *in situ* results suffer from not being able to simulate temperature and insolation well. However, along the Arabian Sea coast, values of 1 to $>5\,mg\,C\,m^{-3}\,h^{-1}$ are obtained.

Table 10.1 Measurements of planktonic primary productivity in various Indian Ocean regions. Measurements were made by ^{14}C method, and with hauls from 200 m deep (from Cushing 1973). Values for Red Sea and its gulfs are from various sources (see text). All units are $g\,C\,m^{-2}\,d^{-1}$.

Region	Southwest Monsoon	Northeast Monsoon
Arabian upwelling zone	1.16	0.23
Arabian Sea	0.76	0.12
Javan upwelling	0.85	0.28
Eastern tropical ocean	0.70	0.26
Equatorial ocean	0.40	0.15
East Africa and Mozambique channel	0.83	0.42
Bay of Bengal		0.21

	Yearly average
Gulf of Aqaba	0.2–0.9
Gulf of Suez	0.22
North Red Sea	0.21–0.50
Central Red Sea	0.39
Southern Red Sea	1.60

(a)

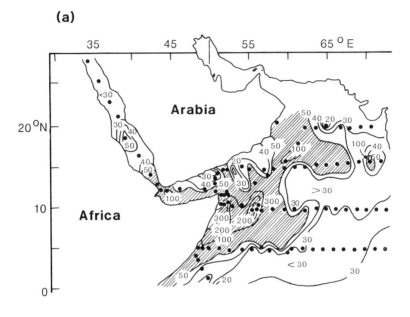

(b)

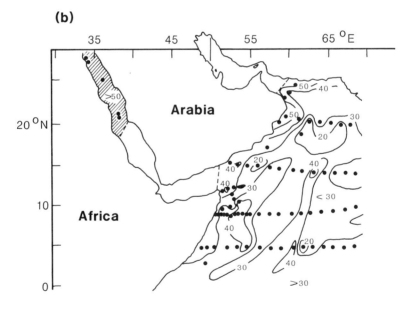

Figure 10.2 *Total pigment concentration (mg m⁻²) for 0–200 m depth in (a) SW Monsoon and (b) NE Monsoon. From Kimor (1973).*

Expressed in different units, El-Sayed and Jitts (1973) showed that the Arabian Sea shore had a primary productivity of $>1\,g\,C\,m^{-2}\,d^{-1}$ along the Oman coast and in a belt from the Horn of Africa to Socotra Island, and that values of 0.5–$1.0\,g\,C\,m^{-2}\,d^{-1}$ occurred elsewhere in the Gulf of Aden. Kimor (1973) reviews data which suggest values of about half of the above. He notes values of 88–$270\,mg\,C\,m^{-2}\,d^{-1}$ in the winter months, and 10–$20\,mg\,C\,m^{-2}\,d^{-1}$ in open oceanic water. Figure 10.2 maps the patterns of total pigments in the region at the peak periods of each monsoon. Unfortunately the collection of comparable data did not extend into the Gulf of Oman, Gulf or Red Sea. It has been shown (El-Sayed and Jitts 1973) that maximum carbon assimilation takes place at depths corresponding to 25% of surface light intensity and that assimilation remains important to about 5% of surface light intensity. In the clear waters of the Arabian Sea this may be 40 m deep, though the transparency declines closer to shore.

Density of phytoplankton cells is reviewed in Kimor (1973). Given that the high primary productivity of the Arabian Sea region is well established, these values are curiously at variance with the productivity pattern. In the Indian Ocean generally, there is a decline in phytoplankton biomass from east to west, and from north to south. Expressed as numbers of cells m^{-3} the Andaman Sea has 6100; Gulf of Aden: 3600; Arabian Sea: 1200. The equatorial Indian Ocean has 400 m^{-3}. The decreasing trend continued through into the Red Sea where concentrations of peridinians in the Gulf of Aden, southern Red Sea and northern Red Sea were 450–950, 200–350 and 130 cells m^{-3} respectively, and comparable figures were obtained for diatoms. At face value, the decrease of phytoplankton density towards the Arabian Sea, where primary productivity is greatest, is not reconcilable without speculation, and illustrates the paucity of data on the pelagic system in this region.

Many species do not survive the transition through the Bab el Mandeb into the Red Sea. Plankton from deeper strata are particularly reduced in the Red Sea because of the lack of a cool deeper layer of normal oceanic salinity. The same decline occurs in the Gulf. Of 452 species of dinoflagellates recorded from the Indian Ocean, only 130 are recorded from the Arabian Sea, 88 from the Red Sea and 57 from the Gulf of Aden. In the Gulf, much fewer still have been reported, though that result is partly due to a lack of research (Kimor 1973). While there is a clear decline westwards, in the Red Sea this is compensated to some degree by the presence of several endemic species, both of dinoflagellates and blue-greens. The former group dominate over diatoms and other phytoplankton in the Red Sea, and whereas blooms of the filamentous blue-green *Oscillatoria erythraeum* also are common, there is an almost complete absence of diatom blooms.

(3) The Red Sea and Gulf phytoplankton

Weikert (1987) notes that primary production in the Red Sea is low because the development of a thermocline and halocline tends to prevent recycling of nutrients from deeper water to the photic zone, and because there is almost no terrestrial run-off. Figure 10.3 shows autumn and spring values of pigments,

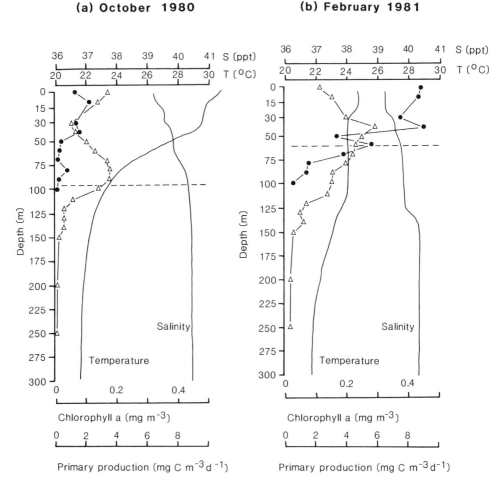

Figure 10.3 *Chlorophyll and pigment concentrations above Atlantis II deep, central Red Sea in October 1980 and February 1981. The broken horizontal line marks the 1% illumination depth. From Weikert (1987). Solid dots, primary production; triangles, chlorophyll a.*

chlorophyll and salinity and temperature in the central Red Sea. In a table showing numbers of cells (phytoplankton and zooplankton), Weikert (1987) also demonstrates the importance of the influence of the Gulf of Aden water on the southern Red Sea. Whereas in the north there were 180 cells m^{-3} and in the central Red Sea there were 300 cells m^{-3}, in the southern region there were 3000 cells m^{-3}. The latter were summer values while winter values were 14,000, 58,000 and 21,000 cells m^{-3} respectively. In the Gulf, densities recorded by Basson *et al.* (1977) were many orders of magnitude greater. Expressed in the same units, namely cells m^{-3}, the numbers were 100–150 × 10^6 in summer and about 150–250 × 10^6 in winter. The sampling location for these results was between the Saudi mainland and Karan Island, located well offshore, with the greater values found near shore and in bays.

Summer populations were dominated by diatoms which were least dense between 5 and 10 m deep, increasing in density below this depth. In winter, high concentrations of dinoflagellates provided the highest values, being greatest of all at about 10 m deep. The Gulf is clearly very productive compared with the Red Sea, though productivity data are difficult to find.

In the total euphotic zone of the Red Sea, productivity was tabulated at 0.21, 0.39 and $1.60 \, \text{g} \, \text{C} \, \text{m}^{-2} \text{d}^{-1}$ in the northern, central and southern regions. In the southern region, about half is contributed by symbiotic dinoflagellates living in the tissues of radiolarian protozoa. Interestingly, Shaikh *et al.* (1986) obtain a rather higher value of just over $1 \, \text{g} \, \text{C} \, \text{m}^{-2} \text{d}^{-1}$ for the central Red Sea, but although they remark on the lack of natural organic input and upwelling, their sample sites are near enough to Jeddah to have possibly been affected by the considerable sewage load discharged from that city. Values of dissolved organic carbon in the coastal waters of Jeddah may be as high as $>100 \, \text{mg} \, \text{C} \, \text{l}^{-1}$ (Behairy and el-Sayed 1984), indicating heavy contamination.

The figures from Weikert (1987) demonstrate the relatively enormous input of nutrients and of plankton too through the Bab el Mandeb, following the seasonal pattern of water exchange described in Chapter 2. Weikert (1987) calculates the contribution of medium sized, mesoplankton, as 96×10^{13} organisms, or 6×10^4 tonnes dry weight per month at peak times of influx. The increasingly harsh conditions of temperature and salinity mean that most die in the southern Red Sea and release nutrients in the process. The fishery of the southern section is the richest of the Red Sea and is supported by this.

In both the Red Sea and Gulf, blooms of *Oscillatoria erythraeum* are common and are of enormous quantitative importance to the pelagic productivity, in large part because of their ability to fix nitrogen in the nutrient-poor waters, especially those of the Red Sea. Shaikh *et al.* (1986) remarked that blooms witnessed by them occurred at the height of seasonal stratification and of lowest nutrient concentration. Within the blooms, concentrations of 9×10^5 cells l^{-1} may be reached, forming floating slicks as well as less dense concentrations deeper in the photic zone. Their effective conversion of dissolved nitrogen into ammonia and amino acids then leads to a succession of other phytoplankton groups, commonly dominated by dinoflagellates in summer and diatoms in winter. The total carbon production is also high: data reviewed in Weikert (1987) suggest that production in the photic zone is $250–500 \, \text{mg} \, \text{C} \, \text{m}^{-2} \text{d}^{-1}$ in the north, rising to over $1000 \, \text{mg} \, \text{C} \, \text{m}^{-2} \text{d}^{-1}$ in the south. In all areas there is also a trend of greater productivity near the shorelines than along the central axis.

In the northern Red Sea, the density-driven turnover of water allows for mixing and a slightly greater planktonic productivity than in the north–central part. In the Gulf of Aqaba, values of $200–900 \, \text{mg} \, \text{C} \, \text{m}^{-2} \text{d}^{-1}$ have been recorded (Weikert 1987), with the greatest values found on the Tiran banks and near coral reefs and seagrass beds. In this region, where water is very clear a significant amount of production takes place between 80 and 200 m deep, even though illumination is greatly reduced, the latter depth having illumination levels of 0.2% of the surface value. In this gulf, considerable detailed work has been carried out. The marked seasonal pattern is such that winter primary productivity is double that of summer when oligotrophic conditions exist; in winter, chlorophyll a standing stock increases to $70\,\mathrm{mg}\,\mathrm{m}^{-2}$, and the dominant organisms appear to be coccolithophores.

In the Gulf of Suez, similar data appear not to exist though a single production value near the entrance in autumn suggested a moderate productivity of $220\,\mathrm{mg}\,\mathrm{C}\,\mathrm{m}^{-2}\,\mathrm{d}^{-1}$.

(4) Zooplankton

Rao (1973) reviews zooplankton studies of the Indian Ocean prior to and including the IIOE, but very little is relevant to the Arabian region. The distribution pattern is similar to that of phytoplankton in that there is a westward decline in diversity, even though this pattern does not match that of total abundance. Peak numbers of all zooplankton both in the south and north of the Red Sea show a lag of a few weeks behind those of phytoplankton. Calanoid copepods, which are the most diverse and abundant group of zooplankton in the Red Sea have >300 species in the Arabian Sea, 60 in the southern Red Sea, 46 in the north and 35 in the two northern gulfs (Weikert 1987). Kimor (1973) presents lower numbers but with a similar pattern, and remarks that many are believed to derive from the inflow of water from the Gulf of Aden in the Northeast Monsoon. Euphausiids are another important group, especially to fisheries, and out of 22 species known for the Indian Ocean, 10 occur in the Red Sea. The latter are mostly surface forms which can pass through the Bab el Mandeb. A very few species increase in abundance towards the north, and because this is against the general pattern it may be assumed that these have developed adaptations to the greater salinity and greater temperature regime of the northern regions, and they experience less competition from other zooplankton also. Weikert (1987) suggests that these are also the older residents of the plankton and points out that one of them, *Creseis acicula* is one of the few known to have survived the late Wurm glaciations.

Total plankton densities, and the substantially higher densities in the southern Red Sea, are noted in the previous section. Higher numbers may be attributed to importation from the Gulf of Aden, an effect which disappears between 16 and $18°N$. In summer, zooplankton immigrants may attempt to descend into slightly cooler deeper water, but generally most die off. Survivors of *Eucalanus*, which is the commonest copepod, show pathological features, and most immigrants appear to be unable to reproduce.

In the Gulf, zooplankton shows marked temporal and geographic variation (Price 1979, 1982b). Diversity may be reduced from that of the Arabian Sea for several groups but appears to be similar to diversity in the Red Sea. Basson *et al.* (1977) reported 33–45 species m^{-3} near offshore islands. Densities may be high too, with up to 3000 individuals m^{-3}. The copepod *Paracalanus* and the tiny pelagic snail *Limacina* were the most abundant in summer. For pteropods and other mesopelagic and bathypelagic species, the shallow depth of the Gulf is a limiting factor, but other zoogeographical considerations also probably affect the zooplankton in a way similar to those which affect larger and benthic groups; these are discussed in a later chapter on biogeography.

(5) Vertical migration patterns in the Red Sea

Within the photic zone, which as noted above may extend to 200 m deep and 0.2% of surface illumination in the clear water of the Red Sea, there is considerable stratification of water bodies and of plankton. There are one and often two thermoclines which are readily noticeable to divers 15–50 m deep in the Red Sea and in such situations a temperature difference of 1 °C is clearly detectable. These may also mark a halocline as well. It is also common to observe a thin but extensive accumulation of seston at the interface, and in addition there may be a markedly different clarity in the water above and below such thermoclines. In winter especially, an upper, turbid layer may extend to about 20 m deep, below which much clearer, cooler water extends to at least 50 m deep. Against this background of water density strata there are diurnal changes in illumination, both of which lead to vertical migration patterns in plankton. In the following brief description, only the pattern in the shallow, most productive part of the photic zone is focused upon.

Because of strong light intensities experienced in waters of the region, photo-inhibition takes place near the surface, typically in the upper 20 m of water. High production values and maximum chlorophyll a concentrations generally occur below this to as deep as 200 m, depending on the degree of vertical mixing and stratification. The most productive part is 50–80 m deep in offshore water (Weikert 1987) though this band is rather shallower nearer land. One exception to this pattern occurs when red tides, blooms of *Oscillatoria*, occur; these floating rafts of filamentous algae exhibit a strong resistance to high illumination and create an additional productivity maximum at the surface. Generally, however, from about 100 m depth, primary productivity from plankton in the Red Sea and Arabian Sea declines strongly deeper than about 100 or 150 m (see Figure 10.3). In the Gulf, maximum depths are never more than about 60 m. While vertical productivity profiles are not available, greater overall turbidity in the Gulf suggests that the plankton productivity maximum will be considerably shallower than the 50–80 m band referred to above. Most water in the Gulf lies within the strongly photosynthesizing zone. The substrate of most of the Gulf also lies within the photic zone and this too supports benthic producers. Both these factors provide reasons why the Gulf is one of the most productive bodies of water anywhere.

As noted in Chapter 2, the Red Sea lacks the cold deep layers of many oceans. Vertical distributions of zooplankton show a greater gradient in the Red Sea than in most tropical oceans (Figure 10.4). Maximum abundance is seen above the shallow thermocline, and while the latter is commonly equated with being 100 m deep this only applies offshore; near shore it can be as shallow as 20 m deep. Below this depth, zooplankton abundance declines markedly. The depth zone between about 300 and 600 m deep in clear water of the Red Sea is the oxygen minimum layer where O_2 concentrations are about 1.3 ml $O_2 l^{-1}$, and in this there is a secondary peak in zooplankton abundance.

Diversity of zooplankton follows the same pattern, namely greatest above 100 m deep which, as explained in Weikert (1987), differs from the situation in other tropical oceans where peak diversity in offshore water is commonly between 500 and 1500 m deep. Hence both peak diversity and density of zooplankton in the Red Sea are within the photic zone, and mostly within the strongly producing zone. Thus it falls also within the region of greater salinity and temperature changes which affect benthic ecosystems too.

Below 100 m deep in offshore water, the system is mostly a consuming one only. In the mesopelagic zone (100–300 m deep), diversity falls rapidly initially. At 400 m lies the oxygen minimum. At these depths, surface environmental changes rarely penetrate, and seasonal stability is greater. There the secondary zooplankton maximum is found. Its zooplankton community is composed of a fairly large group of species, some individuals of which reach a large size (e.g. the relatively abundant calanoid copepods). These species exhibit marked vertical migration, some moving vertically by 800 m or even more each dusk and dawn. This group provides the deep scattering effect commonly seen on sonar, which rises from 600 or even 800 m deep at dusk to 100–200 m deep at night.

The lower edge of the mesopelagial zooplankton zone, or the lowest part of the deep scattering layer, generally marks the start of the bathypelagic zone. In the latter, zooplankton is extremely scarce in the Red Sea. This is explained by the greater temperatures found in the deep water, which accelerates breakdown or utilization of surface production. There is thus a severely curtailed downward passage of food, and Weikert (1987) estimates this to be only 1–3% of that which passes downwards in other warm oceanic systems.

In the Arabian Sea, zooplankton biomass has been recorded as being about 6–25 mg m^{-3} dry weight for the upper 200 m (Lenz 1973). These measurements were taken during the Northeast Monsoon, or during the period when upwelling does not take place, and the values are not very different to Red Sea values of 30–40 mg m^{-3} *wet* weight (Weikert 1987). In the Southwest Monsoon when upwelling is strongest, pelagic productivity values are considerably greater. While details of zooplankton distributions and increases are lacking, an earlier section, and Chapter 2, gave details of the increase in dissolved nutrients and primary productivity.

(6) Reef-associated and "demersal" plankton

A major component of zooplankton near reefs is made up of larvae of the reef dwelling invertebrates. The contribution of this group is probably too variable to provide a useful figure, and depends partly on the season. In winter, the proportion may be very small, and is almost certainly <5%. In spring, or during the reproductive seasons for invertebrates (Shlesinger and Loya 1985) substantial gamete and larval production may turn the usually clear water cloudy over extended areas of coral reef, and thus this component may locally contribute well over 90% of the total zooplankton. Functionally, whether the zooplankton derives from larvae or not may be of little importance and in both cases it is an important food source for the benthic fauna, most of which are suspension or particle feeders.

A group of demersal zooplankton is now believed to be trophically important. This is a group which is wide in its taxonomic composition, and which exhibits

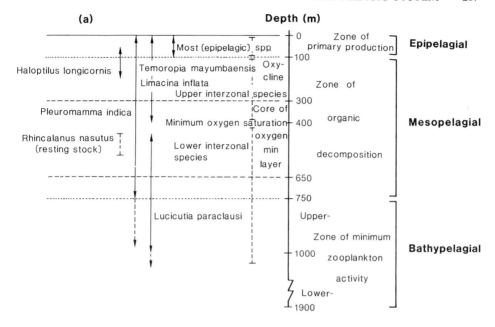

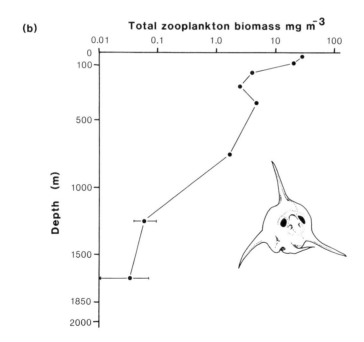

Figure 10.4 *The main vertical zones of zooplankton associations (a) and total zooplankton biomass with depth in the Red Sea (b). The data are from the same location as chlorophyll concentrations in the previous figure. Data from Weikert (1987).*

vertical, diurnal migration from shallow substrates. Different groups show a strong affinity for different kinds of substrate; some originate from interstices in coral reefs or from extensive rubble substrates associated with reefs, while others are associated with sand and seagrass beds (Alldredge and King 1977, Porter *et al.* 1977, Hamner and Carlton 1979). These are commonly very abundant although noticeable only at dusk and dawn. At those periods the shallow sea may become almost cloudy as they rise from the substrate and migrate to the surface layers. Swarms of copepods for example may occur in densities of 1.5 million m^{-3}. Because of their association with distinct substrate types, this group is distributed horizontally in shallow water in addition to their vertical distribution resulting from their diurnal migration. These horizontal distributions over short distances contrast with the oceanic forms which show very little horizontal distribution (other than smaller scale patchiness) over hundreds or even thousands of kilometres (Longhurst and Pauly 1987).

The importance of reef-dwelling zooplankton to the trophic structure of shallow water systems is not adequately described by its quantitative nature (Sheppard 1982a). Because much of the shallow benthos has symbiotic algae, large quantities of zooplankton are not needed for simple energy, but rather it is suggested that they are required more for the supply of minerals and essential nutrients (Johannes *et al.* 1970, Sheppard 1982a). For this, bacteria and smaller elements of the phytoplankton, possibly bound by mucus, are also utilized, especially as such particles may be considerably more abundant than zooplankton. Of this group, very little information specific to the Arabian region is available.

B. Relative importance of planktonic production

Because of the symbiotic association between dominant reef coelenterates and algae, the hard substrate benthos of the Arabian region is itself a primary producer in net terms. Even where reefs do not exist, macroalgae or expanses of blue-green (Cyanophyta) mats similarly produce carbon and fix nitrogen rapidly. In any measurement of relative production which takes into account productivity per unit area, pelagic production is relatively low.

Regarding inshore waters, Lewis (1977) stated: "The consensus of opinion regarding the importance of phytoplankton as primary producers in waters flowing over coral reefs is that it contributes very little to community production of the reef." In his review, Lewis (1977) cites several studies where phytoplankton production near reefs was of the order of $0.03\,g\,C\,m^{-2}\,d^{-1}$. In contrast, production from reefs or tropical benthic algae averages about $7\,g\,C\,m^{-2}\,d^{-1}$ with a range of about $1.5–15\,g\,C\,m^{-2}\,d^{-1}$ (Lewis 1977, 1982). It has been argued that coral reefs are "three-dimensional", a reference to the overtopping and multi-layers of corals which, by giving rise to coral cover values of over 100%, thereby raises the potential productivity of the reef. However, this does not minimize the difference between reef and pelagic system because even overtopping is not as "three-dimensional" as the truly three-dimensional structure of the pelagic water column with its plankton. The figures given above are for gross production, and net production is approximately half. Levels found for planktonic production amongst and over reefs were very similar to those for open water, and these figures suggested that gross planktonic production is indeed far less than that of the benthic system.

Longhurst and Pauly (1987) also use the figures from Lewis' reviews to note that planktonic production is 1–2 orders of magnitude lower than benthic production. However, these values of $0.03\,\mathrm{g\,C\,m^{-2}\,d^{-1}}$ for planktonic production differ substantially from those given in Weikert (1987). Examples from the latter source noted in earlier sections reported values of from 0.2 to over $1\,\mathrm{g\,C\,m^{-2}\,d^{-1}}$ from the pelagic system. These values are ten times greater than planktonic production values given earlier, but they are still over ten times smaller than those given for benthic productivity. Local variation may be partly responsible for the differences, but a broad and general difference of an order of magnitude remains clear.

Some other data from the Gulf (IUCN 1987) support the contention that pelagic production is lower than benthic. Calculations of gross and net production for benthic and pelagic biota in the Gulf concluded that although gross plankton production was high, net production was almost zero. Export of organic matter (excess production, or positive net production) was from benthic habitats.

The difference between the two sets of data on planktonic productivity discussed earlier is tenfold, which could be significant to any assessment of the relative importance of pelagic productivity. There have been several discussions of this relative importance, but it is suggested that the debate is misleading, or at best irrelevant. Certainly, the productivity in terms of fixation per surface area per day is lower in the pelagic, tropical milieu than it is in benthic, tropical systems, perhaps by a factor of 10 or 100. But this misses the main point that the total area of the pelagic system is greater than that of the benthic system by a factor much greater than 10 or even 100. Given the difference in areas, it could be equally argued that the pelagic system is much more important than the benthic one, though this would itself miss another point, namely that the two ecosystems are complementary and are not in some way in competition. The argument of relative importance, as with that of high benthic production per unit area on reefs, should not obscure the significance of high productivity of both. Ratios are not the only consideration either: absolute values of gross and net production are important to aspects such as fishery production as well. In the pelagic case, which is the focus of this chapter, productivity is sub-maximal. Fertilization greatly increases it, as is seen in the Arabian Sea upwelling, as well as in instances of sewage pollution. In contrast, productivity of most benthic systems in the region is maximal, and fertilization can have a markedly deleterious effect on the integrity of the system as well as on its productivity.

Planktonic production is clearly much greater, and much more important in the coastal parts of the Arabian Sea. Here assimilation of carbon was estimated as $1.0\text{–}2.5\,\mathrm{g\,m^{-2}\,d^{-1}}$ (Ryther and Menzel 1965). Important results from their survey include the estimate that total particulate carbon in the water column ranged from about 3 to $8\,\mathrm{g\,C\,m^{-2}}$, and that of this, about 10–20% is living, with the remainder being dead plankton or other detrital material. Thus the living component amounted to $1\text{–}2\,\mathrm{g\,C\,m^{-2}}$. This means that the standing crop of living organisms in the Arabian Sea water column near the coast is equivalent to about one day's primary productivity. (This figure can be contrasted with Red Sea benthic values where standing crop is the result of 3–14 days' production; see Chapter 7.) Both living and dead components increased and decreased in phase, and the turnover rate of total carbon was estimated to be about 7–10 days. This work illustrated the rapid turnover of carbon in the water column of the Arabian Sea as well as its substantial contribution.

The question of transfer coefficients is examined for the Arabian Sea and Indian Ocean by Cushing (1973), who calculated the ratio of yield in one trophic level to that in the one below. The spread of results ranged from about 2% to 34% (Figure 10.5). The interesting results are, firstly, that efficiency in the Southwest Monsoon when upwelling is strong is lower than that in the Northeast Monsoon, and that for both seasons, transfer efficiency becomes greater as primary productivity becomes lower. Cushing (1973) suggests that in areas or times of greater primary production, there is an excess of phytoplankton over the needs of the herbivorous zooplankton. This also suggests that there is a greater inefficiency in more productive areas and seasons, for reasons not yet understood.

The region is important to fisheries as a consequence of this high pelagic primary productivity, and fisheries are covered in detail in Chapter 13. The reef-associated fish are covered in Chapter 4.

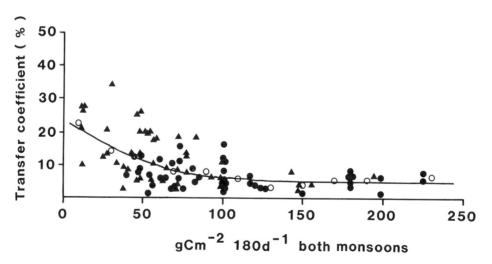

Figure 10.5 *Transfer coefficient from primary to secondary production as a function of primary production. Closed circles SW Monsoon, triangles NE Monsoon, line and open circles are average for both. From Cushing (1973).*

C. Reptiles and mammals

(1) Sea snakes

A recent, major review of snakes of Arabia (Gasperetti 1988) includes the sea snakes. In the Indo-Pacific (there are none in the Atlantic) there are about 55 species in two families. One family, the Hydrophiidae, contains 50 of the species, nine or ten of which occur in some coastal waters of Arabia. These fall into two unequally sized ecological groups. All but one species of sea snake live in shallow coastal water, in conditions which may be turbid and where there is organically rich substrate.

This group are bottom feeders, and are abundant in the Gulf and Gulf of Oman. In the Gulf, most of the substrate is less than 30 m deep. In the Gulf of Oman, deep water occurs, but the shelf is extensive. Known records of all species are documented (Gasperetti 1988) and while these follow clear patterns of human habitation to some degree, records suggest that on the southern side of the Gulf of Oman, sea snakes occur at least as far as Muscat and the Capital Area, but no further. The substrate and oceanographic regime is almost identical southward to the eastern point of Arabia at Ras al Hadd and it is likely that sea snakes would occur to that point at least.

The salinities inhabited by sea snakes may range from brackish and estuarine, through normal oceanic levels, to raised salinities found in many tropical embayments. Records of snakes in water of over 40 ppt are common, and in the Gulf of Salwah some have been observed in water as high as 58 ppt (Gasperetti 1988). Their prey is mostly fish, which explains the need for a remarkably powerful and rapidly acting venom. Like many reptiles, they have glands which excrete excess salt, in this case a posterior sublingual salt gland. In addition they have nasal and cloacal flaps to exclude salt water, and they may also extrude their tongue without opening their mouth. These adaptations permit them to live in a wide range of salinities. Muddy sublittoral substrates, and seagrass beds provide habitat where sea snakes are found from the Gulf to the Far Eastern archipelagoes, and especially along turbid shores of India and Bay of Bengal. Generally this group of sea snakes is found only in water less than 30 m deep, since they are air breathing but feed on or near the bottom.

(Between Arabia and India) a great many water serpents, from 12 to 13 inches in length, are to be seen rising above the surface of the water. When these serpents are seen, they are an indication that the coast is exactly two degrees distant.

Niebuhr 1792

Outside this group is the single species *Pelamis platurus*, of the same family Hydrophiidae. This is a surface feeder which is therefore not restricted to shallow water and ranges across deep ocean. It has a greater tolerance of cold water and as a result it is the widest ranging sea snake, found as far afield as the west coast of America, Japan and New Zealand. In the western Indian Ocean it is the sole sea snake found on the coast of Africa, as far south as the Cape. This species has also been recorded twice in the Gulf of Aden, but it too has never been found in the Red Sea.

Sea snake ecology is little known. They are top level carnivores and their numbers in parts of the Gulf suggest an underestimated importance. However, they lack the appeal of other vertebrates and despite their excellent taxonomic treatment in Gasperetti (1988) there are almost no ecological data available for them in this region. Their distribution is interesting; there are very few groups indeed which have a similar pattern. Reasons for their exclusion from the Red Sea have been discussed fairly extensively in the past, but so far not convincingly, and this possibly biogeographical question is discussed in Chapter 11.

(2) Turtles

Turtle distribution and abundance is summarized very briefly here. For more detailed accounts, see Frazier *et al.* (1987) for the Red Sea, Basson *et al.* (1977) for the Gulf, Miller (1989) for Saudi Arabian shores of both the Gulf and Red Sea, and Clarke *et al.* (1986) for the long Oman coastline of the Arabian Sea. Important details of turtles in each country separately are included in the IUCN Coral Reef Directory for the region (Sheppard and Wells 1988), while the IUCN Red Data Book is also a valuable source (Groombridge 1982). Miller (1989) also provided recent data for parts of Saudi Arabia, where new work is also being conducted by the Wildlife Commission on mainland nesting areas in the Red Sea (Oakley, pers. comm.)

Table 10.2 summarizes distribution details of turtles. The region is now very important for several species of this group. In common with several other areas, there probably once was a much greater abundance in the Gulf and Red Sea, and a significant reduction in numbers from human over-exploitation has clearly taken place, perhaps from as early as Roman times (Frazier *et al.* 1987). Vast piles of turtle bones occur on several islands in the northern and southern Red Sea which suggests both that there were previously large breeding populations there and a correspondingly large exploitation of them. However, their numbers were

Table 10.2 Details of breeding turtle records and counts around Arabia. Sources are numerous. A + is used where an area is known to be important but for which reliable counts are unavailable.

Location	Green	Hawksbill	Loggerhead	Olive Ridley
Gulf islands	+	+		
Daymaniyat Islands	+			
Capital Area, Oman	+	+		
Rass al Hadd	6000			
Masirah Island	+	80	30,000	230
Gulf of Kutch				+
Gulf of Aden	no data			
Red Sea	+	+		

clearly greatly reduced even by the time of the Forskal expedition of the late 18th century when turtles appeared to be very scarce.

In the Arabian seas, we never met with the sea-tortoise; the land-tortoise is more common; the peasants bring the latter, by cart-loads, to the markets of several towns in the east. The eastern Christians eat these animals in Lent, and drink their blood with great relish.   Niebuhr 1792

Today, exploitation in the region appears to be much lower than it is in many other regions where turtles breed, so that remaining Arabian breeding areas assume a correspondingly greater global importance. According to Miller (1989) there is still an artisanal turtle harvest along both Gulf and Red Sea shores of Saudi Arabia, using nets, "rodeo style" chasing, shooting with guns or spearing. There is additional mortality from accidental trawling: in the Farasan Archipelago, about 2000–3000 are caught each year, of which about 25% die, while the remainder are released. When fishermen catch turtles for food, the males are generally selected; the meat is consumed locally, though the penis is considered an aphrodisiac and is readily sold (Miller 1989). In some areas, notably islands in the Gulf, eggs are also collected for consumption.

The most important part of the region for turtles is the Arabian Sea, both in terms of numbers of breeding species and abundance of individuals seen or known to nest there. Whereas most of the pan-tropical species are recorded from the Red Sea and Gulf, some occur very rarely in these seas; generally only the Green, *Chelonia mydas*, and the Hawksbill, *Eretmochelys imbricata*, are at all common, and only these two breed in important numbers.

In the Arabian Sea, Oman contains the most significant populations. The Loggerhead, *Caretta caretta*, is the most numerous turtle, with about 30,000 breeding females nesting on northeastern Masirah Island each year (Ross 1979). This makes Masirah the largest nesting area in the world for this species. Other islands such as the more southerly Kuria Muria group also contain Loggerhead nesting beaches. Oman also supports the largest nesting populations of Green turtles, with 6000 females annually at Ras al Hadd. Numerous other beaches in Oman are also used by Greens, especially on Masirah Island and in the Capital Area where they also feed. Breeding there occurs in the winter. Masirah Island is also used by the Hawksbill, where about 80 females nest annually, though more nest on the Daymaniyat Islands and along other mainland beaches of northern Oman. Also in Masirah Island, the Olive Ridley, *Lepidochelys olivacea*, has a breeding population of about 230 nesting females (Clarke *et al.* 1986), adding to an extraordinary concentration of turtles in and around this island. Elsewhere in the Arabian Sea, the Olive Ridley nests in several places including the Gulf of Kutch and adjacent coasts of India, and on the coast of Pakistan (Groombridge 1984). Greens and Hawksbills are known from similar localities, and from the Gulf of Aden, where the former species at least may move between the Red Sea and Gulf and Aden.

In the Gulf, the Green and Hawksbill are the commonest, and are widely distributed (Basson *et al.* 1977, Miller 1989). Several hundred of both species congregate in the vicinity of the offshore Saudi Arabian islands through the summer, and lay eggs in nests on the islands. Sightings are common of both species along most of the mainland from Kuwait, through Saudi Arabia, to Bahrain and Qatar (Miller 1989) and along the UAE coastline. The extensive seagrass beds in shallow water of this region provide adequate habitat, and turtles have also been seen by the present authors in the Gulf of Salwah and Hawar Archipelago near Qatar in water with a surface salinity of over 47 ppt. In addition to these, the Loggerhead has been recorded in the Gulf, as has the Leatherback *Dermochelys coriacea*, which here, as everywhere throughout the Arabian region, is only an occasional visitor.

In the Red Sea, only the Green and Hawksbill are common species, and both nest throughout its length. Miller's (1989) survey showed that by far the majority of nesting sites are on islands. In a list of the 37 "most critical" sites of Saudi Arabia (Gulf and Red Sea shores), only two are located on mainland shores. Nesting islands include the Tiran Islands, Wedj Bank, the volcanic and reef fringed islands of the south-central Red Sea, and the Farasan and Dahlak Archipelagoes. Of two mainland sites noted in Miller (1989) the promontory at Ras Baridi is the most noteworthy, with the other south of Al Lith.

Sightings of other Red Sea species consist of adults seen offshore with no breeding records. Species recorded are the Olive Ridley, Loggerhead (sighted only off Sinai) and the Leatherback.

In the Arabian region, turtle populations and exploitation of the species have been used to some extent as a barometer of marine conservation measures. As pointed out by Frazier *et al.* (1987), there would seem to be no reason why the Hawksbill is not more common given the extent of coral reefs in the Red Sea and even the Gulf, or why the Leatherback is rare, given their correct observation that the jellyfish food of this turtle is often very abundant. Present levels of turtle populations are clearly reduced from the maximum sustainable, and their value is increasing, not only intrinsically but in terms of acting as a focus for conservation efforts in the region (e.g. Clarke *et al.* 1986).

(3) Dugong

This marine mammal of the order Sirenia occurs in both the Gulf and Red Sea. Along the Arabian shores of the Arabian Sea it is not known to exist, and indeed there would be very few locations where it could find the sheltered expanses of seagrass which it requires. In the eastern Arabian Sea the situation is different, with large herds living between Sri Lanka and India, and with a smaller population probably surviving in the Gulf of Kutch in northern India. In Oman, there are only reports of a carcass washed up on the Batinah coast of the Gulf of Oman (Clarke *et al.* 1986). The only suitable location in Oman with the required habitat would be the Bar el Hickmann coast which is sheltered by a large peninsula and by Masirah Island, but all fishermen questioned in a survey by one of us (C.S., see also Clarke *et al.* 1986) maintained that they had never seen dugong.

The sirens are cetaceous animals of moderate size, whose teeth are prized as ivory, and their thick hide as leather for sandals . . . the ark of the covenant is said to have been made of the hide of this animal.   Klunzinger 1878

A valuable study recently carried out in parts of the Gulf and Red Sea (Preen 1989) revealed considerable numbers of dugong in both areas. Indeed, the largest ever recorded herd of over 600 individuals was observed in the shallow seagrass and *Sargassum*-rich beds between Bahrain and the Qatar peninsula in the southern Gulf. Until the latter survey, estimates of the Gulf dugong population were of only about 50 individuals, and observations of about 37 dugong carcasses following the Nowruz oil spill led to fears that the mortality suffered by these vulnerable mammals could have completely destroyed the total population. However, Preen's (1989) survey revealed a much greater population than had hitherto been suspected. The estimated Gulf population is 7310 ($\pm$ 1300) individuals, making the Gulf the most important area for the species in the western part of its range, and second in global importance only to Australia. They are at their most abundant in the Gulf of Salwah between Bahrain and Qatar, and occur also in significant numbers amongst the shoals and islands west of Abu Dhabi.

In the Red Sea, rich seagrass beds are found around the Tiran Islands and Wedj Bank in the north, and dugong exist there in moderate numbers. They are recorded from the Gulf of Suez also. Similarly they exist around the southern archipelagoes and mainland south of Al Lith (Preen 1989), and in extensive shallows of the Port Sudan and Suakin Archipelago, as well as in the Dahlak Archipelago (Frazier *et al.* 1987). The latter believed that there could not be more than a few hundred individuals in the entire Red Sea due to the shortage of suitable habitat. However, available habitat in the southern and relatively underpopulated Red Sea was severely underestimated, as was the estimated number of dugong, leading Preen (1989) to estimate that there were probably 1820 ($\pm$ 380) for the Saudi Arabian side as far south as Yemen alone, making a total Red Sea population of about 4000.

Dugong have been taken for food in the Gulf, at least until recently. They are both actively hunted and used when caught accidentally in nets (Preen 1989). In most cases, the animals are consumed locally rather than sold in markets, partly because of local regulations. Anecdotal reports in Preen (1989) suggest that in the centre of their range, dugong may be becoming more abundant now because of a decline in their capture and consumption following legal bans on capture. In the case of one report from Bahrain, as recently as 10 years ago several dugong would be sold each day in the main markets, though this number declined to 20 per year in the mid 1980s, and has now ceased.

The flesh of the sea cow, being that of a mammal, rather resembles beef than fish and is readily eaten by the Moslimin. For (according to their law) everything that comes out of the sea is fish and may lawfully be eaten without being slaughtered in the proper way. The flesh of dolphins and turtles is similarly treated.   Klunzinger 1878

In the Red Sea, dugong appear not to have been utilized in an intensive way by man, and possibly their relative, natural scarcity except for rare, large herds has meant that large scale culling has rarely taken place. Fishermen will eat those animals which are accidentally caught, but the demand for dugong meat along the Saudi Arabian coast is apparently declining (Preen 1989). This is in part due to a latter day increase in wealth which allows fishermen to purchase terrestrial animals instead.

(4) Whales and dolphins

There have been no systematic surveys of whales or dolphins in any of the coastal waters of the Arabian region in the manner of those now completed for the turtles and dugong. Sight records abound, however, though inaccurate identification appears to be a particular problem.

The greatest number of records come from the Arabian Sea, where both dolphins (and toothed whales) and several baleen whales have been reported. At least a dozen species of dolphin, and the finless porpoise, have been recorded for the Arabian Sea and its coastal waters. Of the baleen whales, Bryde's, Humpback, Minke, Fin and Blue whales have been recorded there, as has the toothed Sperm whale, Killer whale and False Killer whale. Of these, Frazier et al. (1987) record and tabulate seven species of dolphin, the Killer whale and False Killer whale for the Red Sea, and in some cases these are very abundant, many extending well into the Gulf of Aqaba and Suez.

In the Gulf, Basson et al. (1977) report that "several" species of dolphin occur, some in schools of hundreds of individuals. Also reported are occasional stranded or trapped baleens. Of the latter, few have been positively identified. Positive records include Bryde's whale and the Humpback whale (IUCN/UNEP 1985a). Of the baleens, Bryde's whale is the species which includes fish in its diet to the greatest degree, and perhaps is the most likely to be found in the Red Sea and Gulf. In his survey for dugong, Preen (1989) reported that during a six-week period in September and October 1986, over 500 dead dolphins and a dead whale were observed, killed by unknown causes but possibly because of a red tide.

In the Red Sea, the finless porpoise is apparently absent, and baleens were also not reported by Frazier et al. (1987), though they note that it is probable that they may enter and strand on the coast. Indeed, one of us (C.S.) was invited to observe two of a group of three to four baleen whales which had stranded on the Farasan Archipelago and adjacent Saudi mainland shore in winter of 1989/90. These were unfortunately long dead and in an advanced state of decay, and not readily identified. However, their length (8–12 m) and visible characteristics and remaining patterning suggest they were Bryde's whales. It is likely that the relatively high water temperatures of the southern Red Sea are highly stressful to these large mammals, and that chances of survival by any of them will be slight in the Red Sea.

Summary

Knowledge of the Arabian pelagic system is very patchy, with much conflicting and incomplete data. The contribution from bacteria is probably important. In the Gulf, bacterial productivity is nearly as great as that of algae, and pelagic saprophytic bacteria number from 8–$470\,ml^{-1}$, with greatest accumulations found along thermoclines where organic matter is suspended. Of possibly greater importance may be sulphate reducing bacteria, whose concentrations in deep Red Sea water range from 2.6–$7.3 \times 10^4\,ml^{-1}$.

In the case of phytoplankton, the Arabian Sea is dominated by diatoms as is typical of tropical upwelling areas. It is the most fertile part of the Arabian region, and along the Oman coast primary production is 0.1–$1\,g\,C\,m^{-2}\,d^{-1}$.

In the Arabian region, many species of plankton do not survive. Of 452 known Indian Ocean dinoflagellates, 130 are recorded from the Arabian Sea, 88 from the Red Sea and fewer still exist in the Gulf. The decline westwards in the Red Sea is partly compensated by the presence of several endemics, and by blooms of *Oscillatoria erythraeum*. Cell densities similarly decline westwards. While the trend is similar in winter and summer, winter cell counts are 1–2 orders of magnitude greater. There is a large input of mesoplankton from the Gulf of Aden estimated at 96×10^{13} organisms, or 6×10^4 tonnes dry wt per month at peak times of influx, though most do not survive beyond the central Red Sea. In the Gulf, densities are much greater than in the Red Sea, with cell counts of 100–$150 \times 10^6\,m^{-3}$ in summer and about 150–$250 \times 10^6\,m^{-3}$ in winter, most being diatoms.

Total pelagic primary productivity of the Red Sea is about 0.21–$1.60\,g\,C\,m^{-2}\,d^{-1}$ with the greater values in the south where symbiotic dinoflagellates in Radiolaria are important. In the very clear Gulf of Aqaba, significant production continues to about 200 m deep (compared to 40 m in the Arabian Sea). The chlorophyll a standing stock in winter is up to $70\,mg\,m^{-2}$, contributed largely by coccolitho-phores. Summer values are approximately half of the latter.

Planktonic primary production appears to be less than a tenth of benthic productivity. However, the total area of the pelagic zone is much greater, and previous debates on relative value have obscured the fact that both are important. More important is the fact that pelagic production increases near shores, reaching 1.0–$2.5\,g\,m^{-2}\,d^{-1}$ of carbon assimilation in parts of the Arabian Sea. Pelagic standing crop near the coast is equivalent to about one day's productivity, which is a much faster turnover than occurs in the benthic system.

Zooplankton distribution similarly declines westwards. Peak numbers through-out the Red Sea lag a few weeks behind those of phytoplankton. Calanoid copepods are the most important group with >300 species in the Arabian Sea, 60 in the southern Red Sea and 46 in the north. Euphausiids are important in terms of biomass, though of 22 Indian Ocean species, only 10 occur in the Red Sea. In the Gulf, zooplankton shows marked temporal and geographic variation; diversity is less than in the Arabian Sea but is similar to that in the Red Sea with 33–45 species m^{-3} near offshore islands. Densities are high and up to 3000 individuals m^{-3} have been recorded.

Vertical stratification of zooplankton in the Red Sea is marked, with some species showing vertical, diurnal migrations of 0.5 km. Peak diversity and abundance remains within the photic zone, but a secondary maximum is found near the oxygen minimum layer at 400 m depth. Close to shore, demersal zooplankton and larvae of reproducing invertebrates tend to dominate measurements of numbers and productivity.

Sea snakes are abundant in the Gulf and Arabian Sea, but do not enter the Red Sea. Turtles are found throughout the region which contains some globally important nesting beaches. Of the marine mammals, dugong are relatively numerous in the Gulf, and small groups occur in the Red Sea also. Cetaceans are widespread, though not abundant except occasionally in the Arabian Sea.

SECTION III

Synthesis

Musandam's limestone cliffs.

CHAPTER 11

Marine Biogeography of the Arabian Region

Contents

A. **Coral zoogeography and diversity**   222
   (1) Patterns in the northwest Indian Ocean   222
   (2) Endemism and speciation   226
   (3) Patterns within Arabia   229
B. **Shallow water fish patterns**   230
   (1) Endemism in the Arabian region   230
   (2) Regional clines in species richness   233
C. **Other groups**   234
   (1) Echinoderms   234
   (2) Molluscs   234
   (3) Sea snakes   235
   (4) Marine plants   236
D. **Sub-regions and barriers**   237
   (1) Barriers   238
   (2) Factors of incomplete dispersal   239
Summary   240

The main purposes of this chapter are twofold. Firstly it looks in more detail at geographical distributions of species within the region and between the Arabian region and Indian Ocean. With a knowledge of physical and habitat patterns, this improves the understanding of constraints which act on species and communities. Secondly, marine biogeography is an increasingly active area of research, and the Arabian region previously has been somewhat neglected in this respect, though it offers much of interest. The main requirement for these analyses, namely to have well known biota from several locations, is more difficult than usual to achieve here because the region has only recently been subjected to much study. Several aspects are still sketchy, but this is partly offset by intensive study of particular groups.

   At least three components or attributes make up a biogeographical pattern. All contribute, more or less equally, to creating one, and they have been commonly confused. In order of the most obvious these are: (a) identity or diversity of the species in comparison with other areas, (b) degree of endemism, which has often been equated with local speciation, and (c) patterns of relative abundance. The last is usually ignored though its importance is clear (Rosen 1988). For example, comparing a group with a strong habitat-related distribution from areas with substantially different habitats could lead to conclusions which overlook ecologically based causes of the distribution. A relevant example of this problem is the comparison of fishes from the Red Sea (reef dominated) with those in the Gulf (soft substrate dominated).

A biogeographical pattern must not be confused with a theory. Humphries and Parenti (1986) indicate in *Cladistic Biogeography*: "For biogeographical theories to be meaningful, falsifiable statements must be derivable from them at two different levels. First, What constitutes an area? Second, What are the interrelationships of those areas?" Briggs (1974) had a simple answer to the first part, namely: ". . . if there is evidence that 10 percent or more of the species are endemic to a given area, it is designated as a separate province". His criterion became a widespread assumption embedded in numerous subsequent studies on biogeography. However, it runs into serious problems.

The following addresses questions of whether or not the Arabian region is a different sub-province or realm in the meaning of Briggs (1974), and whether in fact the term has much meaning; in other words does the designation of somewhere as a sub-region aid an understanding of important biological questions? It also discusses the apparently high endemism and its possible causes, and the section on corals is used as a vehicle for discussion of some broader aspects. For general reviews of tropical marine biogeography see Briggs (1974,), Pielou (1979), Rosen (1981, 1988), and Humphries and Parenti (1986).

A. Coral zoogeography and diversity

(1) Patterns in the northwest Indian Ocean

Being a sessile group which is well preserved in the fossil record, corals are well suited to zoogeographical analysis. On the Indo-Pacific scale, Wells (1954) was possibly first to illustrate that coral diversity followed terrestrial patterns; the Indo-Australian region supports the highest numbers of genera and species, and these fall both eastward and westward of it. He worked on genera since species were taxonomically unstable at that time. Stehli and Wells (1971) later produced real and smoothed coral contour maps which showed smoothly descending contours to both east and west. These contours (Figure 11.1 top) crossed vast expanses of open ocean where no reefs existed. It had to be assumed that the higher diversity contours nested simply inside the lower contours. Given the low level of sampling up to 1970 it was not easy to detect, for example, pockets of high diversity in the Indian Ocean at high latitudes, and nor were any particularly expected. Interestingly, their analysis revealed some separation from the main Indo-West Pacific group of sites by species-poor locations in the far south of the Indo-Pacific and in the Gulf.

It was acknowledged by Stehli and Wells (1971) that sampling in the Indian Ocean had a long way to go, which was implicit in their use of genera rather than species anyway. A plot with much improved data by Veron (1985) (Figure 11.1 centre) showed that there was not in fact a smooth decline in diversity across the Indian Ocean, unlike the situation in the Pacific. Later still, and because vast

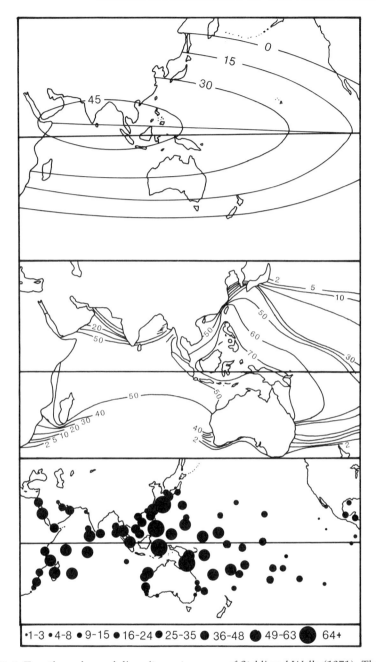

Figure 11.1 *Top: the early coral diversity contour map of Stehli and Wells (1971). The contours are "non-orthogonal polynomial surfaces". Centre: A later generic diversity plot by Veron (1985). Bottom: Spot values of generic diversity in a plot by Rosen (1988).*

areas of Indian Ocean contained no reefs, Rosen (1988) changed to using spot values instead of contours to avoid misleading conclusions (Figure 11.1 bottom). A second value of many of these studies was the graphing of relationships between generic diversity and sea temperature (Figure 11.2). This showed more clearly the decline in diversity with falling temperature, which had been postulated a century earlier.

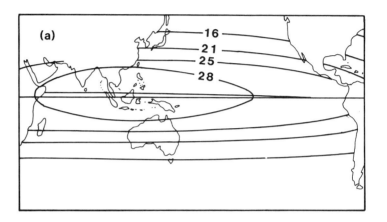

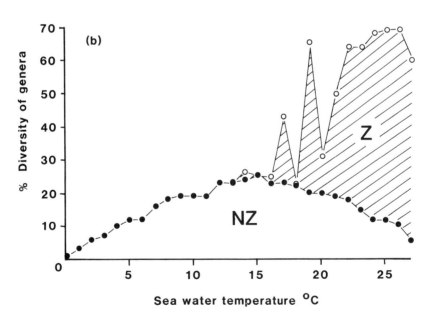

Figure 11.2 *Relationship of coral generic diversity with sea temperature. Material from Stehli and Wells (1971) and Rosen (1981). The contour map (a) uses non-orthogonal polynomial surfaces of sea water temperature values (°C). Graph (b) shows % of total diversity of zooxanthellate (Z) and non-zooxanthellate (NZ) genera against sea water temperature.*

Rosen's (1971) analysis had also suggested that there was a western Indian Ocean sub-province. Similar, updated information by Scheer (1984) supported this hypothesis, which was based strongly on the presence of six (now eight) genera known only from the western Indian Ocean. Generic maps showed a patch of raised diversity at the western end of the central high diversity belt, and showed that in the Indian Ocean, diversity fell more along a north–south axis rather than along a westerly gradient. Another point of interest was that there was an apparent partial separation of the Gulf, which was thought to be mainly a consequence of its much lower diversity; Rosen (1971) stated: "unusual conditions in the Persian Gulf and parts of the Red Sea raise the possibility of temperatures and salinities being too high for some genera . . ."

Most endemic genera of the western Indian Ocean high diversity locus, however, are monotypic: *Horastrea, Gyrosmilia, Astraeosmilia, Siderastrea, Anomastrea, Craterastrea, Erythrastrea* and *Ctenella*. Eight genera in a total of 78 known from the Indo-Pacific (10.2%), compensates numerically for a westwards loss of Far Eastern genera. At species level on the other hand, which is arguably the only biologically meaningful level, eight extra species in the 300–400 known for the Indian Ocean are almost negligible (2%). Endemic species in non-endemic genera occurred, but these were poorly known, and not enough was known about coral species anywhere to judge the importance of these because an earlier proliferation of synonyms had made species level fieldwork all but impossible.

When more taxonomic work on corals had provided a means by which reef ecologists could work at species level, Sheppard (1983b) and Veron (1985) could state that there was a high degree of homogeneity of corals at species level in the Indian Ocean. On a broad scale this is true in so far as there is a large sub-set of Indo-Pacific species spread through both island groups and mainland coasts of the Ocean, but a greater degree of patchiness than expected began to appear. Using all available data, one of us (C.S.) analysed species distributions in the Indian Ocean and adjacent seas (Figure 11.3). Coral species numbers in the Indian Ocean are lower than those for the Far East region, but rise abruptly again in the Red Sea, though not in the Arabian Sea or Gulf. Two or three fairly distinct groups of sites in the Indian Ocean region appeared, one of which comprised the seas around Arabia.

This made a case for a partly separated Arabian sub-province, which supported earlier views based on fish distributions (Briggs 1974). One important point, however, was that at species level there was *not* a clear decline in richness with increasing latitude. As amplified by Rosen (1988), highest species diversity appeared in the Red Sea and southwestern Australia, i.e. two areas with greatest latitude. Species pattern thus does not correlate very well with generic pattern. Generic pattern continues to follow "intuitive" lines of a decline away from the Indonesian region, which may be due to the older age of a genus than of a species (Stehli and Wells 1971).

For corals at least, therefore, a case can be made for a partial Arabian separation, and it is based partly on endemism, and partly on the fact that the Arabian region contains a far greater range of habitat and environment types than do many other coastal and island sites of the Indian Ocean (Sheppard and Sheppard 1991).

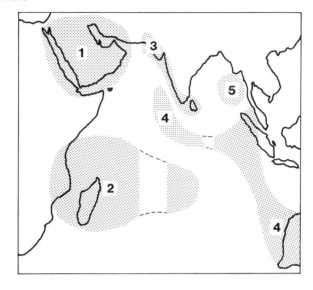

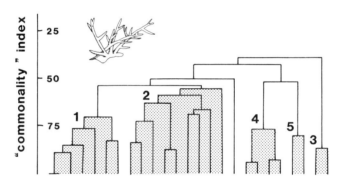

Figure 11.3 *Cluster analysis of coral species showing partial separation of Indian Ocean localities into "sub-regions". Taken from Sheppard and Sheppard (1991).*

(2) Endemism and speciation

While climatic changes in the Pleistocene almost certainly had catastrophic effects on the survival, growth and locations of coral reefs, they did not have a major effect on reef corals generally (Veron and Kelley 1988). In the Indo-Pacific as a whole, there is no major change in identity or diversity of corals over the past 2.3 million years.

The assumption has been made that because glaciations did not significantly increase rates of speciation or extinction, they must have exerted little evolutionary pressure. But this does not necessarily follow. Potts (1983) argues that corals were subjected to strong evolutionary disturbances during the large sea level fluctuations during glaciations, but that even if disturbances did not increase species evolution or extinction they may have caused an increase in intraspecific variation. He points out that during the time of major stress, the average *continuous* time that any given vertical location on a reef remained in the zone of vigorous coral growth was only 3200 years, which would be too short for coral populations to approach evolutionary equilibrium. Large sea level changes would lead to local extinction of populations before any intraspecific variations within them could become expressed as speciation since, Potts argued, over 100,000 years of stable sea level is needed for speciation. One argument he put forward in support of this was the view that there is a general absence of endemic species in isolated peripheral regions of the Indo-Pacific. However, this assertion is not true for the Arabian region under question here.

It is almost as though equable and optimum environmental conditions induce evolutionary instability in reef corals, whereas stabilisation of species occurs during periods of stress.   Goreau 1969

There are 20–23 endemic species of zooxanthellate corals reported for the Arabian region (Sheppard and Sheppard 1991), which constitute 11.3% of the total (Table 11.1). These come from 60 reef building genera, of which only two, or 3%, are endemic to Arabia, (*Erythrastrea* in the Red Sea and Gulf of Aden, and *Parasimplastrea* in the Arabian Sea). Numbers of Arabian genera with restricted distribution increase to five if those extending into the western Indian Ocean are included. Coral species endemism in the Red Sea is about 6.3%; it is 11.3% for the Arabian region, and 16.9% of species are either endemic to Arabia or limited to Arabia and the western Indian Ocean (Table 11.1).

Table 11.1 Distribution of coral species endemic to Arabia or to Arabia and western Indian Ocean. N = north, C = central and S = south.

No. of species	Red Sea N	C	S	Gulf	Arabian Sea	Western Indian Ocean
6	—					
1		—				
1			—			
6	——	——	——			
1	—				—	
1	——	——	——		—	
2	——	——	——	——	——	——
2		——	——	——	——	——
1	—			——	——	
4					—	
1	—				—	
1	——	——	——			——
1	——	——	——		——	——
1	—					——
1				——	——	

If endemism resulted from speciation (a question discussed later), much of it may have arisen in about 15,000 years. Before the Holocene transgression, conditions in the Gulf and Red Sea were probably incompatible with coral growth (Chapter 2). While it is possible that before or during the glaciations the present endemic coral species evolved or continued to survive there, for all of them to have done so is unlikely.

An upper limit can be set, therefore, for a rate of coral speciation in the Arabian region, of about 1.3 species per thousand years. This seems to be unreasonably fast. Veron and Kelley (1988) calculate a rate of 22.4 species per million years from data obtained from the Era fossil reefs in Papua New Guinea and present Australian species. They estimated the average age of coral species to be 20 million years (80 million years for the average generic age). If this upper limit is the true value, species evolution in the Arabian region at 7.5% in 10,000 years is 50 times faster than the value of 4.4% of species per million years in the Far East. The upper theoretical limit therefore seems much too high.

The lower limit to the rate of speciation in the Arabian region remains unknown, and depends on how many species which currently are endemic to Arabia survived somewhere around the Arabian Sea, during Pleistocene low sea level stands. That some of them did is likely. A clear example is the endemic genus *Parasimplastrea simplicitexta* which is extant only in the Arabian Sea but which is known also from the Pliocene fossil beds of Papua New Guinea. Other candidates for a group of survivors must include species which are now "endemic" to both the Red Sea and Gulf but which do not occur in the Indian Ocean; in these cases the existence of the species before the Gulf and Red Sea became habitable again is certainly a possibility. This could be resolved with a discovery of any of the current endemics in raised fossil reefs in the northern Red Sea. Of this group, the shallow water *Stylophora wellsi* should be one of the more readily visible in such fossil reefs, but extensive searching by one of us (C.S.) has not revealed it.

If current Red Sea endemics (6.3% of the total) extend back longer than about 15,000 years, their refuges during the pre-Holocene period would have to have been located outside the Red Sea. If they are now truly Red Sea endemics then they subsequently must have become extinct from their refuges. This is at least possible: one explanation for the pattern shown in Table 11.1 would be to postulate such an external refuge. Recolonization from the refuge after the Holocene transgression by species with varying degrees of success could account for the different ranges. This explanation must still stand along with the other two possibilities that speciation did occur in the Red Sea after the start of the Holocene, or that the Red Sea was never very inhospitable.

A comparison of speciation rates between Arabia and Papua New Guinea/Australia, is not an equitable one as it uses dissimilar approaches. The Arabian rate is calculated from "endemics" while Veron and Kelley (1988) used extensive fossil collections. Neither approach would make much sense, or even be possible, in the other region. The Far East is generally held to be the centre of coral species evolution and so in a sense all corals in this centre of origin which do not occur beyond it are endemic. A corresponding calculation would show a greatly increased rate of perceived evolution in the Far East region, since about 150 species occur in Australia which are absent from Arabian seas. Conversely, there are at present insufficient data on fossil corals from Arabia to provide a comparison to Veron and Kelley's (1988) values.

There is also a possibility that some corals have speciated in the Arabian region and subsequently spread eastwards. These would increase the lower limit to the rate of speciation in the region but would not be recorded as doing so since the corals would no longer be endemic to it. This may have occurred with some corals which exist in the Arabian region and in the western Indian Ocean, but there is no way of knowing whether some radiated out from the Red Sea, or speciated in the Indian Ocean and moved into the Red Sea. As far as corals are concerned, a lower limit cannot accurately be set at present.

It is perhaps unwise to try and do so. Another point, not investigated yet except in a preliminary way by one of us (C.S.), is that given almost any table of sites drawn against species, most sites can be shown to have about 10% of its species unique to itself. This is a sampling or random effect, and while it does not exclude the existence of true endemics, numerical approaches, such as that of Briggs, may have very little value. As Wells (1969) remarked: "I have always thought it very dangerous to derive the phylogeny and paleo-biogeography of a group of organisms solely on the basis of their existing numbers and their distribution, without recourse to the fossil record or in its absence. Neontologists are prone to do this, but the paleontologist, perhaps lacking in imagination, is loath to do so."

(3) *Patterns within Arabia*

While Red Sea and Gulf faunas are fairly well known, those of the Arabian Sea are not, and more species may be expected there, either limited to the area or also occurring in the Gulf or Red Sea. As described in Chapter 3, the Red Sea was initially divided up into four biogeographic sub-zones by Ormond *et al.* (1984c), based not only on corals but on many faunal groups, as well as considerations of "reef quality". Chapter 3 also described more detailed analyses of coral diversity and communities, and determined that overall diversity patterns reflect both more favourable hydrographic conditions in the north and better reef development there.

Special attention was given to the southern mainland section, sub-zone 3. There, clear gradients of coral *community* type were seen to occur, in most cases with several species becoming "extinct" at various points in a southerly direction. In addition, five species stated initially (Sheppard 1985a) to occur in the far south and become extinct in a northerly direction too were shown later (Sheppard and Sheppard 1991) to occur infrequently in turbid parts of the northern or central Red Sea too. This distribution of communities, however, is an ecological distribution caused by factors such as turbidity, salinity and temperature extremes, and depends on relative proportions of the species present. It did not depend on the identity of all the species present, and is an ecological pattern rather than a directly zoogeographical one. The north, central and southern regions all contain most Red Sea species and can therefore not be differentiated in terms of coral zoogeography. It is the proportion in which they contain each species that does vary greatly according to ecological criteria.

Within the main body of the Red Sea, there is at present no reason to suppose that there is any zoogeographic differentiation of reef building corals. Species distribution is not homogeneous, but follows ecological controls. However, one possible exception to this is the position of the Gulf of Aqaba.

In the Gulf of Aqaba or near its entrance, there is a group of seven corals which does not appear to occur elsewhere in the Red Sea. These are *Alveopora viridis*, *Diaseris distorta*, *Cycloseris cyclolites*, *Fungia moluccensis*, *Cynarina lachrymalis*, *Caulastrea tumida*, and *Trachyphyllia geoffroyi*. All are fairly widespread Indo-Pacific corals. Further species of less certain determination have been claimed also, and as noted in a later section, there is some highly localized endemism in other groups too.

The presence of these otherwise widely spread species only at this remotest part of the Arabian region has few possible explanations. A possibility is that they have been misidentified, but several taxonomists have now confirmed most of them. Another is that a refuge did exist in the Gulf of Aqaba, but data from cores in Reiss and Hottinger (1984) indicate that salinity in this region reached about 70 ppt in the last glaciation and thus did not differ from the Red Sea proper. Another is that the species do in fact occur throughout the Red Sea but have not been observed. Many hundreds of hours of searching in suitable and likely habitats by the present authors and several others have failed to find them. Either they are very rare, or they have reached the Gulf of Aqaba and mostly died out in the main body of the Red Sea.

Notable similarities occur in both nature and identity of coral communities in the Gulf of Suez and Gulf. As discussed in Chapter 3, these relate most noticeably to the physiographic form of the corals, but also to the species identity of the common ones. Branching corals predominate in both areas. Important here is that both areas share similar hydrographic regimes: both experience chilling in winter, and both have reefs with high sedimentation on account of their extensive areas of soft substrate. To add weight to the ecological argument, it should be remembered (Chapter 3) that there are also areas of the southern Red Sea along the mainland which share the same coral communities at least in terms of physiography. Stagshorn *Acropora* corals predominate there also. Shared environmental conditions include sedimentation and shallow water, while latitude is clearly a dissimilar criterion.

B. Shallow water fish patterns

(1) Endemism in the Arabian region

Cohen (1973) remarked that for fish ". . . the Indian Ocean does not appear to be a natural zoogeographic unit . . . There are, however, some distinct areas of endemism, certainly the Red Sea and western Indian Ocean . . . It seems probable that the Arabian Sea and south western part of the Indian Ocean are local centres".

Ormond and Edwards (1986) analysed Red Sea endemism in reef fishes, and Smith *et al.* (1987) examined the phenomenon in the Gulf (Table 11.2). Endemism as a whole is about 17%, the same order of magnitude as many other groups. However, there are enormous differences between families. Pelagic, open water fishes which have long larval dispersal phases have very low or no endemism, while small benthic or territorial fishes such as dottybacks (Pseudochromidae) and triplefins (Trypterygiidae) have levels greater than 90%, with their remaining species confined to the western Indian Ocean. Several broadly pan-tropical groups have endemicities of 30–50%. These include butterflyfishes (Chaetodontidae), parrotfishes (Scaridae), blennies (Blenniidae) and pufferfishes (Tetradontidae). However, many families of small fishes have lower degrees of endemism. With this variation, the value of an average percentage is not great though the similarity of the average value with that from fish in the Gulf is interesting.

These figures were derived from analysis of 508 species in 18 families. In the Gulf, 71 species were analysed (Smith *et al.* 1987). Six (10%) are endemic to the Gulf, while an additional five occur in the Gulf of Oman too, with a further seven being found in the Red Sea as well. In addition, once the largest group of truly cosmopolitan species are removed, those Arabian species with a wider distribution divided into two groups of five or six species each, one extending south down the coast of Africa, and the other extending westwards.

Ormond and Edwards (1987) assume that endemism in the Arabian region comes from speciation in the Red Sea. They noted that the Gulf of Aden forms a barrier to easy migration, and that ". . . 27% of the western Indian Ocean species are confined to the extreme northwest of the area and are probably of Red Sea

Table 11.2 Distribution of fish species endemic to Arabia or to Arabia and western Indian Ocean. Includes only species from the families Acanthuridae, Balistidae, Chaetodontidae, Holocentridae, Pomacanthidae, Pomacentridae, Pseudochromidae, Scaridae, Serranidae and Siganidae. N = north, C = central and S = south.

No. of species	Red Sea N	C	S	Gulf of Aden	Gulf	Arabian Sea	Western Indian Ocean
3	—						
6	————						
5	————		—				
3		——	—				
12	————————		—	————			
4	————			————			
3	————			————		————	
1		——		————		————	
4	————			————————			
6	————			————			————
4	————			————————			
1				————	————		
2				————	————		
Total 54							

origin". While it is fair to assume that Red Sea endemics speciated there, it is only an assumption to regard the Red Sea as the source for regionally dispersed species too. If Red Sea fish populations were destroyed in the Pleistocene, then all present Red Sea fishes other than endemics traversed the Gulf of Aden from east to west in the Holocene. For a considerable time, therefore, the barrier in a westerly direction, if it exists, cannot have been insurmountable.

In the account of Gulf fishes by Smith *et al.* (1987) there is a similar tendency to assume that endemism is caused by speciation in the present peripheral sea rather than in the larger Arabian Sea. Again, although speciation in the peripheral location may be likely for species which remain endemic, this is not necessarily so; in the Gulf region more species are confined to the Arabian Sea and northwest Indian Ocean including the Gulf than to the Gulf alone. Speciation in the northwest Indian Ocean and subsequent migration of a proportion of them into the Gulf and/or into the adjacent northern or western Indian Ocean is an equally valid explanation of the pattern found by Smith *et al.* (1987).

The question of what a valid sub-region actually means is related to this. Use of the term has been discussed in relation to fishes as much as with corals. Briggs (1974) tended to separate sub-regions if their endemism exceeded 10% which resulted in his recognition of a major faunal break in the Gulf of Oman separating a coastal, western Indian Ocean region from the east. Briggs (1974) did not go further and separate the Gulf from the western Indian Ocean because of a lack of data, but Smith *et al.* (1987) did. They supported Briggs' contention that a major faunal discontinuity existed in the Gulf of Oman, but added that "the Arabian Gulf reef-fish cannot be strictly related to either the Western Indian Ocean Province or the Indo-Polynesian province." Their view was that the Gulf "may deserve special zoogeographic status". This is a strong conclusion, but is based on an endemism in the Gulf of about 17%. This is high but it also means that 83% of Gulf species overlap to varying extent with adjacent Indian Ocean populations.

Ormond and Edwards (1987) suggest why the Straits of Bab el Mandeb and Gulf of Aden form a barrier which cannot easily be crossed. Firstly species which have evolved to cope with the "peculiar environment" of the Red Sea may find conditions in the Indian Ocean difficult, and secondly species adapted to conditions prevalent in the Indian Ocean and Gulf of Aden may find the Red Sea conditions hostile. In fact, they could have gone further, because the Gulf of Aden's conditions with its cold water upwelling and lack of coral reefs is considerably more different from the Red Sea than is the Indian Ocean. Of particular interest, however, is their comparison of close relatives or pairs of fishes, which they term sibling species, which are split by the Bab el Mandeb, or which at least are found on either side of the presumed Gulf of Aden barrier. They consider 11 pairs from eight families, and a further six pairs of sub-species, which differ significantly from each other. In addition, they point out that there are numerous additional sibling pairs which differ to a lesser degree, having different fin ray counts, colours or body proportions, which may be non-genetic changes induced by different salinity and temperature. In these cases, Ormond and Edwards (1987) assumed that Red Sea species evolved from Indian Ocean forms rather than the other way around.

(2) Regional clines in species richness

Extensive surveys were undertaken throughout the 1980s on reef fishes in the Sinai peninsula and the Arabian side of the Gulf of Aqaba, and along the entire length of the Arabian side of the Red Sea (Chapter 4). These have shown that, as with corals, the Red Sea is not at all homogeneous. Many species are found only in the central and northern Red Sea and not in the south, including numerous endemics, particularly amongst dottybacks, wrasse and damselfishes, while a few others occur in the south but apparently not in the north. Many non-endemic fishes show similar, latitudinally related distributions. As with corals, such distributions have predominantly ecological determinants, either through habitat effects on post-settlement fishes, or water quality influences on survival of larvae. Habitats and food supply for nearshore fishes change markedly down the Red Sea, as does water quality. Numerous other species are common in either the central and northern or southern Red Sea but occur less frequently, or even only very rarely, at its other end.

Regional clines of fishes require a brief comment on the exchange of species with the Mediterranean following the opening of the Suez Canal in 1869. This "Lessepian migration" (Por 1978) has continued for less than 120 years, though two factors qualify this. First is the existence, from early Egyptian and Roman days, of sea level passages linking the Mediterranean and Red Sea via the River Nile, so that a passage of some euryhaline species could considerably predate the activities of M. de Lesseps. Ormond and Edwards (1987) suggest that the killy-fish *Aphanius dispar*, a sparid *Crenidens crenidens*, a herring *Etrumeus teres* and flyingfish *Parexocoetus mento* are likely candidates. Secondly, after the Suez Canal was completed, large salt deposits in the Bitter Lakes kept canal water at salinity levels of over 70 ppt until relatively recently, so that for most species a salt barrier existed until only two to three decades ago. Because mean sea level at the south of the canal is about 0.5 m higher than in the Mediterranean, prevailing water flow in the canal is northwards. An increasingly large selection of Red Sea fish have migrated through, and some are now well established in the waters of the eastern Mediterranean which have only one-third of the number of fish species that the Red Sea has. Several have now become important components of the fisheries there, extending west to at least the central Mediterranean. Currently about 12% of eastern Mediterranean coastal fishes are of Red Sea origin. They appear to be restricted to the upper 100 m of the water because of their intolerance to low water temperatures, but it appears that they have added to the depauperate Mediterranean fish fauna and have not yet displaced any native species.

Because of the prevailing water flow, far fewer species have migrated into the Red Sea, and most of those which have are restricted to northern parts. Only one species of sea-bass has migrated further than the Gulf of Suez (Ormond and Edwards 1987).

C. Other groups

(1) Echinoderms

Echinoderms of Arabian seas are comprehensively covered by Price (1982c, 1983). There are about 350 recorded species from the region. The Red Sea has about 170 species of which 5.3% are endemic, while in the Gulf the value is 12.1%. This group is unusual, however, in that highest endemism values (13.2%), as well as the greatest total numbers, are found in the Arabian Sea. The gulfs of Aqaba and Suez have endemicities of 4.4% and 7.3% respectively. For the former this is a pattern reminiscent of the corals, though the value of over 7% for the Gulf of Suez is certainly larger than is known for other groups. Notably, in the two small areas with high endemism, the Gulf and Gulf of Suez, total echinoderm diversity is lowest, which may be a contributing factor.

Price (1982c) shows that in general the northern Red Sea gulfs associate more closely with each other and with the main body of the Red Sea than they do with the Gulf and Arabian Sea. Within various Red Sea components, similarity based on species presence exceeds 55%, while between the Gulf and the Red Sea it is below 40%. An impression is given of a fairly scattered echinoderm population with, in many cases, no obvious reason why some species are found only where they are. There are, however, clear suggestions that conditions at the Bab el Mandeb act as a barrier preventing many Gulf of Aden species from entering the Red Sea, and that a barrier somewhere along the Arabian coast between the Gulf of Aden and Strait of Hormuz similarly prevents a greater proportion from entering the Gulf.

(2) Molluscs

Approximately 1000 species of molluscs are known for the Arabian region (Sharabati 1984, Mastaller 1987). Of these, only a few are endemic to the Red Sea, and a handful are known only from the Gulf or Arabian Sea (Bosch and Bosch 1982). Within the Red Sea, both Mastaller (1987) and R. McDowell (pers. comm.), who surveyed shallow molluscs along almost the entire length of the Red Sea, concluded that distributions within this body of water were entirely related to local habitat availability and abundance. In other words, no zoogeographical trend has been detected within the Red Sea for this very diverse group. Between the Red Sea and its neighbouring areas, little is known, with the exception that deep water species of the Red Sea are markedly different to deep water species from the Arabian Sea. Mastaller (1987) attributes this to the fact that deep Red Sea water has a temperature closer to that of its shallow water, while the deep Arabian Sea temperature is much colder.

(3) Sea snakes

The distribution of sea snakes (Chapter 10) has not been satisfactorily explained, and seems to add to difficulties of understanding distribution barriers in this region. But this is because in the past certain assumptions have been made without adequate knowledge of sea snake tolerances to environmental extremes. Habitats occupied by sea snakes are known, but there is still inadequate knowledge of their physiology.

Sea snakes are abundant in the Gulf, Gulf of Oman and Arabian Sea, but are completely absent from the Red Sea and, except for the single pelagic species, from the coast of East Africa too. There are no confirmed records at all from north of the Bab el Mandeb despite abundant, apparently suitable habitat. This strongly suggests the existence of a dispersal barrier for animals with their particular behaviour and feeding patterns. It is the nature of this barrier which is unexplained. Here an explanation is put forward, based on assumptions which can only be tested given a better knowledge of sea snake tolerances to environmental extremes.

Several proposed reasons for the absence of sea snakes from the Red Sea despite abundant, shallow and productive habitat are not convincing. It was assumed that warm, salty water excluded the animals. It was also remarked that they might dive to cooler waters to regulate their body temperature. However, the abundance of snakes in the warmest and most saline waters of the Gulf of Salwah suggest that these factors cannot be limiting. As pointed out by Gasperetti (1988), temperature and salinity extremes in the Red Sea are considerably less than in the Gulf. Instead of invoking heat extremes, Gasperetti (1988) postulated a lack of suitable substrate in the Arabian Sea and Gulf of Aden, and suggested that a lack of habitat or food was the barrier. In support he pointed out that deep water east of Java and Borneo separates the sea snakes of the latter from those of New Guinea and Australia. The continental shelf of southeast Arabia is indeed narrow, but only relatively so; there are extensive shallow plains south at least as far as Yemen, and indeed, vast shallow plains and seagrass beds around Masirah Island and the Bar Hickmann areas of Oman would seem to provide ideal habitats, but they are devoid of sea snakes. Also, a lack of shallow substrate cannot be the sole reason for excluding the pelagic *Pelamis platurus*.

What has not been postulated to date is the possible limiting effect of cold water. Recent work on the Arabian Sea upwelling shows that temperatures as low as 16 °C occur annually, while lower temperature limits to migrations of some sea snakes may be around 20 °C (Dunson 1975 in Gasperetti 1988). This is doubtless a barrier to some species at least, and it may be that while there is limited shallow substrate too, the important factor is not the absence of sufficient food on it but that the extent of the shallow substrate is too small to prevent annual pulses of cold water from flowing into and affecting all parts of it. The same control, after all, inhibits corals there.

There is no reason to believe that if sea snakes could penetrate the barrier, they would not flourish very well in shallow environments of the southern Red Sea and Gulf of Suez. Indeed, that other well known group of tropical marine reptiles, the turtles, have greater temperature tolerances and have penetrated into the Red Sea where they do well, at least in the absence of human exploitation.

The barrier to sea snakes is therefore located in the Gulf of Aden and there is no barrier at all into the Gulf. This might seem to contrast with the situation for coastal hard substrate species like corals and reef associated fish which are barred more from the Gulf than from the Red Sea. This is explained, however, by considering the dispersal route likely for sea snakes, which is north along the muddy shorelines from India and Pakistan where they thrive, and along the north coast of the Arabian Sea and into the Gulf. This is all suitable habitat for these reptiles, but is extremely unsuitable for corals and reef fishes which must colonize by crossing the open Arabian Sea from the southeast.

(4) Marine plants

Several species of algae are endemic to the region. In the Red Sea, Walker (1987) separates known species into four geographical blocks of the gulfs, northern, central and southern regions, and shows that between about 8 to 40% of species occur in only one block. In the case of algae, despite the compilation of Papenfuss (1968) and others, considerable undersampling must be regarded as an important confounding factor here. Nevertheless, Walker (1987) points out that many of the southern species are typical of warm, more saline waters of the tropics, while the northern species include members typical of slightly cooler areas, the boundary in this case being drawn approximately through the middle of the Red Sea between Jeddah and Suakin. This boundary complements that observed for fishes and corals.

With algae also, the proportion of species endemic to the Red Sea is about 9%, which is not much lower than that for faunal groups. The major difference, however, appears to be that 64% of species are pan-tropical, occurring over extensive parts of the Indo-Pacific, as well as in the Caribbean or Mediterranean too. No faunal group is known to have such a broad pattern.

There is limited biogeographic information to be obtained from seagrass distribution, since most species found on both island and mainland coasts of the Indian Ocean are found in the Red Sea. However, the occurrence of biogeographical sub-zones within the Red Sea is suggested in seagrass data of Price *et al.* (1988). Habitat availability and extremes of temperature and salinity probably control latitudinal distributions of the abundance of the species, as shown in Chapter 7. In the Gulf, three species are prevalent, *Halodule uninervis*, *Halophila stipulacea* and *H. ovalis*. They are also the three common species of the Gulf of Suez, extending into the Suez Canal, suggesting that they are most tolerant of temperature and salinity extremes. *H. stipulacea* has now successfully colonized the eastern Mediterranean as well. One other hardy species is *Thalassodendron ciliatum* which becomes encountered more frequently and in increasingly dense beds towards the north of the Red Sea, and forms extensive and dense beds around the Tiran Islands. Most other species of seagrass become less frequently encountered and become less abundant with increasing latitude.

Mangrove distribution likewise throws little light onto biogeographic problems. *Avicennia marina* is widespread, but disappears with increasing probability of freezing air temperatures in winter in the highest latitudes of the region. The distributions of *Rhizophora mucronata*, and two other poorly known mangroves reported from the Red Sea, *Bruguiera gymnorrhiza* and *Ceriops tagal*, appear to be related simply to mean salinity. Only *A. marina* exists in the Gulf and along the Oman shoreline today, although *R. mucronata* occurred in the latter area until recently but has become locally extinct due to over-exploitation.

D. Sub-regions and barriers

Within the Arabian region, three different views have been put forward concerning the existence of a biogeographic sub-province: (a) the Gulf may be one, (b) the Red Sea may be one, and (c) the Arabian region as a whole may be one. The foregoing has shown that many groups exhibit endemism of about 10% or more on average, that there appear to be good grounds for recognizing the Arabian region as being to some extent separate from the equatorial high diversity belt of the Indian Ocean, but that differences within the various components of the Arabian region are almost entirely ecologically determined.

Although we have to some degree separated ecological and biogeographical determinants of distribution patterns, any biogeographical barrier must be to a great extent an ecological one too, or from the species perspective, a physiological or behavioural one. An inability to cross a barrier occurs because conditions in the region of the barrier are insurmountable to the species. To some degree the separation is artificial. In this case, the distinction does seem to be reasonable.

When endemism is 90%, the separation is surely justified for that group, but the application has not been consistent. A case has been made on one hand that the species poverty of the Gulf separates it to some degree (Stehli and Wells 1971, Briggs 1974, Smith *et al.* 1987). On the other hand the Red Sea has been designated a sub-region partly because it is very diverse (Briggs 1974, and see Edwards and Head 1987). The Gulf of Suez resembles the Gulf environmentally but it has not been regarded as a sub-province, and nor has the central and southern Arabian Sea coast which differs more substantially still by having a brown algal-based ecosystem of a more temperate than a tropical nature. In the latter case, only echinoderms are somewhat different in this region, and that was because they were more diverse there (Price 1982c).

Also, species poverty should not be used as supporting evidence since it indicates no more than that marine environmental conditions are relatively hostile. If the same peculiar hydrological conditions of shallow water and temperature extremes were reproduced elsewhere in the Indian Ocean, as indeed they are in numerous small lagoons, there would be no reason to call them separate zoogeographical regions. It may be that the peripheral location of the Gulf, and its size, have influenced our view of what constitutes a biogeographical sub-region. This is strictly an ecological sub-region, not a biogeographical one.

For corals, there is a greater degree of endemism in the Red Sea than in any other comparably sized area of the Indian Ocean, but there is little endemism in the Arabian Sea or Gulf, with the important exception of *Parasimplastrea* which proves the point that endemics can be relict populations of once widespread Indo-Pacific corals.

For most of these groups, therefore, it is suggested that any designation of a biogeographical sub-region must lie at the level of the Arabian region and not at the level of its component gulfs and seas. Any designation at a scale smaller than the region cannot be reliable because great habitat and hydrographic differences make any determination impossible. It may be possible to justify a smaller scale designation for some groups of fishes, and for some groups, notably the plants, there is no justification for any sub-provincial designation. Given both differences in organism physiology and dispersal methods, and the different approaches used, it is not surprising that different views have emerged.

(1) Barriers

If the Arabian region is to be regarded as a biogeographical sub-region, it must have some barrier between it and the Indo-Pacific. Factors which could create partial barriers in many other parts of the Indo-Pacific have been extensively discussed in a series of papers by Rosen (1981, 1984, 1988) who considers no less than 13 published hypotheses, theories or mechanisms to account for the marked differences in diversity and species identity across the tropical Indo-Pacific. Barriers, speciation, vicariance and factors connected with larval dispersal have all been invoked and examined, for several marine groups.

In the Arabian region, the main candidate for a barrier, if not the only one, is the upwelling associated with the monsoon system. This has been discussed extensively in various parts of this book, and its effectiveness throughout the Holocene is discussed in Chapter 2. Briefly, Prell (1984) and Prell and Kutzbach (1987) demonstrated that the severity of both monsoon and Arabian upwelling has not been constant. Using planktonic composition of sediments, they showed that before the Holocene there was a stronger monsoon and upwelling than at present, that at the start of the Holocene both monsoon and upwelling dropped sharply and remained weak for about 2000–3000 years, after which both rose steadily to a maximum at 9000 BP. Finally, after the maxima, both fell to today's values (see Figure 2.6). This correlates well also with insolation values, which drive the monsoon. Thus passage of tropical larvae through the Arabian Sea would not have been restricted by upwelling in the early Holocene, and recruitment into the Red Sea might have been much easier.

Even though today the cold upwelling does act as an effective deterrent to coral *reef* growth in the Arabian Sea and Gulf of Aden, and probably acts as a barrier to a range of more mobile fauna such as sea snakes and fishes, this does not necessarily mean that it is a very effective barrier to the passage of larvae. Upwelling occurs only in the summer. In winter, sea temperatures in the Arabian Sea are warmer. Furthermore, during winter months the Northeast Monsoon piles up water in the Arabian Sea from the Indian Ocean, which then flows into the Red Sea at an average speed of 0.5 m per second, and sometimes 1.5 m per second (Edwards 1987). This surface flow is reinforced by southeasterly winds which blow at over 10 m per second for over 50% of the winter months. That these effects are substantial is shown by the fact that winter mean tide level in the Red Sea is 0.5–1 m higher than it is in summer. Thus there is a large transport of warm water in and out of the Red Sea from the Indian Ocean annually. The influx is outside the period of cold upwelling, and it is known to transport a high diversity and abundance of plankton (Weikert 1987).

Despite this indication to the contrary, the bulk of evidence is strong that real barriers are imposed to the passage of several groups. The Arabian Sea may not always have been even a partial barrier, but it seems likely that it has been for the latter part of the Holocene.

Other possible ecological barriers involve soft substrates of the Gulf of Oman and far southern Red Sea. As noted in *Section I*, even where cliffs descend into the sea along much of the Oman coast, the sublittoral is one of very exposed, soft substrates, and the long Battinah coast is likewise soft substrate. The existence of 2000 km of soft substrate with only few areas of hard substrate along it may itself be a deterrent to hard substrate species. Not enough is known of soft substrate species to make a comparison yet, though the pattern exhibited by sea snakes illustrates that these may easily colonize from the northern mainland route.

(2) Factors of incomplete dispersal

Since the major disruptions of the Pleistocene and the stabilization of sea level, there has been only about 15,000 years for marine communities to reach the present condition. For corals at least, this time is probably insufficient for equilibrium to have been attained. Many present day distributions may be no more than chance distributions, as noted by Goreau (1969). He referred to the equilibrium of coral colonies on reefs, but it is likely that his remarks apply equally to ocean wide dispersion. Again with regard to corals, it has been noted that the Indian Ocean shows few clear diversity gradients and that latitudinal gradients about the equator are a more conspicuous feature than any decline from the high diversity Far East region. It has been argued (Sheppard and Sheppard 1991) that the observed partial separation of corals of the Indian Ocean into groups such as the Arabian seas subset could also be partly a result of insufficient "mixing" of the Indian Ocean coral fauna. The cause may be time, but may also be partly a result of the situation whereby there is only widely scattered coral habitat over great distances. Thus chance may play a large part in determining the identity of the subset of the Arabian corals. The fraction of the total fauna found in each island group or stretch of mainland coast shows that there is far from being an even dispersal of corals in the Indian Ocean, and still less is there an equilibrium. Present day differences between coral island groups have no obvious ecological reason in many cases, and a "lottery of location" may well be important.

While the total number of corals and other groups found around Arabia is greater than any of the individual totals found for many other Indian Ocean localities, no location contains all Indian Ocean species, even excluding locally confined endemics. The Arabian region is not richer than the Indian Ocean, but it is much richer in several groups than many other individual sections of coast or island groups in it. It is larger than many, but this need not be important.

The reason for this probably lies in the nature of Indian Ocean sites. Dispersal distances between island groups are considerable, and islands are small. Shallow water in the Indian Ocean provides a small target for larvae. In contrast, once larvae enter the Arabian Sea they may be driven into the Red Sea by each winter's currents, after which the narrow and confined nature of the Red Sea makes it very likely that many of those that do survive the Arabian upwelling and enter alive will find suitable substrate. The distance across the Arabian Sea barrier into the Gulf is much longer, and coupled to that is the fact that survivors may not even find suitable habitat and tolerable environmental conditions when they do penetrate it.

Summary

Although contiguous with the Indian Ocean, the Arabian region contains a fairly discrete faunistic subset. This is based on relatively high degrees of endemism, and on the fact that site clustering tends to separate Arabian sites from others. In the case of corals, diversity across the Indian Ocean is now known not to fall away from the Far East in the manner seen across the Pacific. Rather, there is a high diversity, equatorial belt throughout, and the highest diversity sites within this are high latitude sites, one of which is the Red Sea. While the Red Sea is not richer than the Indian Ocean, it is the richest part of it (west of Australia). Whereas other Indian Ocean sites such as island groups present small or difficult targets for dispersing larvae, because of oceanographic patterns the Red Sea has probably tended to "trap" corals as they recolonized after the Holocene transgression.

There are 20–23 endemic zooxanthellate corals known in Arabia. Endemism for the Red Sea is 6.3%, for Arabia it is 11.3%, and for Arabia plus the western Indian Ocean it is 16.9%. Assuming Red Sea endemism arises from speciation which occurred since the Holocene transgression, an upper limit for the rate of coral speciation might be set at 1.3 species Ka^{-1}, which is 50 times faster than a rate set for the Far East based on quite different criteria. The lower rate cannot be set. Fish endemism in the Red Sea for several groups reaches 50–90%, with lower proportions for the Gulf. Like corals, apparent endemism of fish rises with increasing area, such that Gulf plus Gulf of Oman has nearly double the apparent endemism of the Gulf alone, and it is higher still if the northwest Indian Ocean is added to the area. Although such values, as well as those for other known groups such as echinoderms and molluscs, have supported the case for a discrete, or at least a special, zoogeographic sub-unit of the Indian Ocean, the same arguments can be applied to almost any section of the Indian Ocean. This casts doubt on the value of designating "subregions" or equivalent.

Within the Red Sea, there are many latitudinal gradients for fish, corals and several other known groups. These distributions have predominantly ecological controls by substratum type, salinity, turbidity and temperature, which together adequately explain many of the distributions in the Gulf and Red Sea as well as differences between the two. However, there is probably a dispersal barrier in place in the southern Arabian Sea, caused by the cold upwelling water. This has not been constant. It was strong before the Holocene, weak in the early

Holocene, then became stronger than it is today and peaked about 9000 BP, before falling to present levels. The cold water barrier certainly prohibits significant coral reef development, and might impede passage of many tropical, hard substrate groups. It also explains the absence of sea snakes in the Red Sea.

There has been confusion between ecological controls on species distribution and any underlying biogeographical trends. Examination shows that a designation of biogeographical sub-region must lie at the level of the Arabian region and not at the level of its component seas and gulfs, with a few exceptions where there is exceptionally high endemism in certain groups like butterflyfishes. In addition to the constraint imposed by the Arabian Sea barrier, biogeographical patterns have probably arisen because of insufficient time for complete dispersal after the start of the Holocene, and insufficient "mixing" of fauna throughout the various suitable habitats, or targets, in the Indian Ocean.

CHAPTER 12

Ecosystem Responses to Extreme Natural Stresses

Contents

A. Coral growth and reef growth   243
   (1) Coral growth in severe conditions   243
   (2) Reef growth: the simplest reefs   249
   (3) Reef disappearance   250
B. Community changes and environmental stresses   252
Summary   256

Numerous areas of the subtidal zone in Arabian seas have exposed, hard substrate which are not dominated, or even colonized, by corals. Where this is the case, soft corals or macroalgae generally colonize extensively. The algae which dominate may be greens, but are usually large browns. They may coexist with sparse coral assemblages, even forming canopies over them. Algae may exist in dense zones beside equally dense coral assemblages, or without any significant coral growth at all. The substrata on which this occurs include pre-Pleistocene limestones, limestone which appears to have a recent origin, or a variety of sedimentary or even basaltic rocks.

Generally, such areas are found where environmental conditions are near to, or beyond, the tolerances of most corals. Such areas experience severe stress from both high and low temperature extremes, raised salinity and raised sedimentation. These areas are also the least studied in the tropical marine region, not only around Arabia but in the Indo-Pacific generally. The kelp-dominated area in southern Oman is a notable exception. However, enough is now known of these areas in Arabian seas to identify environmental parameters which inhibit coral or reef growth. Several different parameters appear to be involved simultaneously in most areas, and where these are severe the ecosystem responses can be examined.

The point is first made that most of the following discussion refers to corals; for example, conditions of "increasing stress" refers to a trend towards which corals cannot live, and has usually been inferred in part. Trends of increasing temperature and salinity extremes and raised nutrient levels, for example, may be unimportant, or may even favour, other groups which replace corals under such conditions. However, environmental gradients continue, in several examples, to conditions too severe for the groups which replace corals on hard substrata.

A. Coral growth and reef growth

It is increasingly clear that as environmental conditions become more severe, the processes of coral growth and reef growth become uncoupled. It appears that the processes required for coral reef accretion fail well before corals cease to exist in environmental gradients which become more saline or which experience greater temperature extremes. As noted in Chapter 3, areas within the Arabian region which demonstrate this include: the southern Red Sea, whose reefs are constructed from red algae although they support scattered corals (Sheppard 1985a); several locations in the Gulf such as the Gulf of Salwah; numerous limestone domes off the Saudi Arabian coast where brown and green algae dominate the pre-Holocene limestone (McCain *et al.* 1984, Sheppard 1985c, Coles 1988); and in Oman (Sheppard and Salm 1988). Dominance by macroalgae over corals is also reported to occur at Abd el Kuri Island in the east of the Gulf of Aden (Scheer 1971). There are an increasing number of examples from locations further east, such as southwestern Australia and Japan (see Veron 1986 for illustration).

To date, there has been little interpretation of coral communities growing on non-reef substrata, or of why their substrates do not accrete, other than vague remarks about inhospitable conditions or water temperature being too low. The following addresses these two questions of coral and reef growth in tandem.

(1) Coral growth in severe conditions

Most work on coral growth in severe conditions considers salinity and temperature, and less commonly turbidity and sedimentation as well. In the United Arab Emirates, Kinsman (1964) recorded 11 species of corals on substrata which extended to 15 m depth. His survey was one of the earliest in the Gulf and has been much cited as a study which "extended" known salinity and temperature tolerances for hermatypic coral growth. The extension he made was from views of Dana (1843) and Vaughan (1916) who determined that the lowest temperature tolerance values of corals were about 18 °C. In Kinsman's Gulf locality, water temperature ranged from 20 to 30 °C below 4 m depth, while surface layers fluctuated more widely from 16 to over 36 °C. The salinities tolerated by corals which Kinsman found reached 45 ppt, with an unconfirmed *Porites* species tolerating as much as 48 ppt.

Work in Bahrain (Sheppard 1988) considerably extended the number of corals known to be able to withstand such salinities, and it also extended the known salinity range tolerated by some. These are shown in Table 12.1. In addition to the 10 Gulf coral species indicated, about 40 more thrive in salinities of up to 41 ppt (and even more species than this survive in such salinities in the northern Red Sea). Three species, *Siderastrea savignyana*, *Porites nodifera* and *Cyphastrea microphthalma*, tolerate salinities up to 50 ppt, and are found living under such conditions at least for many weeks during summer, if not continuously. There was a significant negative correlation of salinity with species number ($r = -0.512$, $p < 0.05$) within this study in which salinity ranged from 42 to 50 ppt and coral

Table 12.1 Coral species which survive high salinity (lasting 1–3 months at least) in the Gulf of Salwah and Bahrain region. Data from Sheppard (1988).

Salinity (parts per thousand)	46	48	50	>50
Cyphastrea serailia	⟶			
Porites compressa	⟶ ? ⟶			
Platygyra daedalea	⟶			
Favia pallida	⟶			
Favites chinensis	⟶			*Sargassum* only
Leptastrea purpurea	⟶			
Porites nodifera	⟶		⟶	
Cyphastrea microphthalma	⟶		⟶	
Siderastrea savignyana	⟶		⟶	

diversity from 3 to 20 species. Least squares regression showed a drop of 1.1 coral species for each 1 ppt increase in salinity. This represents the drop in already very saline and depauperate conditions; the drop in species diversity from "normal" salinities (about 34–36 ppt) to 42 ppt will be considerably greater than this.

The above study considered salinity only, but it recognized that a combination of salinity and temperature is probably more important than either factor acting alone. The study did not monitor water temperature, other than to note that temperatures of 30–34 °C were common in summer on some reefs off northern Bahrain both at the surface and depths to 8 m deep. In the latter case, warm, saline wedges of dense water appeared to flow off nearby shallow coastal shelves towards the reef slopes. Although there is no way at present of assessing the duration of such streams, the fact that they are warm yet still flowed beneath cooler water strongly suggests that their salinity was high.

Lowest winter temperatures for the Bahrain sites are not known. Coles and Fadlallah (1991) provide an assessment of effects of low water temperatures from areas close by in the Gulf, at Manifa at about 27° 40′ N and at Tarut Bay at 26° 30′ N. These authors considered temperature only, possibly because salinities were not excessively high at their study sites. For the first time they were also able to record not only extreme low temperature values but, very importantly, the durations of the lowered values. They point out that the duration for which a particular temperature prevails is as important as the temperature value itself (Figure 12.1). The more northerly site at Manifa was subjected to two months of continuous exposure to temperatures of 11.5–16 °C, and temperature remained below 13 °C for one entire month during early 1989. This is the lowest recorded water temperature for coral reefs, and as a result of it several species of corals were killed. Mortality was most pronounced near the surface and on the reef flat. Towards the reef base, at 7 m depth, most other species showed signs of sub-lethal stress including loss of colouration. Significantly, however, survivors included the important reef builder *Porites compressa* and several faviid coral species, which appeared to have recovered six months later.

At the more southerly site at Tarut Bay, temperatures fell less markedly and remained below 14 °C for only a day or so at a time. However, they remained

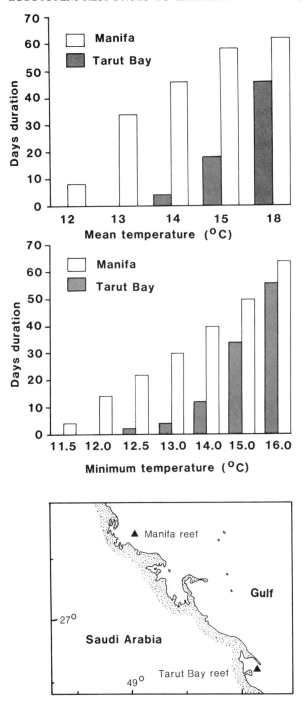

Figure 12.1 *Duration of mean and minimum temperatures on coral reefs in the Saudi Arabian Gulf from December 1988 to March 1989. These exposures to sustained low temperatures are the lowest recorded over coral reefs. Map shows localities. Data from Coles and Fadlallah (1991).*

between about 14 and 16 °C for a month. In the latter site, little evidence of coral stress was observed, and no mortality appeared to have occurred. Slight bleaching amongst several coral genera was noticed. It is noteworthy that the Tarut Bay temperatures recorded were below those traditionally regarded as lethal to corals; Coles and Fadlallah (1991) include a historical review of opinions regarding lower limiting temperatures for corals.

Table 12.2 shows recorded water temperature extremes for several Arabian sites. Very few world regions other than the Arabian region show annual ranges of as much as 11–16 °C, with a notable exception from Heron Island where a range of 19 °C has been recorded (see Coles and Fadlallah 1991). Although range data are not available, Macintyre and Pilkey (1969) observed corals in temperatures as low as 10.6 °C off North Carolina too. These existed as scattered colonies on non-reef substrate. Sites in the Gulf, however, clearly have greatest temperature ranges. Our own observations together with data in Edwards (1987) also suggest that the northernmost reefs of the Gulf of Suez experience cold winter lows and marked summer highs, providing a greater range of temperature than is seen over the same latitude in the Gulf of Aqaba where water is deep. However, the range is not nearly as great as in the Gulf.

Coles and Fadlallah (1991) name several species which are susceptible to temperature extremes. Important reef builders such as *Acropora* species and *Platygyra* were largely killed in Manifa, while *Porites lutea*, three *Favia* species, a *Favites, Cyphastrea, Coscinaraea, Leptastrea* and *Turbinaria crater* generally survived. However, they showed symptoms of sub-lethal stress. In addition, Sheppard and Sheppard (1991) name a few more species which are particularly tolerant. These are shown in Table 12.3. While this table gives a good indication of which species are most temperature tolerant, the bands are deliberately broad because insufficient data are available at present to give a more accurate picture.

None of the values tabulated for temperature ranges are from the Arabian Sea. Although the Arabian upwelling produces cold water, this rarely falls below about 17 °C. High summer temperatures of the low to mid 30s are not recorded from the Arabian Sea, and indeed the upwelling itself is a summer phenomenon. The low upwelling temperatures have often in the past been cited as lethal to corals, as indeed they are in many oceanic areas for example. It is now clear that many of the same species living in the Arabian region have adapted to them.

Table 12.2 Temperature extremes recorded from Arabian coral reefs or from limestone platforms supporting coral communities.

Location	Lat.	Min. (°C)	Max. (°C)	Range	Source
Saudi Arabia, Gulf	27° N	11.4	36.2	24.8	Coles and Fadlallah 1991
Qatar, Gulf	24° N	14.1	36.0	21.9	Shinn 1976
Abu Dhabi, Gulf	25° N	16.0	36.0	20.0	Kinsman 1964
Kuwait, Gulf	29° N	13.2	31.5	18.3	Downing 1985
Suez, Red Sea	29.5° N	17.5	30	12.5	Edwards 1987 and unpublished observations
Aqaba, Red Sea	29° N	20.0	28.0	8.0	Fishelson 1973b

Table 12.3 Coral species which survive temperature fluctuations of the range indicated. Data from Sheppard (1988), Sheppard and Sheppard (1991), Coles and Fadlallah (1991).

Temperature fluctuation (°C)	8–15	15–20	20–28	30
Acropora horrida	———→			
Stylophora pistillata	——————————→			
Porites nodifera	——————————→			*Sargassum* only
Cyphastrea microphthalma	- - - - - - - - - - - - - - - -→			
Siderastrea savignyana	————————————————→			
Porites compressa	————————————————→			
Platygyra daedalea	————————————————→			
Porites lutea	————————————————→			
Psammocora contigua	————————————————→			
Pavona varians	————————————————→			
Coscinaraea monile	————————————————→			
Leptastrea purpurea	————————————————→			
Favia pallida	————————————————→			
Favia speciosa	————————————————→			
Favia favus	————————————————→			
Favites pentagona	————————————————→			
Turbinaria crater	————————————————→			

Temperatures of this order may have an inhibitory effect on reef building itself, which is discussed later.

In Oman, greatest coral diversity was found in the north, where there are substantial reefs in Musandam and in the Capital Area. About half of the 91 species recorded from Oman show a north–south distribution split. Thirty-one species have been found only in the north and do not appear to penetrate the south. It was concluded that these were excluded from the south by the cold summer upwelling (Sheppard and Salm 1988). Fifteen species occur in the south of Oman only and do not appear to occur further north. These are typical Indian Ocean species and there is no obvious reason why their range does not extend further north. They are all still abundant at the Yemen border, and it is not known whether scattered coral assemblages of a lower diversity than this might be found further south. It may be that this distribution, which involves only 15% of Oman's corals, is a random distribution, or it may be attributable to a lack of time since the start of the Holocene for migration north (see Chapter 11). Closer to Aden, diversity begins to rise substantially again (Sheppard and Sheppard 1991).

The three species noted above from Bahrain as present in salinities of up to 50 ppt are common in all such areas. They are most conspicuous, not surprisingly, when all other species have disappeared. Of the three, *Siderastrea* appears most tolerant of all. In the southern Red Sea, *Siderastrea* was again one of only a few coral species found on algal reefs. Generally it is not a common component of reefs at all but is found more often in sandy chutes and sedimented areas associated with reefs. In the southern algal reefs it reaches its greatest known abundance in the region with colonies present every 5–10 m^2. Other coral species which were fairly common on algal reefs include several faviids, but there was an almost complete absence of *Acropora* and other species commonly associated with providing coral framework. In these reefs, calcareous red algae provided over 95% of the reef fabric (Sheppard 1985a).

Kinsman (1964) did not make distinctions on whether the 11 coral species he investigated in the central and eastern part of the Gulf were reef forming or not. However, his assumption appears to have been that the corals, being zooxanthellate or hermatypic, were reef building. In fact, almost all coral species considered by researchers in Arabian seas have been zooxanthellate forms. Exceptions to this have been lists of corals without algal symbionts by Scheer and Pillai (1983) and Sheppard and Sheppard (1991), and a comment on unusually high abundances of some of these in Musandam (Sheppard and Salm 1988). The equation of hermatypic with "reef building" has been modified recently (Schuhmacher and Zibrowius 1985) due mainly to the fact that many hermatypic, or symbiotic, forms contribute only marginally if at all to reef construction. Proposed alternative terms which overlap in meaning but which are not synonymous with each other are zooxanthellate, constructional and hermatypic. However, these definitions by Schuhmacher and Zibrowius (1985) are also not adequate to address the condition found in Arabian areas, and this aspect is discussed later.

Working in the Farasans in the southern Red Sea, Wainwright (1965) was probably first to notice the difference between corals on coral reefs and coral communities on non-reefal substrata. He observed the distinction in areas composed of raised reefs and limestone blocks which have been uplifted by rising underlying salt domes (see Chapters 1 and 3). They lie in a fairly sedimented area due to the fact that the southern Red Sea has a broad, sedimented shelf and has many sheltered sites. This is also demonstrated by the fact that the area is well populated by mangroves. Additionally, water of the southern Red Sea is very much richer in nutrients than the central and northern parts (see Chapter 2), and has summer water temperatures which rise close to the upper limits tolerated by many coral species. Sheppard (1985a) observed that fossil reefs fringing the mainland in this area contained a much higher generic diversity of corals than was seen on adjacent living reefs. However, without dates for the former no interpretation can be made of whether the present unfavourable conditions (for corals) have existed throughout or only in the latter part of the Holocene.

The distinction between reef and coral growth has also been used by Pichon (1971), Heydorn (1972) working in marginal areas of southern Africa, Hopley (1982) and by Sheppard and Salm (1988) for the Arabian Sea. In southern Africa, it was remarked that the coral communities were superficially indistinguishable from rich coral communities on true reefs and that this applied also to associated invertebrates such as echinoderms as well as to fish. Further south in Africa, brown algae dominate, and this condition therefore parallels that found off Oman. In the latter, many areas support coral communities on non-accreting substrata which are indistinguishable from those on growing reefs. In Oman the gradation continues further to a condition of scattered colonies coexisting amongst green algae or under a canopy of browns.

Acropora species provide framework on reefs which are apparently accreting normally as well as on non-limestone substrate, but framework species of this genus are not amongst the most tolerant group. As conditions increasingly favour large brown algae, framework corals become rare in southern Oman. The most tolerant corals include some faviids, together with *Siderastrea* and *Turbinaria*. These gradations are amplified later.

It is possible that similar coral communities grow on uplifted limestone domes in the Gulf of Suez. As noted in Chapter 3, these experience conditions similar to those of the Gulf and also have a low coral diversity. However, observations there have been few, and there is no reason yet to question the assumption that the northern part of the Gulf of Suez has true coral patch and fringing reefs.

(2) Reef growth: the simplest reefs

A comment is required first on the present view of what exactly causes reef growth. This subject is more complex than was first thought and still requires considerable further research. Whereas it was once universally thought that growth of reefs equated with growth of corals, it later became clear that this picture of "corals growing on corals" was far too simplistic. Most of the matrix or fabric of living and raised fossil reefs was seen, for example, not to consist of coral skeletons at all, even though numerous skeletons might be found embedded in it, even in their positions of growth. The concept of "framework" was thus introduced, together with an improved understanding of the role of compacting and solidifying sediments within the framework. "Framework" has had different definitions according to the viewpoint of different authors. However, whether it is used in reference to a particular group of organisms or to larger masses of material, it refers to matrix which traps sediments, allowing the latter to consolidate into the bulk of the reef fabric.

Earlier chapters covered corals, other groups of benthic invertebrates, and algae which bind sediments and rubble sufficiently to permit consolidation. Mechanisms by which they facilitate sediment consolidation appear to be chemical and biochemical, but are still largely unidentified. Obviously, limestone producing organisms are required for reef growth (corals, algae etc.), and equally clearly carbonate sediments accumulating within spaces are important (Hopley 1982). However, without the consolidation processes referred to, the result would remain merely coral skeletons with sediments; not a durable reef.

There is a special case where coral reefs are formed in a simpler way: in fact by the "corals-on-corals" scheme of old. This is seen in several parts around Arabia when small scale "reefs" of the order of metres or a few tens of metres develop, usually in areas where coral diversity is low. These reefs may result simply from aggregations of enormous coral colonies, mainly *Porites*, and such aggregations mark an important transition in the Arabian gradients discussed here. Such "corals-on-corals" aggregations are also found in many back reef areas amongst more developed reefs in the Arabian seas and elsewhere (which has possibly led to the durability of this idea of reef construction). Where this occurs, massive colonies develop on shallow substrata (such as in bays) where numerous adjacent colonies tend to fuse together forming a structure whose cross-sectional profile is identical to that of a small reef. These appear similar in nature to the "proto-reefs" of Reunion (Pichon 1971), and some in the Arabian Sea have been referred to as "incipient reefs" by Glynn (1983). No sediment binding processes appear to be required for their formation. In their simplest form they may be merely a large, even isolated, coral colony growing close to shore and limited in upward growth by the water surface. Such cases are common in the Arabian Sea and in the Gulf. In their more elaborate form they consist of numerous such colonies, fused or partly fused together, and supporting other coral species. Examples of this are common in the Capital Area of Oman. They undoubtedly trap carbonate sediments too, and it might be supposed that this form grades into typical fringing or patch reefs.

Schuhmacher and Zibrowius (1985) define the three terms constructional, hermatypic and zooxanthellate for different properties or characteristics of corals in a paper which discussed what reef building meant. The simplest Arabian reefs just described are "constructional" in the sense that they form a bioherm or "an elevated durable carbonate structure". They are also hermatypic in that they "significantly contribute to the framework of reefs . . .". They are, in the case of *Porites*, zooxanthellate too. Schuhmacher and Zibrowius (1985) go on to say that "hermatypic forms are always constructional, as the reef represents one type of bioherm". This is true in the simplest corals-on-corals case, in which the entire reef is constructed from large skeletons of single colonies or of groups of colonies. But it is not true in many Arabian examples where thriving "hermatypic" forms (where hermatypic is defined as a species "significantly contributing to the framework") do not lead to any reef construction at all. Substantial framework may become well established on non-limestone or directly on pre-Pleistocene limestone in many places and provide very high coral cover without leading to any reef or bioherm construction. Somehow sediments trapped by the framework do not consolidate or solidify. The framework itself does not build up either in the case of *Acropora*, though the massive *Porites* skeletons persist. On steep slopes such as in parts of Musandam, sediments are not trapped at all, and once they are dead the *Acropora* growths do not appear to remain on the substrate. Corals may thrive, but growth is no longer coupled to reef construction in marginal conditions such as these.

(3) Reef disappearance

The pattern of reef disappearance into conditions of increasing severity for corals can now be traced for several parts of the Arabian region.

In the Gulf, one area where reef decline and disappearance has been documented is the Gulf of Salwah, between Bahrain and Qatar (Sheppard 1985c, 1988). There is possibly a second similar gradient off the Saudi Arabian shore documented by McCain *et al.* (1984) and Coles (1988). They did not examine the gradient in the same light as this, but reported on the gradual change from coral-dominated communities to macroalgal-dominated communities on limestone as proximity to shore increased.

In the Gulf, offshore reefs have the greatest diversity of corals and reef associated invertebrates. While the reef fabric in some of these is more friable than is usual in the Indian Ocean and Red Sea (Sheppard 1985d) the structures are clearly accreting. In many instances they are several hectares in size. *Acropora* provides primary framework in shallow water to about 3–5 m deep, together with substantial growths of *Porites compressa* in deeper water. Moving south into the Gulf of Salwah, changes in the coral community occur (discussed later), the most important of which in the present context is an increasing dominance of *Porites nodifera*. By the point where salinity exceeds about 45 ppt, this species dominates, and is increasingly seen as huge, solitary colonies or as groups of more or less fused colonies, arising from extensive areas of limestone. This limestone is occasionally bare but is more commonly covered by a thin veneer of mobile sand. *P. nodifera* also commonly appears as groups of colonies rising to the surface from extensive beds of seagrasses, initially with several other coral species growing on their dead portions, but increasingly with only the two or three highly tolerant species.

These *Porites* reefs are the best examples in the Gulf of the special case of corals-on-corals reef growth noted above. The examples noted here are visually remarkable, since they attract high densities of diverse invertebrate and fish species and provide a striking contrast to vast expanses of seagrasses. In locations where the recorded salinity in August exceeded about 47–48 ppt, *Porites nodifera* does not form colonies substantial enough to be termed reefs. Instead it is found as smaller, scattered colonies on outcrops of limestone which are not scoured by sand. At salinities above 48 ppt it is no more abundant than the remaining *Siderastrea* and *Cyphastrea*, the other two salinity-tolerant species. Green and brown macroalgae replace corals completely at these salinities in the Gulf of Salwah, and the same broad pattern appears to take place on reefs and limestone domes off the Saudi Arabian coast.

In Oman, reef disappearance has been documented in more detail and over a geographically longer gradient (Sheppard and Salm 1988). Three categories of reef or coral population were identified. They are arranged loosely in geographical order from north to south. Firstly, clear coral reef growth occurs mainly in northern areas, where original bedrock is overlaid or obscured by a characteristic reef topography of a horizontal reef flat at low tide level and a steeper reef slope. These are in most cases true accreting coral reefs, and they exist in Musandam and the Capital Area.

Secondly, many examples are seen where coral framework is present, with coral cover values of 25–75% and sometimes more, but there is no characteristic reef topography. The substrate may be limestone or other rock types, and in either case is commonly almost completely obscured by the corals. The other benthic and fish fauna are also those typical of a coral reef. Both *Porites* and *Acropora* communities occur commonly. However, *Montipora* and *Pocillopora* may also form substantial, monospecific stands as well, in almost all cases in water less than about 5 m deep. These coral communities are typical of the framework condition. In many locations, large growths of *Porites lutea* and *P. solida* also appear, forming the kind of simple reefs referred to above. In parts of the Capital Area the stoutly branching coral *Pocillopora damicornis* forms monospecific stands which form coral mounds 2–3 m above the substrate. Rich coral communities such as these are common from Musandam to the Capital Area of Oman, with others recorded in southern Oman also. In the latter area, *Acropora* framework may be dense on a range of rock types in embayments. Presumably, in the latter, water temperatures remain warm in the summer upwelling, and in such locations massive *Porites* forms the single-species or even single-colony reefs noted earlier.

These conditions are not restricted to one clearly defined locality but occur throughout Oman, albeit more commonly in the south. The important criterion appears to be exposure to low summer temperatures. Embayments in the south escape the worst temperature drops, probably because water in them is warmed sufficiently by the intense summer insolation. Conversely, the Capital Area has many localities where cold water wedges have been reported by divers, even though it lies on the Tropic of Cancer and is well west of the Ras al Hadd promontory. The common feature of all these areas is that despite good coral growth, reef growth (except for the single-colony reef condition) does not occur. Dense growths of zooxanthellate, hermatypic, framework-constructing corals continue to provide high cover well into adverse conditions, but reef construction does not accompany it.

There are at present insufficient data to determine exactly what causes reef development to cease. The present state of knowledge concerning this problem is little better than in the 1940s or even earlier when low temperature was correctly invoked as the limitation to coral reef growth at high latitudes. Exact combinations of temperature and duration below an optimum value (the "degree days" of Coles and Fadlallah (1991)) remain to be determined, and clearly differ among species. But unlike the view of the 1940s, it is clear that the point marking the termination of true reef growth comes before the point marking the termination of many species of corals (including many species important in framework construction and reef building). In other words, corals (such as a high cover carpet of *Acropora*) may be termed hermatypic or constructional when present on growing reefs, but the same species may not be so termed when found growing equally abundantly on, for example, basalt.

In the third category seen in Oman, in more severe conditions still, corals provide up to 15% cover. Reef growth of course does not occur, and this cover is also too low to be considered as framework, especially when algal cover greatly exceeds coral cover. Initially, corals remain diverse but they do not include more than rare examples of the main framework genera, *Pocillopora*, *Porites*, and *Acropora*. The change to this category from the second one is not usually a gradation but is sharp. It appears to result from a sudden demise of the high cover species, leaving only hardy species which are widespread but rarely abundant either in these habitats or on reefs. This final category appears to continue until greater diversity and growth recommences in the Gulf of Aden.

B. Community changes and environmental stresses

For many major groups and habitats there are clear north–south gradients in terms of both species presence and absence and of assemblage or community type. These have been described in earlier chapters, where they were shown to match the physical controls outlined in Chapter 2 in a predictable way in most cases. However, as noted in the latter, the Arabian region has a greater range of extreme habitats than do most regions of comparable size, and the following summarizes changes seen at the most extreme ends of the species' or community's ranges. This may be defined as their terminal condition, as they either disappear leaving a void, or are replaced by another group.

Figure 12.2 summarizes the main patterns. Whether the environmental gradient is one of increasing turbidity, high salinity or high or low temperature extremes, the pattern has certain basic similarities. Progressing from the most amenable conditions, changes of coral community take place, this being evident at the level not just of dominant species but at dominant genus level, especially near increasingly turbid sites where mangrove development increases. Concurrently, there is an increase in brown macroalgae as well, especially along the crest of the reef. At first there is an increase in abundance of stagshorn corals, *Acropora*, in most places but of particularly elongated *Stylophora* in the Gulf of Suez. Where there is high turbidity and high salinity such as in the Gulf, this then changes to *Porites* dominated coral communities, which leads in turn to simple coral reefs of single species as described above. Where there are severe temperature fluctuations, this continues with a sudden elimination of the main framework species, which causes a sharp and substantial loss of coral cover.

It seems likely given existing data, that as conditions become harsher for corals, coral species drop out in a more or less smooth manner, or at least in a continuous one. Those which later on (under harsher conditions) become dominant are present in low densities in more favourable conditions. However,

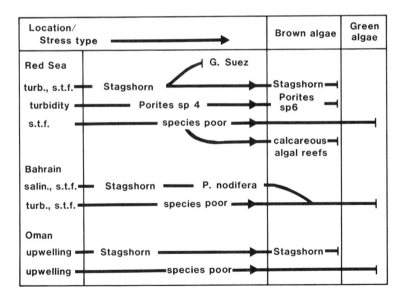

Figure 12.2 *Schematic representation of the main trends in dominant hard substrate communities under conditions diverging from those optimum for corals, and from typical oceanic conditions. s.t.f., severe temperature fluctuations; turb., turbidity; salin., salinity. Data extended from Sheppard (1988).*

possibly due to competitive effects in the higher diversity environment, they only achieve dominance when many other species have disappeared. The increase in cover of these species (e.g. *Porites nodifera* in the Gulf) continues as diversity and cover of other species falls, keeping overall coral cover high. In fact, cover may even increase as diversity falls, and in the Gulf cover of 65% or more is maintained by *Porites nodifera* as diversity falls to no more than 10 or even three species. Finally the dominant species cannot continue, and its disappearance leaves a void which no other coral fills. This appears to be the reason why the change between conditions of abundant coral framework and very low coral cover is sudden, both in the Gulf and in adjacent parts of southern Oman.

Before framework coral species disappear, they are no longer found on accreting coral reefs, as discussed earlier. The void left by the disappearance of the coral framework is filled by algae. In the southern Red Sea this includes calcareous red algae. Throughout the Red Sea, the latter occur in abundance in shallow water on reef crests, but in the far south they become the principal reef constructors. In the southern Red Sea and in the Gulf, brown algae of the *Sargassum* group replace corals, while in the Arabian Sea the kelp *Ecklonia* becomes important on rocky substrate too. These at first coexist with corals, and then occur in their absence.

In the Gulf and Arabian Sea, several examples exist where brown algae are themselves eventually replaced by green algae. This is seen in salinities of 50 ppt or more in the Gulf. Such values do not occur in the Arabian Sea where the cause of greens replacing browns remains unknown. Indeed, the locations where this is observed in Oman fall geographically between areas of moderate and limited reef development and areas of brown algal dominance. The extensive area does, however, lie in the eastern Gulf of Oman where cold summer upwelling may inhibit corals but where there may be insufficient nutrients to encourage much brown algal growth. This explanation remains speculative though.

Severe conditions exist in the Hawar Archipelago near Qatar. Biota on the sublittoral rock grades in this small chain of islands from browns to greens, with increasing amounts of space uncolonized by macro-biota. Summer temperatures exceed 37 °C and salinity exceeds 50 and rises to 80 ppt. Increasingly shallow substrata finally grade into sabkha-like conditions, which are dominated by blue-green algae in the manner described in Chapter 9.

Chapter 6 tried to identify the importance of competition between large seaweeds and corals in areas where one community replaced the other, but pointed out that work designed to separate effects of competition from the response to increasing environmental parameters has been minimal. For the Arabian region, Coles (1988) examined inshore Gulf reefs with this in mind but reached no clear conclusion, except to postulate that corals suffered both from competition with algae and from high temperatures closer to shore. Similar gradients of corals to algae seen in other regions of lowered temperatures and nutrient enrichment may also involve competitive effects. The latter has usually been asserted without any controlled experiments to distinguish competition effects from the response without competition. What is clear is that in the case of the Arabian Sea upwelling system, the response is similar in nature to that experienced by a few other high latitude reef systems. In all of these falling sea temperatures and rising dissolved nutrients leads to a gradation from coral to algal dominance. In southwestern Australia and the Atlantic several examples are documented (Crossland 1983, Johannes *et al.* 1983, Lighty 1983), in some of which competition by shading is regarded as important. The Oman response occurs in decreasing latitudes because of the geography of the upwelling, but the response is the same; the term "pseudo-high latitude" was suggested to allow for this inconvenient geographical difference. Oman is not the only location where this applies, however; Chuang (1977) reported a similar trend from Malaysia to Singapore.

Limits to mangroves are much less complex. Only one natural parameter appears to restrict *Avicennia* growth, and that is low winter temperatures. This is an effective control in both the western part of the Gulf and in the Gulf of Aqaba. In the latter case, high salinity which rises to over 40 ppt has been cited as a limiting factor also, but this is clearly not the case since the species survives similar salinities in the Gulf. Controls on seagrasses have not been investigated, but it appears likely that salinity in the range experienced in parts of the region limit some. Three very hardy species noted in Chapter 7 do not appear to be limited by salinities up to about 48–50 ppt. They occur in the Gulf of Salwah as well as throughout the Gulf of Suez as far as, and into, the Suez Canal.

Much less work has been done in the Arabian region on the effects of environmental stress on other taxa. Price (1982c) showed that few echinoderms were able to inhabit salinities above 50 ppt in the Gulf. In the same study populations of two asteroid species living under conditions of high salinity (52–60 ppt) showed evidence of dwarfism, suggesting physiological stress.

Fishes follow similar patterns of decreasing species richness with increasingly extreme environmental conditions as documented for corals above. For example, number of species drops markedly from south to north in the Gulf of Aqaba (Figure 4.3). Such patterns are difficult to interpret since it is not possible at present to separate direct effects of stress (physiological) from ecological effects. Furthermore, the data do not support a similar effect in the Gulf of Suez where environmental conditions become more extreme than those of the Gulf of Aqaba. Decreases in the availability of reef habitat or food, as other species drop out, might explain declining diversity better than conditions simply exceeding environmental tolerances of species. Nevertheless, extreme conditions almost certainly do limit the penetration of fishes into some habitats. For example the shallow reef-flat zone may experience very wide temperature and salinity extremes which few species are able to tolerate.

There are still considerable questions to be addressed on limitations of species and habitats by natural environmental conditions, and with the increase of human-induced stresses on marine ecosystems there is an increasing need to discover causes to the underlying trends, in case the one can help to solve the other.

Summary

Numerous hard substrata in the Arabian seas are not dominated, or even colonized, by corals. These may be old limestones or mixtures of sedimentary or other rocks. As environmental conditions become increasingly saline or incur greater temperature extremes, the dual processes of coral growth and reef growth become uncoupled. Processes required for reef accretion fail well before corals cease to provide high cover. The Arabian region has numerous examples, especially in the Arabian Sea, Gulf of Salwah and southern Red Sea. Temperatures below 11 °C in winter, combined with annual offshore temperature ranges of 11–16 °C and salinity over 42 ppt all greatly reduce coral diversity and lead to increased cover by macroalgae. A gradient in Oman showing a coral decline towards the southern upwelling area reciprocal to the increase of algae, strongly implicates increased nutrients and reduced temperatures.

As conditions become more extreme from the coral point of view, diversity of corals declines but total coral cover initially remains high at 40–60%, and reef construction continues. Under increasing severity of conditions, diversity falls further, until cover suddenly rises to 80% or more, due to an increase of "framework" building species such as *Acropora* and *Porites*. Reef construction continues and typical reef profiles of reef flat, crest and slope persist, overlying the basement rock. At this stage also, the simplest reefs are formed by aggregations of large colonies of species such as *Porites*. Such reefs, made by accumulations of "corals-on-corals", require no additional processes to result in formation of typical reef profiles.

At a later stage, however, when cover is still high, trapped sediments cease to consolidate or solidify. Coral framework itself does not accrete, and corals attach to basement rock, whether old limestone or non-limestone material. Once they die, coral colonies detach from the substrate. As conditions become more extreme, coral diversity falls progressively and eventually includes the high cover, framework species. Only at this point is there a sharp, marked fall in coral cover to <10%. Algal cover increases with the fall in coral cover.

Examples are described from the Gulf of Salwah and southern Gulf, the coast of Oman and the southern Red Sea. In all cases, where salinity rises above about 42 ppt or where temperatures fluctuate through more than about 15 °C, then reef growth fails, coral growth becomes severely reduced, and macroalgae replace the corals as dominant biota. Where nutrient enrichment occurs, algal replacement occurs with much smaller temperature or salinity extremes. In most areas the algae are phaeophytes, though in the southern Red Sea calcareous rhodophytes also appear with a vigour not seen elsewhere, to the extent that these form red algal bioherms, or algal reefs. Competition to corals from algae may be one cause of the change, though whether the decline in corals is caused by competition or whether algae merely benefit from increasing space is not determined. In some cases, such as in shallow water in southern Oman, or where salinity exceeds 50 ppt and summer temperatures exceed 37 °C in the Gulf, phaeophytes are themselves replaced by chlorophytes. In such conditions, corals cover <1% of the substrate.

For many major groups and habitats there are clear north–south gradients in terms of both species presence and absence and of assemblage or community type. Most if not all of these may be related to increasing environmental fluctuations. Mangroves are limited by winter temperatures, and for this reason are not found in the north of the Gulf.

Use and Management

Artisanal fishermen, central Oman.

CHAPTER 13

Arabian Fisheries

Contents

A. The bases of fishery production   261
B. Interdependence among different systems   263
C. Demersal fin-fisheries   264
   (1) Reef associated   264
   (2) Deep water fishes   266
   (3) Soft-bottom habitats   267
D. Pelagic fishes   269
   (1) Purse seining   269
   (2) Gill and drift netting   271
   (3) Reef-associated pelagics   272
   (4) Levels of production   272
   (5) Mesopelagic fishes   272
E. Crustacean fisheries   273
   (1) Shrimps   273
   (2) Lobster   279
F. Mollusc fisheries   280
   (1) Artisanal fisheries   281
   (2) Commercial fisheries   281
   (3) Cephalopods   283
G. Other fisheries   284
   (1) Aquarium fish   284
   (2) Ornamental shells and corals   284
H. Prospects for the future   285
Summary   286

The fisheries of tropical regions are typically very diversified due to the broad spread of species available for exploitation. The Arabian region is no exception, having well developed fisheries for a wide range of species. However, like fisheries elsewhere in the world, they can be classed into three main groups: fin-fish, crustaceans and molluscs. In the following account these groups will be described separately.

A. The bases of fishery production

The productivity of fisheries varies widely around the Arabian peninsula. Figure 13.1 shows the locations of the most productive fishing grounds. Fisheries

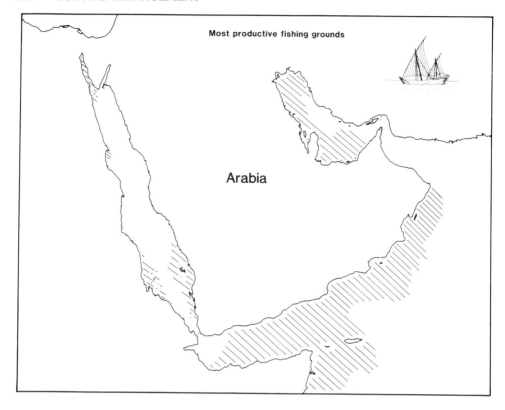

Figure 13.1 *Locations of the most productive fishing grounds in the Arabian region.*

production depends on several factors but the most important are level of primary production, which is largely dependent on nutrient input, and the ease with which resources may be exploited, which depends heavily on the nature of the resource and where it occurs.

Wherever there are high levels of nutrient input, productive fisheries tend to occur. Four main sources of nutrient input are present in the region: upwelling, terrestrial run-off, resuspension of bottom sediments, and nitrogen fixation by blue-green algae on reefs and mud flats.

Off the southern coast of Arabia, from south Oman deep into the Gulf of Aden, there is seasonal upwelling of cool, nutrient-rich water. This area provides high yields of pelagic fishes, as well as supporting productive demersal fisheries, mainly for crustaceans and molluscs.

Evaporative losses and a density-driven outflowing current from the Red Sea result in continuous input of surface water from the Gulf of Aden through the Strait of Bab el Mandeb. Inflowing water has a high nutrient level due to upwelling, and this results in relatively high water-column (i.e. planktonic) productivity in the southern Red Sea. In addition, the broad, shallow shelf areas of this region allow resuspension of bottom sediment by turbulent mixing, which also contributes to the raised nutrient levels. These shelves support some of the most important fisheries of the Red Sea. The other main sources of fishery production are the Gulf of Suez and Foul Bay on the Egyptian–Sudanese border. Here also nutrient availability is increased by resuspension of bottom sediments from the shallow seabed.

Throughout most other areas of the Red Sea and the Gulf of Aqaba, the continental shelf is narrow and very deep water found close inshore. Nutrients are rapidly lost below the thermocline and consequently plankton productivity and abundance of pelagic fishes are low.

In the Gulf, levels of nutrients in the water column are high due again to resuspension of bottom sediments. There is also some input carried by freshwater run-off from the Shatt al Arab and other rivers along the Iranian coast.

Fisheries yield is also heavily dependent on the ease with which resources can be exploited. Trawling, purse seining, gill netting and long-lining are among the most efficient techniques for commercial exploitation, but their use is limited to the open sea, or areas of level, relatively featureless bottom. Trawling for demersal species is possible throughout much of the Gulf, the shallow shelves of the southern Red Sea, Gulf of Suez, and some parts of the Gulf of Aden. However, use of this method is greatly restricted elsewhere by a lack of suitable grounds. This dependence on particular types of bottom is one reason why shallow areas of the region support greater fisheries production.

Although coral reefs are highly productive systems, their exploitation on a commercial scale is limited by the difficulty of fishing them. Their irregular and shallow topography prevents the use of trawls, seines or large nets, and makes access by large boats very difficult. Consequently, although important, fisheries on reefs are mainly artisanal.

B. Interdependence among different systems

Transfer of nutrients among systems is also important to fisheries production. For example, areas of mangrove forest are thought to export significant amounts of detrital material and dissolved organic carbon to adjacent systems, although the importance of this input is disputed, and in any case is likely to vary substantially depending on region (Ogden 1988). Detrital food webs support many crustaceans and other invertebrates, including shrimps (Basson *et al*. 1977). Such invertebrates are themselves the basis of much demersal fish production. There are also a number of fishes caught commercially which feed directly on detritus, such as the mojarras *Gerres* spp. (Gerreidae). The shores of the southern Red Sea support relatively large areas of mangrove forest, as do parts of the Gulf. Linkage between mangroves and adjacent soft-bottom habitats, such as seagrass beds and areas of sand and mud, is likely to contribute to the productivity of fisheries in these areas.

A number of exploited organisms also complete their life cycles in more than one biotope. For example, mangroves and seagrass beds have also been suggested to be important nursery areas for many commercially important species (Quinn and Kojis 1985). The life histories of several such species are described below.

C. Demersal fin-fisheries

(1) Reef associated

Fisheries for species associated with coral reefs are of greatest importance in the Red Sea, due to the abundance of reefs in this region. However, the Gulfs of Aden and Oman and the eastern Gulf also support limited reef fisheries. Fishing grounds follow the distribution of reefs closely for artisanal fisheries (see Chapter 3 for reef distribution), whilst the main grounds for commercial fishing lie within the Gulf of Suez, the southern Red Sea and eastern Gulf. Reef-associated fish contribute only a small amount to overall production in the Gulf as reefs there cover only a very small area.

A) METHODS USED

The main methods used to catch reef-associated fishes are hook and line, traps, trammel nets, gill nets and spears. Throw nets may also be used, often to collect bait fish, and bottom-set longlines are employed in some areas. As noted above catching methods are constrained by the irregular topography of the reef, and limit the scale of fishing activities. Commercial fisheries generally operate at a small scale, and products are usually sold locally (Neve and Al Aiidy 1973, Barrania 1981). The bulk of exploitation is artisanal, and the main importance of the fishery is in supplying dietary protein for coastal people (Head 1987c). Nevertheless, there have been moves to greater commercialization of the fishery, for example in Saudi Arabia. The fishing community of Tuwwal, for instance, supplied around 40% of the reef fish reaching the market in Jeddah in the mid 1970s (Kedidi 1984a).

B) SPECIES CAUGHT

Catch composition is dependent on fishing method. Hook and line fishing targets mainly predatory species such as groupers (Serranidae), snappers (Lutjanidae), emperors (Lethrinidae) and bream (Sparidae). Traps, trammel and gill nets are less selective, catching a wide variety of species from several trophic levels, including herbivores (e.g. parrotfishes [Scaridae] and rabbitfishes [Siganidae]), detritivores (e.g. mullet [Mugilidae]), and planktivores (e.g. fusiliers [Caesionidae]), as well as predatory piscivores. Spear fishing is not widely used, but is strongly size-selective with large fishes caught most readily (Oakley 1984). They include mainly parrotfishes, surgeonfishes (Acanthuridae) and groupers.

Despite the huge variety of species available to fishermen, a remarkably small number make up the bulk of the catch. This appears to be largely due to differences in catchability rather than differences in abundance. For example, there are 13 species of parrotfishes in the Red Sea but only three comprise the majority of landings: *Hipposcarus harid*, *Scarus ghobban* and *S. psittacus*. All are species which regularly feed over sandy bottoms adjacent to reefs and so are easily caught in trammel nets and small gill nets. Similar trends are apparent for predatory species. Landings of reef-associated predators from Egypt were dominated by four species of grouper, *Epinephelus summana*, *E. areolatus*, *E. tauvina* and *Variola louti* (50% of catches) and two species of emperor, *Lethrinus nebulosus* and *L. mahsena* (25% of catches) (Sanders and Kedidi 1981a). The three species of *Epinephelus* and *L. nebulosus* are all most common among sandy areas with patchy reef, more easily accessible to boats.

C) SEASONALITY OF THE FISHERY

Reefs support year-round fisheries, although spawning and recruitment of fishes may be highly seasonal. The main spawning season in the Red Sea is from April to September (C.R., pers. obs.), and appears to cover roughly the same period in the Gulf (Coles and Tarr, 1990). Catch statistics generally lump species into family or other groupings and so seasonality of catches for individual species is difficult to demonstrate. However, several of the fishes commonly caught in the Red Sea show marked seasonality. This is usually associated with the spawning season, or onshore–offshore migration. Perhaps the most notable is the emperor *Lethrinus nebulosus*, which was the subject of a detailed study in the Gulf of Suez and adjacent areas (Sanders and Kedidi 1984a). This is a relatively long-lived species which first reaches sexual maturity between the ages of four and six years. It spawns between April and July and during this period is particularly vulnerable to capture. From 50% to 80% of annual catches were made in these months in the northern Red Sea. Recruitment to the fishery does not take place until the age of first spawning with current fishing methods (hook and line, bottom longline, gill and trammel nets).

Success of capture during the spawning season appears at least partly due to the migrations to mass spawning sites which the species undertakes. These areas are well-known to local fishermen and are targeted seasonally. For example, the area adjacent to the western tip of the Sinai peninsula, close to Ras Mohammed, is an important spawning ground (C.R., pers. obs.) and Sanders and Kedidi (1984a) report that this area is fished almost exclusively between April and July.

D) PRODUCTIVITY

The main basis of both the artisanal and commercial fisheries on reefs is carnivorous species. Detailed analysis of the artisanal fishery at Tuwwal on the Red Sea coast of Saudi Arabia showed that catches were predominantly composed (over 50% by weight) of predatory piscivores (Kedidi 1984a). Similar findings were obtained from commercial Sudanese reef fisheries. Groupers and emperors combined constituted an average of 64% of the catch by weight for two ports over two sampling years (Kedidi 1984b). Similar trends are apparent for most of the other countries in the region (Sanders and Kedidi 1981, Morgan 1985).

Overall productivity of these fisheries is relatively low. Kedidi (1984a) estimated an average production of 0.4 tonnes $km^{-2}y^{-1}$ for the Tuwwal fishing area. On the basis of catch and effort data, Kedidi concluded that this fishery was fully exploited, and that production could not be increased above these levels. He further suggested, using catch statistics and the areas of fishing grounds available, that reef fisheries along the whole Saudi Arabian Red Sea coast north of

$20°N$ were fully exploited. A similar productivity of 0.31 tonnes $km^{-2}y^{-1}$ was reported from the Gulf of Suez by Sanders and Kedidi (1984b). Studies elsewhere have shown that reefs can be much more productive than this. For example, yields from Philippine reefs have been estimated between 8 and 20 tonnes $km^{-2}y^{-1}$ (Alcala 1981, Munro and Williams 1985, Bellwood 1988), and those from American Samoa up to 44 tonnes $km^{-2}y^{-1}$ (Wass 1982). What is the basis for the lower Red Sea production?

Part of the discrepancy lies in the fact that calculations for the Philippines and American Samoa excluded non-reefal shelf area, whilst those from the Red Sea included low productivity habitats other than reefs within the fishing grounds. However, perhaps the main reason for low productivity in the Arabian region is the fact that most of the production comes from high in the food chain. By contrast, Philippine and Samoan reefs have been subject to very high levels of exploitation (e.g. Norte *et al.* 1989) and support only low populations of predatory fishes. Most of their production is based on herbivores, such as surgeonfish and rabbitfish, or planktivores such as fusiliers (Bellwood 1988). Fisheries targeted lower down the food chain will be substantially more productive than those based on large piscivores or other predators.

E) POTENTIAL FOR INCREASED PRODUCTION

There is probably only a small potential to increase production of the species currently fished. Some parts of the region are at present only lightly exploited, such as the Gulf of Aqaba, parts of the Sudanese coast and the southern Red Sea off Ethiopia. Production from these areas is limited by their inaccessibility and distance from markets (and civil war in the case of Ethiopia). There have been some moves to increase production from such areas by gear improvements and provision of better onshore facilities. Sanders and Kedidi (1981) estimated that catches from the Red Sea using artisanal methods (the majority of them being of reef-associated fish) could potentially be increased by 35,000–45,000 tonnes y^{-1} from a level of approximately 50,000 tonnes y^{-1}. Substantially greater increases in effort would be necessary to realize this potential.

There is, however, a much greater potential for increasing yields with shifts in the species targeted. Harvesting lower down the food chain, catching greater quantities of herbivorous and planktivorous fishes, would lead to increases in production. This would arise because of the greater proportion of primary production being incorporated into fish biomass at these levels. However, the majority of such species are currently not widely eaten in the region and so this approach would have to be combined with a campaign to increase their acceptability.

(2) Deep water fishes

In many parts of the world, such as the Seychelles, there are well-established fisheries for deep water species found on slopes off the continental-shelf edge. These are usually exploited using hook and line and consequently catches are dominated by predatory fishes, mainly snappers and groupers. However, this kind of fishing is little used in the Arabian region, although deep water fishes such as the grouper *Epinephelus morrhua* occasionally turn up in markets (Randall and Ben-Tuvia 1983). It is likely that substantial potential exists in some areas of the southern Red Sea and Gulf of Aden for development of a deep water fishery. Exploratory fishing at the shelf edge in depths of around 200 m has revealed stocks of deep water snapper, especially *Pristipomoides typus*, which may support commercial exploitation (Peacock 1979).

(3) Soft-bottom habitats

Soft-bottom habitats, including sand and mud bottoms and seagrass beds, support the bulk of fisheries for demersal fin-fishes in the region. They also support major fisheries for other species, such as shrimp (the latter will be covered separately below). These habitats are not as productive as coral reefs, and estimates suggest that sand and mud bottoms may produce $200 \, g \, C \, m^{-2} y^{-1}$, seagrass beds around $1000–3000 \, g \, C \, m^{-2} y^{-1}$, and coral reefs in the region of $2000–4000 \, g \, C \, m^{-2} y^{-1}$ (Basson *et al.* 1977, IUCN 1987). The higher fisheries productivity of these habitats is mainly due to the fact that fishes can be more easily harvested from them on a commercial scale, and because they cover relatively large areas by comparison with coral reefs.

A) METHODS USED

Fishes from soft-bottom habitats are caught mainly by trawling, although limited quantities are captured by gill and trammel nets set in shallow water, traps and beach seines. In addition bottom longlines are used in some areas. Figure 13.2 shows the main trawling grounds, which largely correspond to areas of shallow shelf free of major obstructions. Trawling is conducted both by artisanal fishermen working from dhows (15–20 m long) and steel-hulled industrial vessels (20–25 m long) (FAO 1989a). Dhows generally fish closer inshore than the industrial vessels (IUCN 1987). Beach seining is in widespread use on the Red Sea coast of Yemen (A.R. Dawson Shepherd, pers. comm.), Ethiopia and in the Gulf (IUCN 1987) but is restricted to artisanal fishermen.

B) SPECIES CAUGHT

The species caught vary depending on area and fishing method. However, the majority of the catch comes from the following groups: lizardfishes (Synodontidae), goatfish (Mullidae), snappers, threadfin bream (Nemipteridae), sweetlips (Haemulidae), and mojarras (Sanders and Kedidi 1981, 1984c). In addition, trawling catches a number of species of pelagic fish of which the most important are horse mackerels and jacks (Carangidae) and mackerels (Scombridae) (Sanders and Kedidi 1981, 1984d,e). Bottom longlines catch primarily predatory species such as the emperor, *Lethrinus nebulosus* and the grouper *Epinephelus chlorostigma*.

Trawling for fishes takes place during the day and so catches are dominated by

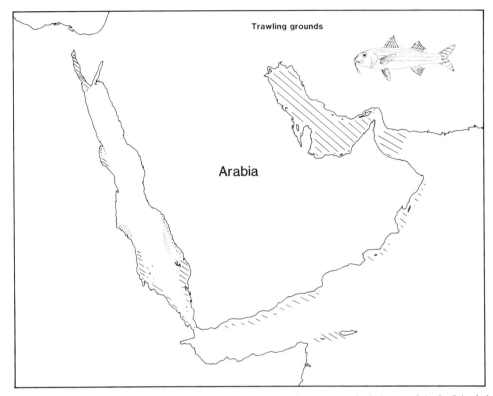

Figure 13.2 *Main trawling grounds. Hatched areas are those currently being exploited. Stippled areas represent potential grounds. Data are from Sanders (1981a), Sanders and Kedidi (1981), Venema (1984), IUCN (1987) and FAO (1989a) and unpublished data.*

diurnal species. By contrast, trawling for shrimp is done at night (Basson *et al.* 1977). Much of the commercial fleet in Foul Bay on the Egyptian coast operate as trawlers by day and purse seiners at night (Sanders and Kedidi 1984f).

Composition of the catch varies between fishing areas. For example, in the Gulf of Suez trawlers caught twice as many of the bigeye snapper, *Lutjanus lineolatus*, as they did in grounds close to the Sudanese border over a three year period. Catches landed were dominated by only a few species or species-groups, reflecting both relative abundance and marketability. In Egyptian fishing grounds lizardfish (predominantly *Saurida undosquamis*) and *L. lineolatus* together comprise over 55% of catches (Sanders and Kedidi 1984g). In the Gulf, the other main trawl fishing area in the region, catches consist primarily of croakers (*Otolithes ruber*; Sciaenidae), grouper, bream and mullet. Together these constituted 27 and 40% of the fin-fish catch landed by Kuwaiti trawlers in 1987 (FAO 1989a).

Unlike reef fisheries, trawling is usually seasonal. The best information available is for the Egyptian Red Sea (Sanders and Kedidi 1984f), where trawling is highly seasonal with vessels operating from September to May.

C) CATCH TRENDS

Sanders and Kedidi (1984f) estimated that the trawl fishery in the Gulf of Suez was fully exploited or possibly over-exploited for some species. They argued that reducing the size of the trawl fleet would increase profitability whilst only reducing total catches by a small amount. For example, a 50% reduction in effort was calculated to lead to a reduction in catch of only 22% (Sanders and Kedidi 1984c). Fishing effort is generally high in this area, with 74 vessels operated on the Egyptian coastline in 1983 (Sanders and Kedidi 1984g). During the nine month season each vessel undertakes trips of an average of eight days' duration in which there are around seven trawl shots per day, each of approximately three hours. Sanders (1984) suggested that the trawl fishing fleet in the Gulf of Suez should be reduced by about 10% to limit fishing effort.

D) POTENTIAL FOR INCREASED PRODUCTION

Although the fishing grounds presently used are probably fully exploited, there are some areas which are not at present fished commercially (Figure 13.2). For example, the shelf areas of the southern Red Sea provide extensive grounds suitable for trawling, estimated at 19,000 km^2 in the waters of Ethiopia, Saudi Arabia and Yemen. Surveys in the late 1970s identified substantial potential for exploitation (Walczak 1977, Peacock 1979) which, for Saudi Arabia at least, is only just beginning to be realized (Oakley and Bakhsh in press). Although a trawl fishery existed in Ethiopian waters which was quite productive during the 1960s and early 1970s (Sanders and Kedidi 1981), production subsequently declined substantially due to disruption by the war in Eritrea. Recent data for this area are not available. Surveys of the shelf area of the Sudanese coast have failed to reveal commercially viable stocks to support a trawling industry (Sanders and Kedidi 1981).

D. Pelagic fishes

There are important fisheries for pelagic species throughout the Arabian region. Fishing is by a variety of methods including purse seining, gill and drift netting, trawling, longlining and trolling. Although the majority of the exploitation of pelagic species is in the open sea, artisanal fishing close to reefs also results in substantial catches. Fisheries for these reef-associated pelagics are also covered below.

(1) Purse seining

This method is used widely in parts of the Red Sea, off the south coast of the Arabian peninsula and in the Gulf. The main fishing grounds are shown in Figure 13.3. By far the most important in the Red Sea lie in the Gulf of Suez and Foul Bay close to the Sudanese border, with 73 and 11 vessels operated in each area

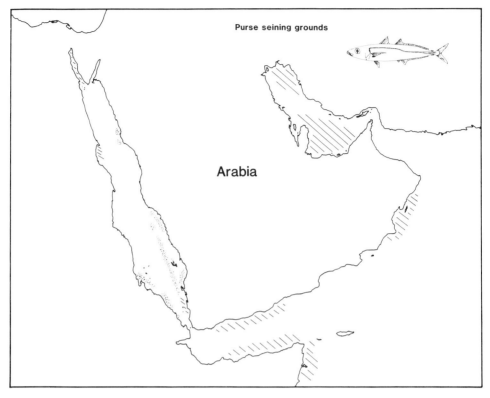

Figure 13.3 *Main purse seining grounds. Hatched areas are those currently being exploited. Stippled areas represent potential grounds. Data are from Sanders and Kedidi (1981), Venema (1984) and FAO (1989a).*

respectively by 1983 (Sanders and Kedidi 1984g). Fishing takes place at night and fishes are concentrated prior to capture by lights placed on small boats. Each boat carries 9–10 paraffin lamps and is left for a period of a few hours before the net is set around it. Two or three shots are made from each vessel per night of fishing. Fishing does not take place during a period of around ten days every month when light from the moon reduces catches. Purse seining is seasonal in the northern Red Sea, taking place from October or November through to May (Sanders and Kedidi 1984g) although with a slightly extended season in Foul Bay (Sanders and Kedidi 1984h).

Fishing off the southern coast of Oman is also seasonal, with peak catches occurring following the period of upwelling brought on by the Southwest Monsoon (L. Barratt, pers. comm.).

A) SPECIES CAUGHT

Catches consist of two main groups of species: predatory piscivores and zooplanktivores, and herbivores which depend mainly on phytoplankton for food. In the former category the dominant species from Egyptian fisheries include horse mackerels (*Trachurus indicus* and *Decapterus maruadsi*) and mackerel (*Rastrelliger kanagurta*). In the latter category, sardinellas (e.g. *Sardinella gibbosa* and *S. sirm*; Clupeidae) and round herring (*Etrumeus teres*; Clupeidae) are most important. By weight, catches are roughly evenly split between these two categories (Sanders and Kedidi 1981).

B) POTENTIAL FOR INCREASED PRODUCTION

Purse seining was estimated to yield around $16,500\,t\,y^{-1}$ from the Red Sea by 1980 (Sanders and Kedidi 1981). Fisheries for pelagic species are fully-exploited in most parts of the Red Sea and Gulf, although there are potentially under-exploited stocks off the Saudi Arabian and Yemen coasts of the Red Sea (Sanders 1984). However, it is likely that substantially more productive fisheries could be supported on the southern Arabian coast, due to upwelling. Early estimates of productivity for this area were almost certainly over-optimistic. For example, Shomura estimated a potential productivity of $800,000$–$850,000\,t\,y^{-1}$ for waters off the southern coast of Yemen and Oman and the northern coast of Somalia (Gulland 1971). More detailed surveys of the area have suggested potential yields of $450,000$–$650,000\,t\,y^{-1}$ (Venema 1984). The southern coast of Oman appears to be especially productive, with an estimated standing stock of 600,000 tonnes of small pelagic fish in Masirah Bay alone (Strömme 1983).

Maskat Cove (Oman) some years ago was so abundantly filled with the small fish called sardinas in the Mediterranean, that they might be obtained in any quantity . . . but about two years ago they wholly deserted their former quarters . . .

Wellsted 1838

---

(2) Gill and drift netting

Gill netting has not been established on a large commercial scale anywhere in the region. The great majority of landings are from small-scale fishermen operating close inshore in depths of less than 200 m (FAO 1989b). Gill netting is primarily directed at fast-swimming pelagic predators, such as mackerel (e.g. *Scomberomorus commerson* and *S. guttatus*), tuna (e.g. *Thunnus albacares*; Scombridae) and jacks (e.g. *Carangoides bajad*). At present, drift netting is used primarily off the Omani coast, and is mainly to capture tuna and mackerel (FAO 1989b).

There appears to be some potential for increased use of gill nets in deeper, offshore waters off the south coast of the Arabian peninsula, where the seasonal upwelling results in relatively high productivity. These waters are presently exploited by Japanese and Korean boats catching tuna using longlines. Nevertheless, there appear to be greater resources available for exploitation further south in the Indian Ocean, for example around the Seychelles (FAO 1989b).

(3) Reef-associated pelagics

Pelagic fishes, especially predators, occur in relatively high densities in the waters close to coral reefs. Most important among them are jacks, mackerel, tuna and barracuda (Sphyraenidae). As noted above, these resources are virtually impossible to harvest using large-scale commercial fishing operations due to the difficulty of fishing close in to reefs. Consequently, they are exploited almost exclusively by small-scale fishermen.

Three main methods of capture are used: gill netting (described above), trolling and handlines. Reef-associated pelagic fishes are of great importance within the catches of artisanal fishermen operating throughout the region. For example, in the fishery at Tuwwal studied by Kedidi (1984a), jacks, mackerel and barracuda combined constituted 23–33% of the catch, and were similarly important in Sudanese reef fisheries (Kedidi 1984b).

(4) Levels of production

Estimated total landings of tuna, mackerel and billfish (Istiophoridae) from the Red Sea, southern Arabia and Gulf were 100,000 tonnes in 1987 (FAO 1989b). Four countries, Oman, Iran, Saudi Arabia and the United Arab Emirates, caught over 90% of this quantity with Oman alone contributing approximately 50% of the total. In general it has been concluded that tuna and mackerel are currently heavily exploited (FAO 1989b). However, it was suggested that production of neritic (shallow shelf) tuna could be increased within the Red Sea, from the low level of 1987 (1300 tonnes).

(5) Mesopelagic fishes

Surveys of the Arabian Sea area during the 1970s and early 1980s were undertaken to determine whether there were significant stocks of fishes present which were under- or unexploited. It was expected that there would be extensive fishery resources in this region based on seasonal upwelling (Venema 1984). Although important stocks of shallow-water pelagic fishes were discovered, mesopelagic fishes were found to have an even greater biomass (Gjøsaeter 1984). Mesopelagic fishes are usually defined as those occurring between depths of 200 and 1000 m.

The estimated total standing stock of mesopelagic species for the region bordering southern Arabia, including the Gulfs of Aden and Oman, was approximately 94 million tonnes (Gjøsaeter 1984). The greatest densities were found in the Gulfs of Aden and Oman. Most of the biomass was concentrated in the range 250–350 m deep, but there was upward migration at night to depths of 100–150 m.

The dominant species present include the lanternfishes *Benthosoma pterotum* and *B. fibulatum* (Myctophidae) although over twenty other species occur (Gjøsaeter 1981). *B. pterotum* was estimated to occur at densities of 100–300 individuals m^{-2} surface area in the Gulf of Oman. It is a small, fast-growing species which feeds predominantly on copepods, crustacean larvae and amphipods. Individuals are estimated to reach a length of 4 cm in one year and at this size they are believed to spawn and subsequently die. Spawning takes place throughout the year but with two seasonal peaks in March–June and September–November.

Mesopelagic fish can be caught using pelagic trawls. Trials in the Gulf of Oman yielded average catch rates of approximately 200 kg per hour during a survey in 1979 and approximately 1000 kg per hour in 1981 (Gjøsaeter 1984). However, the small size (most are under 10 cm long) and unusual morphology of the species present mean that they are unsuitable for human consumption, although they could be processed into fishmeal. Even so, Gjøsaeter (1984) concluded that the stock density was probably too low over most of the area, except the Gulf of Aden, to support a commercial fishery. He believed that catching methods would have to be improved before stocks would be commercially viable.

E. Crustacean fisheries

Shrimps and lobsters are the most important crustacean resources harvested in the area and will be discussed separately below. The swimming crab, *Portunus pelagicus*, is also fished but is caught mainly by artisanal fishermen, sometimes using spears (Head 1987c), or is landed as part of the by-catch from other fisheries such as those for shrimps (Basson *et al.* 1977).

(1) Shrimps

The main fishing grounds for shrimp are shown in Figure 13.4. The Gulf is the most important area, with shrimps one of the prime fishery resources, especially along the east coast of the Arabian peninsula (Farmer 1981, IUCN 1987). They currently make up around 15–20% of total fishery landings (FAO 1989a) from the Gulf. There are also substantial shrimp resources in the Red Sea, primarily in the Gulf of Suez, but also off the coasts of Sudan, Yemen and Ethiopia (Sanders and Kedidi 1981). However, total Gulf landings are seven to nine times higher than those from the Red Sea (Venema 1984). Shrimps are also fished in the lakes of the Suez Canal (Gab-Alla *et al.* 1990).

Shrimps have been exploited in the region since historic times, early fishermen mainly catching them in traps (Price 1976). Commercial exploitation is a recent phenomenon. An industrial fishery for shrimp only began to develop in 1963 in Saudi Arabia (Lewis *et al.* 1973) and around 1959 in Kuwait for example (FAO 1980). This fishery is of considerable economic importance and is served by both industrial and artisanal fleets (IUCN 1987). In 1985, shrimp landings in Kuwait were valued at US$10 million and constituted the country's second largest export earner after oil (Pauly and Mathews 1986).

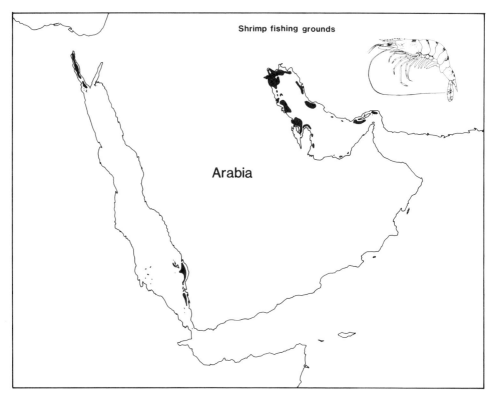

Figure 13.4 *Main shrimp fishing grounds. Data are from Sanders and Kedidi (1981), Farmer (1981), IUCN (1987) and Oakley and Bakhsh (in press).*

A) CATCHING METHODS

Shrimp are caught primarily by trawling at night since they are nocturnally active (Basson *et al.* 1977), although night trawling has now been banned in some Gulf states due to security restrictions. Industrial trawlers are typically around 20–25 m long whilst artisanal vessels are somewhat smaller, the latter generally operating close inshore. In addition, shrimp are taken on a very limited scale in barrier traps made of palm fronds or nets set in intertidal areas, and by manually operated nets (FAO 1989a). The species caught depend on the area fished and the methods used. The largest and most valuable shrimps are *Penaeus semisulcatus*, *P. latisulcatus* and *P. japonicus* in both the Gulf and the Red Sea (Price and Jones 1975, Sanders 1984). Smaller shrimps are also important in catches and include *Metapenaeus affinis* and *M. stebbingi* (Farmer 1981, Gab-Alla *et al.* 1990).

Penaeus semisulcatus forms the backbone of the industrial fishery and constitutes over 90% of industrial landings in the Gulf (Price and Jones 1975, IUCN 1987) and over 60% in the Gulf of Suez (Sanders and Kedidi 1981). *Metapenaeus* spp. are also of considerable commercial importance, particularly to artisanal fishermen. *M. affinis* are abundant in the catches of the artisanal fleet in Kuwait (Van Zalinge 1984) and exceeded catches of *P. semisulcatus* there for the first time in 1986/87 for the combined industrial and artisanal landings (FAO 1989a). *M. stebbingi* is the most important species in landings from the lakes of the Suez Canal (Gab-Alla *et al.* 1990).

B) LIFE HISTORIES

The life cycle of shrimps has three main stages: a pelagic phase, settlement and early development in inshore nursery grounds, followed by later movement offshore as they grow. Basson *et al.* (1977) describe the life cycle of *Penaeus semisulcatus* from the Gulf in detail. Spawning takes place at night, mainly in early spring (February–March) on the Saudi coast (Price and Jones 1975), and a little later in Kuwait (FAO 1980). *P. semisulcatus* are able to spawn several times throughout their lifetime, and females produce an average of approximately 200,000 eggs per spawning (FAO 1980). Eggs are pelagic and hatch in less than 24 hours. Larvae then undergo several developmental stages before settling into shallow water close inshore. The planktonic stage is completed within 2–3 weeks. Juveniles initially inhabit clumps of macroalgae, such as *Sargassum* and *Hormophysa*, and settlement coincides with the period of maximum algal development (Coles 1988) in March to early April. As they develop they move to seagrass beds, at least in Saudi Gulf waters. Subsequently they migrate offshore into sandy and muddy habitats where they are fished by industrial trawlers, although part of the population remains inshore.

Other species common in the area have somewhat different life histories. For example, early development of *Penaeus latisulcatus* appears to take place over shallow sandy bottoms immediately below the wave-wash zone, whilst *Metapenaeus stebbingi* settle into tidal creeks of intertidal mud flats (Basson *et al.* 1977). Adults of the latter species are found in these creeks and in shallow, muddy habitats. They appear to be tolerant of relatively high salinities which is perhaps why they have been successful in the lakes of the Suez Canal, where salinities reach 45 ppt (Por 1978, Gab-Alla *et al.* 1990). By contrast, *M. affinis* appear tolerant of relatively low salinities, and are common in estuarine habitats. This may explain their greater abundance in the northern Gulf around the mouth of the Shatt al Arab where salinities are seasonally reduced (Farmer 1981).

Shrimps are carnivorous or omnivorous, feeding predominantly on small invertebrates and detritus. They grow rapidly, at rates of 2–3 mm per month for several species in Kuwaiti waters (FAO 1980). Recruitment into the fishery takes place as little as three months following settlement. *P. semisulcatus* have been estimated to increase their body weight by around 2000 times within one season, reaching adult size after eight months' development within the seagrass biotope (Basson *et al.* 1977). Although *P. semisulcatus* have been estimated to reach 4–5 years old, heavy fishing mortality results in virtually an annual life cycle. Basson *et al.* (1977) report that few survive for much longer than one year. Consequently, the fishery is based primarily on 0 and 1 year age classes and so production is heavily dependent on spawning success and larval survival in a given year (FAO 1980).

C) CATCH TRENDS IN THE GULF

In the early days of expansion of the shrimp fishery, catches were initially very high. Figure 13.5 shows total Gulf catches over the period 1961–1987. Peak catches were obtained in the late 1960s to early 1970s but following this declined markedly. Over-exploitation and consequent severe depletion of stocks was perceived to be the cause of this trend, with some areas being more seriously affected than others. Landings from Kuwait and northern Saudi Arabia suffered a particularly large decline, whilst those of the south appeared less badly affected (Farmer 1981, IUCN 1987). Decreases in catch were sufficient to put three major companies out of business in 1979 alone (Farmer 1981).

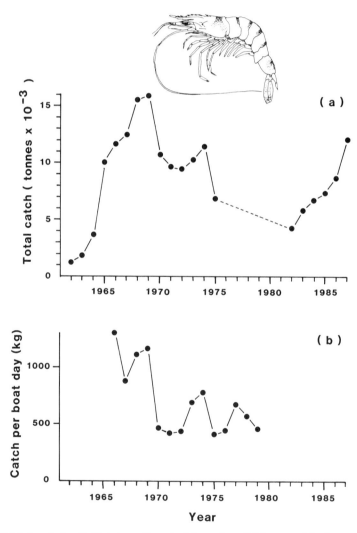

Figure 13.5 *Catch trends for Gulf shrimp. Data in (a) are from FAO, quoted in Farmer (1981) and FAO (1989a). Data in (b) are from FAO, quoted in Oakley (in press b).*

The reasons for decline in Gulf catches have been controversial. Undoubtedly, there has been heavy exploitation. However, other contributing factors have also been suggested. In particular, coastal infilling of shallow areas of the western Gulf coast may have caused substantial loss of juvenile habitat (IUCN 1987, Oakley in press a). Some of the main nursery grounds are also areas of rapid urban and industrial development, for example Tarut Bay on the Saudi coast and Kuwait Bay further north (Al-Attar 1979).

D) MANAGEMENT

The steep decline in catches experienced indicated the need for management of the fishery. Consequently, shrimps were one of the first resources in the Arabian region to receive formal management with the implementation of closed seasons in the Gulf. Initially, fishing for shrimp was prohibited in eastern Saudi waters between June and July (Lewis *et al.* 1973). However, later this period was extended from February to July (FAO 1989a), and in 1989 the fishery was closed for eight months from January to August (Oakley in press b). In Kuwait, the closed season has been variable, operating from July to February or September to April, the exact season depending on the expected level of recruitment (Morgan 1985). The United Arab Emirates have instituted an indefinite, year-round closure of the fishery to allow stocks to recover (FAO 1989a).

Effort has also been reduced within the fishery by the banning of foreign vessels from Saudi waters in 1979 (Van Zalinge 1984). Kuwait implemented a policy of decommissioning vessels and fixing the number of industrial and artisanal boats licensed to fish. Thirty-three vessels had been removed by 1988 (FAO 1989a). In addition, there has been legislation to close some nursery grounds to fishing within Kuwait Bay, and mesh size restrictions for trawlers have been introduced (FAO 1989a).

Measures to limit effort of the Saudi fishery have been less comprehensive. There is no limitation on vessels licensed to fish and numbers obtaining licences have been increasing, from 342 in 1980/81 to 1245 in 1987/88 (Oakley in press b).

Reduction of fishing effort appears to have resulted in some recovery in catches (Figure 13.5) from the Gulf overall. However, after a recovery in the early 1980s, catches from the Saudi Arabian fishery have again been declining (Oakley in press b), despite increases in length of the closed season. There is evidence that the industrial sector have been landing shrimp during the closed season in defiance of the law, whilst regulations have been strictly enforced on artisanal vessels (Oakley in press b, Pilcher *et al.* in press). This and the continued infilling of nursery areas may have contributed to declining catches. Furthermore, trawling in seagrass beds does substantial damage, and intense fishing activity may have considerably reduced their productivity.

A recent report (Siddeek *et al.* 1990) showed catches from Kuwaiti waters at record levels for 1988–1989. This was attributed to reduction in effort and a favourable environment for recruitment.

E) RED SEA SHRIMP FISHERIES

These are concentrated in the Gulf of Suez and off the southern Saudi Arabian coast, although shrimp are also landed in Sudan, Yemen and Ethiopia. The Gulf of Suez has been exploited commercially for much longer than the southern Saudi grounds. Over 600 tonnes were landed from the Gulf of Suez in 1979/80 with shrimp constituting approximately 12% of trawl catches by weight (Sanders and Kedidi 1984b). Stocks appeared to be fully exploited at this level of catch (Sanders 1984). By contrast, the southern Saudi fishery is still actively expanding with catches dominated by two species: *P. semisulcatus* (83% by weight) and *Metapenaeus monoceros* (14%) (Oakley and Bakhsh in press). However, the grounds are less productive than those of the Gulf with less than half the catch per unit effort. Little information is available as yet to manage this fishery but a closed season has been implemented from April to July, and an area closed to fishing to conserve stocks (Oakley and Bakhsh in press).

F) BY-CATCH

Shrimps comprise only a small proportion of the total weight of catches from trawlers. The rest of the catch, which may be five to ten times the weight of the shrimp, consists of fish, other crustaceans and molluscs (Price and Jones 1975, Basson *et al.* 1977). Morgan (1985) estimated the weight of by-catch in the Gulf at 30,000 to 40,000 tonnes annually. Dominant species or species-groups within the by-catch include mojarras, pony fishes (Leiognathidae), tripod fishes (*Pseudotria-canthus* sp.; Triacanthidae), goatfishes (Mullidae), catfish (*Arius thalassinus*; Ariidae), and sharks and rays (Pauly and Mathews 1986, IUCN 1987). Among crustaceans caught are slipper lobster (*Thenus orientalis*) and swimming crabs (*Portunis* spp.). Most of the by-catch is not marketable or has only a low value, especially when compared with shrimp. Consequently, the majority is discarded at sea, fishermen preferring to save ice for high value catch. Only a few prime species from the by-catch are generally retained, including groupers, snappers, jacks and barracuda. However, some could potentially form the basis of commercial fisheries in their own right. Pauly and Mathews (1986) estimated the maximum sustainable yield of *A. thalassinus* in Kuwaiti waters as 15,000–17,000 t y^{-1}, over ten times the yield of the shrimp fishery. However, most was discarded because this species is rarely consumed there. Substantial potential exists, therefore, for increasing fishery production by more rational use of the by-catch.

This pattern of by-catch wastage is aggravated on the Saudi Arabian coast where the long closed-season for shrimp fishing favours industrial trawlers over the artisanal sector. By the time the season opens, shrimp have moved well offshore and so are difficult to exploit by small artisanal vessels which have a smaller range. As the amount of catch they can store is substantially lower than the industrial boats, the necessity to work further offshore is likely to result in increased dumping of by-catch, even of relatively marketable species (Oakley in press b). Some other countries of the region, such as Iran and Kuwait, have designated areas close inshore for exclusive use by artisanal craft to help overcome this problem.

The by-catch may actually have some positive effects even though discarded.

Pauly and Mathews (1986) estimated that fin-fishes consumed three times more shrimp from Kuwaiti waters than were caught by the fishing fleet, estimated at approximately 6000 tonnes of shrimp eaten by fish per year. High mortality of shrimp predators caught incidental to shrimp trawling may reduce predation rates and thus increase the maximum sustainable yield. Furthermore, discarded by-catch may provide food sources for the prey of shrimps.

(2) Lobster

Lobster are typically associated with coral reefs or areas of hard substrata. There are substantial fisheries for them within the Red Sea and around the coast of southern Arabia. However, they have little importance within the Gulf. The main fishing grounds are shown in Figure 13.6. Two types of lobster are important, spiny lobsters of the genus *Panulirus,* and deep sea lobsters of the genus *Puerulus.* The former are caught extensively by artisanal fishermen on reefs throughout the Red Sea, Djibouti and on the Omani coast, whilst the latter form the basis of a commercial fishery in Yemen (Sanders 1981a, Crossland *et al.* 1987).

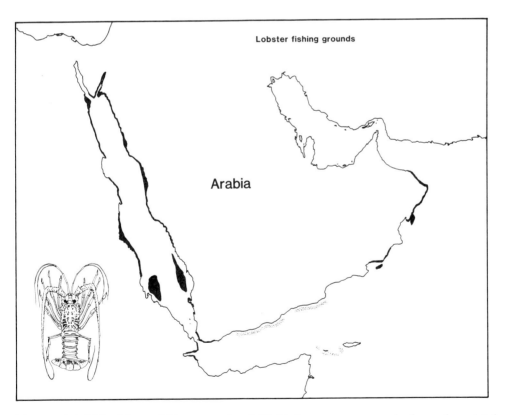

Figure 13.6 *Main lobster fishing grounds. Solid black areas are grounds for reef-associated* Panulirus *species, and stippled areas for the deep sea lobster* Puerulus sewelli. *Data for the latter are from Sanders (1981a).*

Shallow-water reef-associated species common in the Red Sea include the double-spined rock lobster *Panulirus penicillatus* which is common in clear, shallow areas, especially the reef-flat. This species is caught mainly at night by fishermen wading on the reef-flat with paraffin lamps. The painted rock lobster, *P. versicolor* is also common in water less than 10 m deep and is caught using the above method and by trapping. A third species, the ornate rock lobster *P. ornatus*, is rare but commonest in turbid coastal waters. There may be commercial stocks present especially in the shelf areas of the southern Red Sea. However, these have so far been little exploited (Crossland *et al.* 1987).

Productivity of reef-associated species is generally low throughout the region. Total catches reported from the Saudi Arabian coast came to only a few tonnes annually in the early 1970s (Neve and Al-Aiidy 1973) and there has been little expansion since. Red Sea populations have probably been heavily over-exploited in some areas, particularly in northern waters, even by relatively low levels of fishing effort. For example, in Sinai there has been recent expansion of the fishery by Bedouins to supply the burgeoning tourist industry. Demand is outstripping supply and the average sizes of lobsters appearing in hotels declined markedly during the late 1980s. Immature lobsters now constitute much of the catch, and females with eggs are often taken. Furthermore, small lobsters are often discarded on the beach rather than returned to the sea to grow further. This is compounded by hotels buying even very undersized lobster to decorate seafood dishes. Similar problems exist in Oman where hotel development and tourism have also contributed to heavy fishing of lobster (Sheppard 1986b).

At present the artisanal fishery is completely unmanaged and in Sinai lobster are probably recruitment overfished (fished to such low numbers that levels of recruitment are reduced), although data are completely lacking. In three years diving around the reefs of southern Sinai one of us (C.R.) observed virtually no lobster. Most of the catches now appear to come from the reefs surrounding offshore islands, with an important fishing area among the islands adjacent to the southern Gulf of Suez.

Fisheries for deep sea lobster, *Puerulus sewelli*, are much more productive. This species inhabits the continental slope at depths ranging from 200 to 600 m and is caught by trawling. It has been the subject of a detailed study by Sanders (1981a). There are two spawning seasons per year leading to two separate periods of recruitment. Growth is rapid and recruitment to the fishery takes place at around one year old. The average life-span is roughly four years. Fishing is seasonal, taking place from September or October to April, the rest of the year vessels being engaged in catching cuttlefish. Between 100 and $400\,t\,y^{-1}$ of tails have been landed from the fishery, and Sanders (1981a) predicted a maximum sustainable yield of approximately $200\,t\,y^{-1}$.

F. Mollusc fisheries

Fisheries for molluscs can be divided into two groups: demersal and pelagic. The former, based mainly on bivalves and gastropods, are exploited at an artisanal level, whilst the latter, based on cephalopods support industrial fisheries in some areas.

(1) Artisanal fisheries

Little published information exists on fisheries for demersal molluscs apart from pearl oysters. This is probably because their importance is primarily to local fishermen, with little commercial exploitation. Species of four genera are commonly fished within the region, mostly in coral reef areas: giant clams (*Tridacna* spp.), spider conch (*Lambis truncata*), *Strombus tricornis*, and top shell (*Tectus dentatus*). In some areas, other molluscs such as *Murex* spp. and oysters are fished, and the bivalve *Tapes* supports a modest commercial fishery in lakes of the Suez Canal. In addition, octopus are often important in catches from reefs. All of these are caught for food and the majority are either consumed by the fishermen and their families or used as bait. However, *Tectus dentatus* are also valued for mother of pearl, and are collected at a commercial scale at least in the Sudanese Red Sea (Head 1987c). Head also notes that the operculae of *S. tricornis* are used as a base for perfume manufacture. Although most of the exports of operculae are purported to come from the Gulf, average sales in the Sudan markets were estimated at the production from perhaps 18,000 shells per day (Schroeder 1981).

Most of the above are very easy to harvest since they are abundant in shallow water and are easy to find. Genera such as *Tridacna* and *Strombus* have been heavily over-exploited in other parts of the world, such as the Caribbean for *Strombus* and the Pacific for *Tridacna*. There is some evidence for depletion in the Red Sea, primarily of populations of *Tridacna* living on the reef flat in the vicinity of fishing villages. However, there are very healthy populations, at least of *Tridacna*, at depths of greater than 3 m even close to such villages. This is because fishermen as yet rarely dive to harvest them. Collection of molluscs is mainly a task for women and children, and has been restricted to wading depths.

(2) Commercial fisheries

Abalone and pearl oyster are the targets of commercial fisheries. Abalone are present in abundance only along the southeast coast of Oman (although there may be unexploited stocks further south), whilst pearl oysters are fished in limited areas of the Red Sea and the western Gulf.

A) ABALONE

Sanders (1982) made a preliminary assessment of the abalone fishery of Oman in 1981. Abalone, *Haliotis* sp. support a significant fishery with annual catches of around 200 tonnes in-shell weight. This was worth around US $1.3 million in 1981 and most was exported dried to the Far East after removal from the shell. Abalone inhabit rocky reef habitats in depths of up to 20 m. However, they are caught by divers, usually using masks, in depths to 7 or 9 m, by prising them from rocks with metal bars. The Omani coast has the only extensive area of suitable habitat for abalone in the Arabian region, these being more usually fished in temperate and warm-temperate waters (e.g. Beinssen 1979). Abalone feed on benthic algae, including large macroalgae, the tips of whose fronds are trapped under the shell prior to being grazed (Barratt 1984). Such algae are seasonal on the coast of Oman, with a massive increase in productivity from September to November following the upwelling brought on by the Southwest Monsoon (Barratt 1984).

The fishing season is limited to the period of calm weather from September to April; at other times the Southwest Monsoon causes rough seas. This has the effect of protecting stocks for a period of around four months, and divers indicate that abalone are around three times as abundant at the beginning of the season as at the end. Abalone recruit to the fishery at an age of approximately 16 months old. Prior to this they remain in crevices and fissures inaccessible to divers. With seasonal limitations, this appears to have so far prevented over-exploitation of stocks.

There have been suggestions to increase catches in the fishery by introduction of diving gear such as the hookah. Sanders (1982) estimated that the current fishery yield could be realized with about one-sixtieth of the manpower used in 1981 with such gear. However, he warned that gear improvements would have serious socio-economic impacts. Equipping all divers with improved gear would result in excessive fishing effort and declining yields. Equipping only a proportion would bring the two groups into conflict since those with diving equipment may be expected to obtain considerably greater fishing revenue than those using artisanal methods. Given the several thousand people engaged in the fishery it was recommended that no improvements be made and that legislation be passed to prevent the use of diving gear, at least until a more detailed assessment had been made of the potential of the fishery.

B) PEARL OYSTERS

Pearls have been fished in the Arabian area since ancient times, and in the past were an important part of the economy of the areas in which they occurred. Wellsted (1838) estimated the value of the pearl fishery at £400,000 sterling at the time (perhaps £16 million today), employing over 30,000 people at the height of the season. However, competition from the cultured pearl industry in the Orient led to the decline of the fishery towards the middle of this century. The main fishing grounds were in Sudan and the western Gulf, although there was some harvest from other areas, such as the Farasan Bank of the southern Red Sea. Wellsted (1838) calculated that on average Bahrain produced 3500 pearls per season, the Persian coast (Kuwait, Iraq and Iran) only 100 and the area between Bahrain and Musandam 700. Although there are several species present in the region, two represent the mainstay of commercial fishing in both the Red Sea and Gulf: *Pinctada margaritifera* and *P. radiata*.

Pearl oysters have an interesting life history. Unlike most hard-substrate bivalves, they are capable of some degree of movement. Basson *et al.* (1977) have described the life history of *Pinctada* in the western Gulf. Settlement takes place mainly into seagrass beds in late spring (April) although also on algae. Juvenile oysters attach to macrophytes, sometimes in huge numbers, where they develop for a period of several months until around 5 mm diameter. As the seagrass blades begin to detach in October, the young oysters move to other blades which are still attached, leading to very dense concentrations. Eventually even these blades detach and the oysters are carried down-slope into deeper water. They attach to hard substrata wherever this is encountered.

In the Red Sea, the main fishing area was in Dunganab Bay on the Sudanese coast. The English biologist Cyril Crossland established a pearl oyster farm there in 1905 which operated successfully until 1969 when a mass mortality wiped out most of the stock (Head 1987c). There have been recent attempts to revive the pearl fishery, especially in Bahrain. This has partly been supported by an increase in popularity of natural pearls but the revival has also had a cultural basis. Pearl fishing is perceived to be part of the heritage of Gulf States and moves have been made to prevent its disappearance.

(3) Cephalopods

There are significant fisheries for cephalopods, such as squid and cuttlefish, in the region. These animals usually live close to the bottom and are mainly captured by trawling. Consequently, the fishing grounds are located in areas suitable for trawling, including the Gulf of Suez, the shelf areas of the southern Red Sea (eastern coast), the south Yemen coast in the Gulf of Aden, and in the Gulf. Cephalopods are also caught on a small scale in the lakes of the Suez Canal. In most of these areas, they are captured primarily as a by-catch from fishing operations directed at other species, such as shrimp. However, on the south Yemen coast there is a well-developed fishery for cuttlefish, *Sepia pharaonis* (Sanders and Bouhlel 1982).

S. pharaonis occur to depths of 110 m but are most abundant in water less than 40 m deep. Off the Yemen coast there are two peaks in spawning, one in spring (March to April), the other in autumn (August to October) (Sanders 1981b). Spawning takes place in shallow water close inshore following migration from offshore feeding areas. The eggs are laid on the bottom in clusters (Roper *et al.* 1984). They feed on crustaceans and small demersal fishes. This species once supported catches of 3000–10,000 t y^{-1} from the Yemen grounds during the 1970s (Sanders 1981b). Sanders and Bouhlel (1982) estimated a fishery of 5000 tonnes in 1982 but stocks collapsed at about this time with the 1981 catch being estimated at only 900 tonnes (Roper *et al.* 1984). This was probably due to overfishing.

As a result of the decline in stocks, Sanders (1984) recommended that there be a 50% reduction in fishing effort. This was partly achieved by the banning of Japanese trawlers from Yemen waters in 1981 (Roper *et al.* 1984). Sanders also recommended that the size at first capture be increased and that trawling should not be permitted in depths of less than 30 m during the autumn spawning period. Since then there has been some stock recovery.

G. Other fisheries

(1) Aquarium fish

The region holds considerable potential for the establishment of aquarium fisheries but at present there are very few facilities operating. Egypt has a fishery based at El Tur on the eastern Gulf of Suez coast, representing one of the only sources of Red Sea fish for the world market. Unfortunately, it has been poorly managed to date. The collecting methods used there are crude and cause habitat destruction. Techniques used include driving fish into barrier nets using weighted scare lines and removal of coral heads containing fish. Most fish arrive at the holding tanks in poor condition leading to high mortalities. Hence there is much scope for improvement and so increased profitability.

There have been recent moves to establish another collection facility at Sharm-el-Sheikh in southern Sinai by a joint Egyptian–German venture. The proposed methods of capture and holding of fish were a big advance on those used at El Tur. However, the proposals called for too large a harvest, and there were potential conflicts with the rapidly expanding tourist industry in the area and so plans were shelved. Plans were also made for a Saudi aquarium trade, but this did not progress, partly because such a trade was not considered to be in the country's best interests, or in keeping with its wildlife policy.

In 1987 an assessment was made by FAO of the potential for development of an aquarium fishery in Djibouti (Barratt and Medley 1990). At the time Djibouti had no exports and such a fishery was perceived as a way of developing the economy. The study recommended that there was significant potential for an aquarium fish industry there, providing that it was set up under strict management guidelines to protect the resource base. Barratt and Medley (1990) concluded that a fishery would stand the best chance of competing with established fisheries such as those in the Maldives or Philippines by gaining a reputation for supply of high quality fish. Many other countries of the region could benefit from the expanding trade in aquarium fishes by using a similar approach.

(2) Ornamental shells and corals

Corals and shells are not at present fished commercially in the region, although supporting significant export industries in other parts of the world (Wood and Wells 1989). The only area in which they are harvested in any quantity is Hurghada, a tourist centre on the Egyptian Red Sea coast. Here marine curios are collected and sold locally, primarily to Egyptian tourists. Such a cottage industry has been operating there since early in the century. A similar fishery has not become established anywhere else in Egypt, despite the recent expansion of coastal tourism. This has probably been mainly due to increased environmental awareness by hotel and diving companies operating within the country, and among the tourists visiting resorts.

H. Prospects for the future

Fishery production from the Arabian region has expanded enormously in the last 30 years. There have been problems, however, stemming predominantly from over-exploitation of certain stocks, for example shrimps in the Gulf and cuttlefish in the Gulf of Aden. Over-exploitation has tended to arise where development of the fishery has proceeded more rapidly than legislation to control it. A further problem has been a lack of adequate data on the status of stocks, or the biology of fished species, with which to design and implement proper management controls. Countries throughout the region have been developing fisheries programmes for collection of the necessary data and to manage fisheries. However, regulatory structures are still poorly developed in some, such as Ethiopia and Iran.

The fishermen maintain that there is not the same abundance of fish as formerly.
Klunzinger 1878

Many of the fish stocks of the region are shared among a number of countries, particularly those of migratory pelagic species, such as tuna and mackerel. Consequently, management must be coordinated among states using such resources. Multi-national coordination of data collection and fishery management are still at an early stage of development. However, cooperation among states is essential to safeguard future yields.

Two further problems are of much concern: loss or degradation of habitat and pollution. It was mentioned earlier that loss of habitat, due to coastal infilling on the western Gulf coast may have contributed to the decline in yields of shrimp. There are many other species which may be dependent to a greater or lesser extent on similar critical habitats now under threat from development. Extensive landfill of shallow-water biotopes has already taken place on the Gulf coast, with perhaps 40% of the Saudi Gulf coastline affected in this way (IUCN 1987). Further loss of habitat may be expected to result in deleterious effects on fisheries production (Oakley in press a).

Pollution is a particularly serious threat to fisheries of the region. The enclosed nature of the Red Sea and Gulf lead to very low rates of turnover of water. The turnover time of Gulf waters is estimated at 3–5 years (Hunter 1982) whilst that of the Red Sea is likely to be much longer. Consequently, pollutants released into the marine environment will tend to be retained. The massive input of oil into the Gulf during the Iran–Iraq war caused significant increases in the levels of hydrocarbon present in the water column and in the sediments of the seabed. Following this war, Gulf waters had the highest concentrations of petroleum hydrocarbons of any sea in the world (Emara 1990).

The recent massive inputs of oil into the Gulf during the Gulf war will have further increased pollution levels. Shallow inshore habitats suffered the greatest amount of damage due to this oil spill, with large areas of the western coast badly oiled. The spill occurred during the spawning season for the most important shrimp species, *Penaeus semisulcatus,* and some important nursery areas have been seriously polluted. At the time of writing it is too early to say whether shrimp stocks will suffer a decline due to the oil, but the possibility certainly arouses concern.

Oil pollution on this scale will not only have toxic effects on the species themselves. It may also reduce the palatability of marine organisms harvested. In addition, it is likely to interfere with the harvesting process itself. Much of the Gulf fishery, especially in inshore waters, is based on trawling; sunken oil may cause serious fouling of nets and traps.

Levels of oil pollution are also high in the Gulf of Suez, one of the most productive fishing areas within the Red Sea (Dicks 1987). Continued inputs of oil and other pollutants into the seas of the region, especially the Gulf, may compromise the future of fisheries.

Summary

Overall levels of fisheries productivity vary greatly throughout the Arabian region. This is a result of differences in levels of primary production, catchability of stocks and fishing effort. Primary productivity depends heavily on levels of nutrients within the photic zone. There are four main sources of nutrient input in the region: upwelling off southern Arabia, resuspension of bottom sediments in shallow basins, nitrogen fixation by blue-green algae on reefs and mud-flats and terrestrial runoff (mostly in the Gulf).

A wide variety of fin-fishes, crustaceans and molluscs are exploited in Arabian waters. Fisheries production within the region has expanded greatly over the past 30 years. The most productive areas lie in the Gulf of Suez and Foul Bay in the northern Red Sea where there are important purse seine and trawl fisheries; in the southern Red Sea where shallow shelf areas support trawling and beach seine fisheries for demersal species; in the Arabian Sea, where seasonal upwelling results in high productivity of pelagic fishes; and in the western Gulf where there are important shrimp fisheries. Soft bottom habitats, including areas of sand, mud and seagrass bed support the bulk of demersal fisheries production in the region. Although highly productive in terms of carbon fixation, coral reefs of the region do not support very productive fisheries. This is largely a result of targeting predatory species high in the food chain and of difficulties in large-scale harvesting and transport of catch to markets. However, reef fisheries do provide an important source of protein for coastal communities, especially in remote areas. Compared with many other parts of the world, the reefs of Arabia are lightly exploited.

Prospects for increasing overall fishery production in the region are limited Yield of reef fisheries could be substantially increased by targeting lower down the food chain. In the Arabian Sea, large stocks of mesopelagic fishes have been identified but harvesting is not commercially viable using present methods. There is still some potential for developing trawl fisheries in the southern Red Sea particularly off the Ethiopian coast. Greater opportunities exist for developing fisheries for ornamental species. At present only small-scale operations exist in two or three countries.

Fisheries management in the region is in its early stages. A sharp decline in yield of Gulf shrimp fisheries during the 1970s, thought to have been due to a combination of overexploitation and destruction of nursery areas, indicated the need for management. Management efforts began in the early 1970s with imposition of closed seasons by countries bordering the western Gulf area. Catches have subsequently improved substantially. Similarly, collapse of the Yemen cuttlefish fishery in the early 1980s led to the introduction of management measures to reduce fishing effort, again with subsequent recovery in catches.

Fisheries management in the Arabian region in general is seriously hampered by lack of data on most stocks. Stocks of some species are being heavily overexploited but still have no form of management, for example spiny lobsters in the northern Red Sea and Oman. Since many stocks are shared among several countries, multinational coordination of data collection and management are essential to safeguard future yields.

Two potentially serious problems facing countries bordering Arabian seas are increasing pollution and degradation or loss of nursery areas. Both could cause declines in yield. The recent oil spillage during the 1991 Gulf war may have seriously damaged inshore nursery areas of shrimp during their peak spawning period. However, reductions in fishing effort during and subsequent to the war may have helped protect stocks from serious declines as a result.

CHAPTER 14

Human Uses and Environmental Pressures

Contents

A. **Human activities and uses**   288
  (1) Introduction   288
  (2) Major coastal and marine uses   289
B. **Environmental pressures and effects**   289
  (1) Short-term to medium-term impacts   290
  (2) Medium-term to long-term impacts   296
  (3) Possible longer-term impacts   300
  (4) The 1991 Gulf war   302
Summary   306

A. Human activities and uses

(1) Introduction

For millennia people have been attracted to the shores of the Arabian peninsula. An early maritime civilization, Dilmun, encompassing what is now Bahrain and the eastern coast of Saudi Arabia, prospered 4000–5000 years ago. Even by the tenth century AD, the Arabs had established a trade network extending from the Gulf as far eastwards as China. This was possible through their sophisticated knowledge of astronomy and navigation. Cargoes of textiles and spices were accompanied by the exchange of new ideas, science and religion.

The foundations of life and development in the region remain firmly based on renewable and non-renewable natural resources. These contribute to the food, transport, industrial, recreational and other needs of local people. The Gulf and the Red Sea play a particularly vital role in providing most of the population with fresh water via desalination plants. The fisheries are a multi-million dollar industry, and the artisanal sector, in particular, is socially significant (IUCN 1987). Discovery of oil in the Gulf during the 1930s and 1940s was largely responsible for the immense economic wealth and strategic importance associated with much of the Arabian region today.

Table 14.1 Coastal and marine uses and major environmental pressures.

Coastal and marine uses	Actual or potential environmental pressures
Shipping and transport	
Shipping	Oil spills; anchor damage
Ports	Coastal reclamation and habitat loss; dredging, sedimentation; oil and other pollution
Residential and commercial developments	Coastal reclamation and habitat loss; dredging, sedimentation; sewage, fertilizer and other effluents; eutrophication; solid waste disposal
Industrial development	
Oil and petrochemical industry	Oil, refinery and other effluents containing heavy metals; drilling muds and tailings; air pollution
Mining	Sedimentation and elevated heavy metal levels
Desalination and sea water treatment plants	Effluents with elevated temperatures, salinities and sometimes heavy metals and other chemicals
Power plants	Various effluents; air pollution, increasing greenhouse gases and global warming; acid deposition
Fishing and collecting (including dynamiting)	Population decline of target and non-target species, habitat degradation, and changed species composition of fish, shrimp and other biota
Recreation and tourism	Decline in "curio" species; reef degradation; anchor damage; eutrophication (e.g. from resorts and hotels)
War-related activities	Multiple pressures and ecological effects (direct marine and atmospheric)

(2) Major coastal and marine uses

Major coastal and marine uses of the Arabian region are summarized in Table 14.1. Also shown are some of the resulting environmental pressures, which are considered in the next section. Particular pressures may arise from a variety of different human uses. Uses and environmental pressures are sometimes considered together, since the two clearly are highly interconnected. Agriculture and related uses are mainly terrestrial rather than coastal in origin, and are not shown in Table 14.1. Nevertheless, effects from these activities have already significantly impinged on some coastal environments of the Arabian region (section B2h). Uses of the coastal zone and offshore areas have increased dramatically over recent years, and heavy use is also likely in future.

B. Environmental pressures and effects

The effects of prolonged human use of the coastal and marine environment of the Arabian region are multiple, complex and in many cases not well understood

(Table 14.1). For convenience these have been divided, qualitatively, into short-term, medium-term and long-term impacts, according to the known or perceived immediacy of effect, duration of effect, and likelihood of successful ecological recovery under appropriate management. Hence, an impact such as nutrient enrichment of a bay, unless the source is continuous, is likely to have less far-reaching ecological effects than the irreversible loss from coastal infilling of a seagrass bed, or the prolonged effects of global warming. However, in some instances the ecological effects of impacts may well operate over both the short-term and long-term time. For instance, an oil spill that is chronic, or acute and heavy (as in the case of the 1991 Gulf war), may have effects extending beyond the short-term. Nevertheless, the divisions provide a convenient framework into which the array of impacts affecting the coastal and marine environment may be placed.

Most human activities and impacts have clearly resulted in environmental degradation, or even complete loss of coastal habitats. However, in some cases creation of new structures (e.g. platforms and piers) and new sedimentary environments may actually have ecological benefits. Environmental pressures in the Gulf are generally much greater than elsewhere in the Arabian region.

(1) Short-term to medium-term impacts

A) OIL POLLUTION

i) Overall extent and impact of oil operations

The Middle East harbours more than half of the world's proven oil reserves, thereby ranking as the world's largest oil production area (30%) (Dicks 1987). Also, the Safaniya area of the Saudi Gulf is the largest known offshore oil field in the world. Most oil produced is exported, via sea or pipeline, while local refining and consumption constitutes less than 10% of total production (Dicks 1987). The Arabian seas are therefore among the busiest tanker routes. For instance, at the Strait of Hormuz there are 20,000–35,000 individual tanker passages annually (Linden *et al.* 1990).

For these reasons, reports of widepread pollution in the Red Sea (Wennink and Nelson Smith 1977, 1979, Dicks 1987), and particularly in the Gulf region (Wennink and Nelson Smith 1977, 1979, Fowler 1985, Price *et al.* 1987c, Coles and Gunay 1989), are not surprising. Summarized data (Table 14.2) suggest that whereas in the Gulf most inputs of oil to the marine environment originate from tanker and ship traffic, inputs in the rest of the world's seas originate principally from refining, industrial and urban sources. Comparable data are not available for the Red Sea, although the pattern for this region is probably similar to that for the Gulf (Dicks 1987). Inputs from tanker and ship traffic originate primarily from the discharge of dirty ballast water and other oily water, which during 1986 ranged from 400,000 to 750,000 tonnes in 1986 in the Gulf (Linden *et al.* 1990). Well blow-outs during drilling, and as a consequence of war activities, can also be a major source of oil pollution. The Nowruz blow-out in 1983, resulting from hostilities between Iraq and Iran, was the longest recorded in the region. It lasted for nearly

Table 14.2 Sources of oil pollution and estimated magnitude (as percentage of total) for the Gulf vs. rest of the world.

Source of oil	Gulf Neumann (1979)	World Golob (1980)	Cowell (1978)
Tanker and ship traffic (spills and routine discharges)	86%	58%	23%
Offshore production and natural seepages (including drilling muds)	14%	22%	14%
Refining, industrial and urban	<0.12%	20%	63%

eight months, during which an estimated half a million barrels leaked into the Gulf (Linden *et al.* 1990). Oil and other pollution arising from the 1991 Gulf war are discussed below (section B4). Oil also reaches the environment from terrestrial sources and natural seepages.

Outside the Gulf, the marine environment is also subjected to oil spills (Burns *et al.* 1982, Salm *et al.* 1988), and in the Gulf of Oman surface tar balls are common (Price and Nelson Smith 1986). Even in the Red Sea, oil pollution has been widespread for at least 10–15 years, dispelling the notion that the region as a whole is relatively unpolluted (Wennink and Nelson Smith 1977, 1979). The Gulf of Suez has been most affected (Dicks 1987).

ii) Oil in water, sediments and biota

Elevated concentrations of petroleum hydrocarbons in sediments (up to $3950 \, \mu g \, g^{-1}$ dry wt), and bivalves ($>500 \, \mu g \, g^{-1}$ dry wt) have been reported around some industrial areas, although comparison between different localities is problematic (see Linden *et al.* 1990). Relatively high levels in sea water ($546.4 \, \mu g \, l^{-1}$) occur in some industrial areas (Samra *et al.* 1986). Nevertheless, petroleum hydrocarbon levels in water, sediments or biota are not as high as might be expected in and outside the Gulf, considering the frequency of oil spillages (see Coles and McCain 1990, Linden *et al.* 1990). Compared with beaches, the generally low levels of petroleum hydrocarbons in the offshore environment might be explained by their rapid breakdown, as a result of intense solar radiation and high summer temperatures. This is undoubtedly enhanced by biological processes, since large numbers of bacteria which degrade oil hydrocarbons have been found in sea water samples from the region (Diab and Al-Ghonaime 1985, Linden *et al.* 1990). In the Red Sea, it appears that petroleum hydrocarbons may be relatively high in the Gulf of Suez, and high levels in biota are also apparent in some areas (UNEP 1987).

iii) Beach oil

Beach oil is widespread, often in high concentrations, throughout much of the Arabian region (e.g. Wennink and Nelson Smith 1977, 1979, Burns *et al.* 1982, Dicks 1987, Linden *et al.* 1990). This contrasts with levels generally found in the water column, sediments and biota (section ii above). During a recent survey of the western Gulf, tar balls were encountered at 77% of 53 coastal sites (Price *et al.* 1987c). Using a semi-quantitative scale of 0 (no oil) to 6 (extensive tar pavements), the modal value recorded was 3, denoting moderate amounts of tar at most sites. During a quantitative survey, tar concentrations of $1–10\,kg\,m^{-1}$ of the Gulf shoreline were frequent, and concentrations exceeding $10\,kg\,m^{-1}$ also commonly occurred (Coles and Gunay 1989). Tar levels ranged up to 100 times the upper values reported for other world regions, and up to 10 times values previously reported for other areas in the Gulf. Beach tar is also widespread in the Red Sea. Along the coast of North Yemen it was recorded at 58% of 131 coastal sites (Barratt *et al.* 1987).

iv) Ecological effects of oil

The effects of oil on biota arise from mechanical smothering and from the presence of toxic substances (see Nelson Smith 1984, Clark 1989). The overall severity of effects depends on the nature and quantity of oil spilled, in conjunction with factors such as wind, water movements, temperature and probably also salinity. Animals at particular risk include surface swimmers and feeders, such as waders and seabirds, marine reptiles and marine mammals. Although there is still controversy about ecological effects of oil in the marine environment, many of the world's major oil spills have had long-term consequences (Vandermeulen 1982). On the other hand, recovery after oil spillages can begin remarkably quickly. Even major population declines can be compensated for by increased reproduction during subsequent years (Clark 1989, Baker *et al.* 1990).

Despite the oil industry in the Gulf, the response of marine ecosystems to oil pollution is still not well understood. Gross effects of the Ras Tanura oil spill and Hasbah 6 blow-out in 1980 included heavy casualties on birds, such as the Socotra cormorant which is endemic to the Arabian peninsula, herons and other waders (see Linden *et al.* 1990). In Bahrain alone several thousand birds were estimated to have died. Up to 100% mortality of littoral invertebrates in Bahrain and Qatar also occurred from this event, and mortality of marine turtles and sea snakes was reported. Reef corals and other marine ecosystems in the Gulf may be affected as much by chronic oiling as by infrequent, heavy spillages. Experiments using oil and dispersant on Gulf corals (Legore *et al.* 1983) suggested some behavioural changes and delayed effects, but growth was not affected. Assessing the effects of oil (and other impacts) in the Gulf is problematic (Nelson Smith 1984, Price and Sheppard 1991). Whether or not the widespread animal deaths reported on Gulf beaches at the time of the Nowruz blow-out (1983) resulted from oil pollution has never been resolved satisfactorily (Linden *et al.* 1990). Similarly, the approximate synchrony between a fractured pipeline in the Gulf (Spooner 1970a,b) and decline in shrimp fisheries was apparently coincidental (Nelson Smith 1984).

Few assessments of the effects of oil on community structure in the Gulf have been undertaken. In a recent study along the western Gulf, only one organism, the polychaete *Perielectroma zebra*, showed a significant relationship with sediment petroleum hydrocarbons (Coles and McCain 1990). These authors showed that its abundance increased significantly with sediment hydrocarbons within the relatively low concentrations recorded. This species was also primarily reponsible for the observed relationships between numbers of individuals, species diversity and petroleum hydrocarbons. It is suggested that oil residues may reach the shore without major contamination of the offshore benthos (Coles and McCain 1990).

The effects of oil pollution on ecosystems and species groups are better understood in the Red Sea (Dicks 1987). In an extreme instance of oiling on a reef in the Gulf of Aqaba in 1970, a colonization failure (e.g. by *Stylophora pistillata*) was attributed principally to oiling. However, the additional influence of sewage and phosphate eutrophication in a nearby lagoon complicated this. Subsequent research in the Red Sea has shown more clearly the negative effects of oil on reproduction (Loya 1975, Rinkevich and Loya 1979). Oil-based drilling muds are known to affect corals and other ecosystems, and may constitute an increasing problem in both the Red Sea (Dicks 1987) and the Gulf (Linden *et al.* 1990). Drilling also produces rock cuttings (sediments) which, together with drilling muds, can produce various biological responses (Szmant-Froelich *et al.* 1981, Dicks 1987). In general, however, there is little clear evidence for significant deterioration of reefs due to oil pollution in the Red Sea (IUCN/UNEP 1985b). Studies on oiled mangroves (*Avicennia marina*) along the Egyptian Red Sea coast revealed that mangrove survival can be high in coarse, well-drained, oxygenated sediments, even with the pneumatophores and substratum completely coated with oil, but not in fine-grained anaerobic muds (Dicks 1986, 1987).

Ecological effects of oil on habitats such as seagrasses, seabed sediments and open waters of the Red Sea are not known in detail. Results of laboratory tests on several macrofaunal species of coelenterates, molluscs, crustaceans, echinoderms and fish (Eisler 1973, 1975, Dicks 1987) indicate a variety of lethal and sub-lethal effects. However, these findings cannot readily be used to predict effects in the wild (Dicks 1987). Little information is available on the effects of oil pollution on the fisheries of the Red Sea or other parts of the Arabian region. However, concern arises from the known vulnerability of egg, larval and juvenile stages of crustaceans and fish. Spawning and nursery areas undoubtedly occur in many shallow coastal areas. This is certainly true for penaeid shrimp (e.g. *Penaeus semisulcatus*) in the Gulf, where spawning and nursery areas have been located in shallow waters of less than 8 m depth (Basson *et al.* 1977, Price 1979, 1982b, Sheppard and Price 1991). It is often these shallow coastal waters that are exposed to the oil pollution and other coastal pressures.

On the coast of Oman oil is threatening breeding seabirds on the Kuria Muria Islands, aside from contamination of recreational beaches (Salm and Jensen 1989). However, effects of oil on marine fauna of the region are not known in detail.

B) WASTEWATER POLLUTION

Numerous sources contribute to wastewater pollution in the region, which is probably greatest in the Gulf (see IUCN/UNEP 1985a, IUCN 1987) and in some parts of the Red Sea (IUCN/UNEP 1985b). Future problems might also arise from increasing development in other parts of the region (Salm *et al.* 1988). Of domestic and urban sources, untreated sewage and abattoir wastes are discharged at several localities. For instance, in the Saudi Arabian towns of Al-Khobar, Al-Hasa and Al-Qatif daily inputs of raw sewage into the Gulf may be as high as $40,000\,m^3$, equivalent to production by 175,000 persons (IUCN 1987). Effects include unsightly solids and grease mats, local eutrophication, increasing biological oxygen demand (BOD) and algal blooms. Sewage treatment plants are helping to alleviate some of these problems. Domestic wastes and pollution are becoming an increasing problem in Oman (Salm *et al.* 1988).

Among the industrial wastewater inputs are desalination effluents, wastewater from fertilizer plants (e.g. urea) and refinery and other industrial effluents containing heavy metals. Elevated concentrations of lead, mercury and copper in bivalves and fish have been reported around some ports and industrial areas. High levels of cadmium, zinc and vanadium have been found in sediments (see Linden *et al.* 1990). However, a recent study revealed that copper and vanadium levels, although comparable to values elsewhere in the Gulf, were an order of magnitude lower than heavily polluted areas in some other parts of the world (Coles and McCain 1990). High concentrations of heavy metals are reported near some industrial areas of the Red Sea, such as desalination plants (Dicks 1987, UNEP 1987), whose effluents also are normally above ambient temperature (+5°C) and salinity (+3 ppt: Linden *et al.* 1990). Chlorine is also used as an antifouling agent in various industrial plants using sea water. Long-term ecological effects of effluents may be considerable locally, but probably only minor over larger areas, at least in the Gulf (Linden *et al.* 1990).

While the occurrence and consequences of wastewater pollution, such as coastal algal blooms, are documented for some areas, there are still major gaps in knowledge. It is not certain whether the increasing prevalence of offshore algal blooms, jellyfish and "red tides" in the Gulf are primarily signals of large-scale eutrophication, or more the effects of natural cycles (Linden *et al.* 1990). Studies during the late 1970s revealed significant increasing diversity of diatoms (Basson and Hardy, unpublished), but significant decrease in benthic faunal diversity (Burchard, unpublished) towards an oil refinery outfall. Other studies in the Gulf (Coles and McCain 1990) do not suggest major effects of heavy metals on marine biota. However, further sampling is needed, particularly in view of the extreme spatial and temporal heterogeneity in biophysical conditions in many regions (Price 1979, 1982b).

C) SOLID WASTE POLLUTION

Throughout much of the region the coastal zone has become a repository for large quantities of industrial, commercial and residential trash and other solid waste (IUCN 1987). Often this takes the form of plastics, metal containers, wood, tyres, and even entire scrapped automobiles at some localities (IUCN 1982, 1987). In the Gulf, oil sludges constitute the most important type, in terms of quantity, of solid waste (Linden *et al.* 1990). Much of the lighter debris has become spread along widespread tracts of shoreline through wind and water movements. During a recent survey, solid waste was encountered at 87% of 53 sites inspected along the Gulf coast (Price *et al.* 1987c, Price 1990). Using a semi-quantitative scale of 0 (no impact) to 6 (greatest impact), the modal value was 3, denoting moderate amounts at most sites. In the southern Red Sea, such refuse was reported at 67% of 131 coastal sites (Barratt *et al.* 1987).

In recreational areas, solid waste can have ecological as well as aesthetic consequences. In areas containing extensive metal and industrial debris, the potential exists for toxic substances to leach into the marine environment. Wooden pallets and driftwood may form a physical barricade to female turtles crawling up beaches to nest. Further, if such debris becomes impacted by an oil slick, the problem becomes compounded, and also increases dramatically the cost of any future oil clean-up operations. For this reason, several offshore coral islands in the Gulf, where turtles and birds nest in high densities, were cleared of debris shortly after the 1991 war (see Chapter 15). On the Red Sea coast, Ras Baridi is one of the few mainland nesting sites for the Green turtle (*Chelonia mydas*). It is also the site of a cement factory and dumping ground for various wastes (UNEP 1987). In Oman, solid wastes are also becoming an ecological problem on habitats such as reefs, lagoons, and mangroves (Salm and Dobbin 1986, Salm and Jensen 1989, IUCN 1987).

D) MINING AND OTHER INDUSTRIAL INPUTS

Deep trenches (c. 2000 m) of the Red Sea are associated with rich deposits of metalliferous muds in association with hot brines. The brines are greatly enriched with manganese, iron, zinc, cadmium and copper, compared with overlying waters. Pilot schemes for extraction have been undertaken. If extensive deep sea mining takes place, the potential clearly exists for environmental problems, particularly concerning heavy metals and waste sediments (e.g. silts and clays). It is predicted that a commercial venture might produce c. 270,000 tonnes of waste tailings, mixtures of mud and warm brine, as well as various treatment chemicals (Dicks 1987). Additional information on resource and environmental aspects of deep sea brines is given by Thiel *et al.* (1986), Dicks (1987) and Karbe (1987). In parts of Oman, sand mining is reducing the suitability of beaches for nesting turtles, and also increases the likelihood of coastal erosion (Salm *et al.* 1988).

In parts of the northern Red Sea inputs of phosphate, manganese and bauxite minerals, through loading onto ships, are major pollutants (Fishelson 1973a, Walker and Ormond 1982, Dicks 1987). In the Gulf of Aqaba, death of corals (*Stylophora pistillata*) was observed to be four to five times greater in an area of phosphate spillage as in a control area (Walker and Ormond 1982). However, the picture was complicated by the influence of additional factors such as sewage input. It appeared that reduced light, inhibition of calcification (by phosphate) and increasing sedimentation together contributed to degradation of corals in the polluted area. Enhanced algal growth from nutrient enrichment was probably not a direct cause of coral mortality, but appeared to increase sediment loads in the areas through increased sediment trapping.

(2) Medium-term to long-term impacts

A) COASTAL RECLAMATION AND HABITAT LOSS

Together with dredging, coastal reclamation probably represents one of the most significant impacts on the coastal and marine environment of the Arabian region (Linden *et al.* 1990). Reclamation has been undertaken for residential developments, ports, bridges, causeways, corniche roads and other purposes. Favoured areas often have included intertidal flats often with mangroves, shallow embayments and other biologically productive areas, whose true bioeconomic value is seldom recognized by developers. Coastal development and infilling has been far greater along the Gulf coast than in the Red Sea or other parts of the Arabian region, where its occurrence is more localized (IUCN 1987). Approximately 40% of the Saudi Arabian Gulf coast has been developed (IUCN 1987, Sheppard and Price 1991), involving extensive infilling and reclamation. Only an estimated 4 km^2 of mangroves now remain along these shores (Price *et al.* 1987a). A similar situation has arisen in other Gulf states, such as Kuwait and Bahrain. More than 30 km^2 (3036 ha) of Bahrain is now reclaimed or artificial land (Madany *et al.* 1987). This has also involved much loss of original intertidal ecosystems such as mangroves. Further, there have been plans for infilling on an area of almost 200 km^2 in Bahrain (Linden *et al.* 1990). In the Red Sea, infilling has taken place place principally around industrial centres, such as Jeddah, Yanbu, Rabigh and Jizan on the Saudi Arabian coast.

Apart from the direct and permanent loss of habitat, landfill usually increases sedimentation. This may directly smother habitats, or may limit photosynthesis of communities such as algal mats, seagrasses and coral reefs (IUCN/UNEP 1985a,b, Vousden and Price 1985, IUCN 1987, Ormond 1987). Whether this has had measurable effects on the fisheries is not known, but nevertheless has caused concern (IUCN/UNEP 1985a, IUCN 1987, Sheppard and Price 1991). Aside from ecological concerns over habitat degradation, urban growth is encroaching on some archaeological sites in Oman, which date back 6000–7000 years (Salm and Dobbin 1986, 1987, Salm *et al.* 1988).

B) DREDGING AND SEDIMENTATION

Dredging provides much of the infill material needed for coastal reclamation, hence the two activities often occur simultaneously. The former also takes place to deepen shipping channels and harbours. Like landfill, dredging has taken place most extensively in the Gulf. As a result of projects in Jubail and Dammam, Saudi Arabia, an estimated 46.5 km^2 of coastal habitats have been dredged; and for landfilling the residential and industrial areas of modern Jubail City, more than 200 million m^3 of sediments adjacent to the development site (IUCN, 1987) were removed. Dredging has also been extensive in Gulf States such as Bahrain, where $15–20 \text{ km}^2$ of shallow coastal waters and habitats have been dredged during recent decades (Linden *et al.* 1990). During construction of the Saudi–Bahrain causeway, nearly 60 million m³ of mud and sand were dredged (Linden *et al.* 1990). In the Red Sea, dredging has also been most prevalent in commercial and industrial areas, in particular Jeddah and Yanbu in Saudi Arabia. Most other parts of SE Arabia have been affected less, but dredging is becoming an environmental problem in some areas (Salm *et al.* 1988). Sedimentation can also result from other factors including propeller action of large ships over shallow, sedimentary areas.

The ecological effects of dredging are similar to those described above for coastal reclamation, and include both direct habitat loss and various secondary effects (Price *et al.* 1983, IUCN/UNEP 1985a,b, IUCN 1987, Vousden and Price 1985, Madany *et al.* 1987, Linden *el al.* 1990). In parts of the Gulf, sedimentation stimulated by dredging has created new soft substrate feeding areas for some species such as waders (IUCN 1987). Some fauna (e.g birds) may have benefited from new habitats created by sedimentation, for instance in Tarut Bay, Saudi Arabia (IUCN 1987). In general, however, ecological effects are undoubtedly more adverse than beneficial.

C) POSITIONING OF RESIDENTIAL AND INDUSTRIAL STRUCTURES

Construction of jetties, bridges, piers, shore residences, oil platforms and other artificial structures has often involved some loss of natural habitat, as in the case of the Saudi–Bahrain causeway. Secondary effects include increased sedimentation (Price *et al.* 1983). On the other hand, the presence of a solid structure in an area otherwise often devoid of hard substratum can be beneficial ecologically (Vousden and Price 1985). Such structures rapidly become colonized by sponges, hydroids, bryozoans and other encrusting biota, which in turn provide food or refuges for larger organisms such as fish (Basson *et al.* 1977). For species with only a short larval duration, the presence of oil platforms can aid dispersion and hence increase distributional range. However, growth of fouling organisms on hard surfaces, such as ships' hulls and intake pipes of industrial pumps and plants, can be a problem.

D) AIR POLLUTION

Major sources of air pollution in the Arabian region, particularly the more industrialized centres, include oil refineries, oil wells and oil platforms; petrochemical and fertilizer plants; asphalt plants; and motor traffic (Baker and Dicks 1982). Burning of solid wastes (e.g. open pit burning) also occurs extensively in some countries, but the practice seems to be declining because of air pollution problems (Linden *et al.* 1990). It is of ecological significance that atmospheric pollutants often end up in the marine environment. For instance, it has been shown that hydrocarbons adsorbed on particles of dust from desert regions in Iraq are transported in the atmosphere to the Gulf (Al Mudaffar *et al.* 1990).

Major air pollutants arising from industrial and other operations are probably universal and are described by Baker and Dicks (1982) for the Gulf. It may be added that smoke from the incomplete combustion of fuels is currently a serious environmental problem in the Gulf (see section 4 below). Dust arises from several sources including cement factories, one of which on the Red Sea coast at Yanbu has caused greater impact on nearby turtle beaches than on coral reefs.

E) RECREATION AND TOURISM

Primary effects of coastal tourism include, in particular, inadvertent coral breakage by divers, other types of habitat damage, and the collection of souvenir species. Spearfishing is discussed in section f (below). Secondary effects include a range of impacts, for instance from hotels and other infrastructures needed to support an expanding tourism industry.

Pressures from recreation and tourism are higher in the Gulf of Aqaba than other areas of the Arabian region. The popularity of the Red Sea is largely a reflection of its magnificent coral reefs and its relative close proximity to Europe. However, the central Red Sea actually contains more highly developed and diverse reefs, and diving is intense in limited areas. Recreational diving also occurs on reefs throughout much of the Arabian region.

Collecting of "curio" or souvenir species is generally limited. In the Red Sea there is a small market for corals for display, and reef molluscs (e.g. *Trochus* sp.) are also collected. Some export of reef fish for the aquarium trade occurs from the Red Sea (Ormond 1987). Collection of porcupine fish (*Diodon hystrix*) and pufferfish (*Arothron hispidus*) takes place, particularly in Egypt, for use as ornamental lampshades. This fish is a predator of crown of thorns starfish (*Acanthaster planci*) and also needle urchins (*Diadema setosum*). Reef damage by abundant urchins reported in Hurghada (Egypt) may be an effect of predator elimination (Ormond 1987). Near major industrial and residential centres, such as Jeddah and Yanbu, pressure from collection is now increasing. In some areas reef flats are systematically combed for corals, shells, urchins and other species, such that large areas are now quite bare on some fringing reefs (IUCN 1987). Elsewhere in the Arabian region some coral, shell and other collecting occurs. For instance, there is a small trade in Hawksbill turtle curios in Oman (and possibly other areas) but, despite constant taking of eggs, there is no systematic exploitation (Ross and Barwani 1979).

Damage to coastal and other vegetation from the use of off-road vehicles is also evident in the Red Sea (IUCN 1987) and in some other areas. This not only reduces vegetation available for birds, grazing mammals and other wildlife, but also the loss of halophytes can destabilize sand dunes that border many shore areas. Similar problems of perhaps greater magnitude will have resulted from tanks and other military vehicles during the 1991 Gulf war (section B4).

F) FISHING AND OTHER DIRECT RESOURCE EXPLOITATION

Fishing, like many other uses, has many environmental effects. Apart from the obvious removal of individuals of target species from the stock (Chapter 13), there can be indirect effects which generally are not well understood. These include changes in genetic stock of target species (Smith *et al.* 1991, Law 1991), capture of non-target species, including turtles, and disturbance of the benthos when trawls drag over seagrass beds (Basson *et al.* 1977: Fig. 81). Fishing by dynamiting remains uncommon, although it is reported from the shores of the Gulf of Suez, particularly around military installations. Spearfishing takes place in a number of areas, although the practice is officially banned in several countries (Ormond 1987). It can seriously deplete populations of large predatory fishes, even at low intensities.

Collecting of marine wildlife and eggs takes place throughout the Arabian region. People on Masirah Island and nearby areas consume a mimimum of 1000 Green turtles annually from nearby feeding pastures (Ross and Barwani 1979). Turtles and eggs are also taken on the mainland Dhofhar coast, but the extent and effects are unknown (Salm and Jensen 1989). Consumption of turtles and eggs also takes place elsewhere in the Arabian region, but mostly at low intensity.

Bird eggs are also taken, for example from the coral islands of the Saudi Arabian Gulf. In one instance during 1986, all tern eggs were taken from Harqus Island by fishermen, apparently to be sold. It is suggested from egg counts during the summer of 1986 that this island supports a total of 1750 breeding lesser crested terns (IUCN 1987). The Gulf supports a significant proportion of the world population of this species (IUCN 1987). Egg collection may be an important factor limiting their population there. Human predation of seabirds and their eggs on one of the Kuria Muria Islands in Oman has caused the extermination of breeding colonies of at least two species of seabirds (Salm and Jensen 1989). Large-scale trapping, although not for marine birds, is reported along the Red Sea coast, where more than 100,000 turtle doves are captured annually at three trapping stations (Büttiker 1988). The practice represents a threat to this migrating species (Büttiker 1988).

Systematic harvesting of algae for fish bait has taken place in Bahrain for many generations. Species used include *Ulva lactuca, U. reticulata* and *Enteromorpha intestinalis* (Basson 1989). It is assumed that this is a sustainable practice which has not adversely affected the coastal ecology.

G) CORAL DISEASES

Two coral diseases, white-band and black-band, are common in the Caribbean, and both exist in the Red Sea (Antonius 1988). Black-band disease appears to be caused by Cynanophyta. The cause of white-band disease is not known. Increasing pollution might be a contributing factor, although the aetiology of coral diseases is complex and not well understood. Reefs that are already heavily degraded (e.g. from heavy sedimentation or pollution) may be rendered more susceptible to the effects of disease.

H) EFFECTS OF AGRICULTURE AND RELATED ACTIVITIES

Although less acute than in many other parts of the world, agricultural and related activities appear to have had significant effects in some coastal environments of the Arabian region (Koeman 1982). Agricultural chemicals (e.g. urea fertilizer) dissolved in run-off contribute to eutrophication in some coastal areas, for instance in the western Gulf. However, levels of chlorinated hydrocarbons such as aldrin, lindane, dieldrin and DDT in marine sediments and biota are generally low, at least in the Gulf, compared to other regional seas (Linden *et al*. 1990). Little information appears to be available for the Red Sea (UNEP 1987).

Grazing of mangroves and other vegetation by livestock such as camels has been a longstanding practice in the Red Sea. Unless managed, it can be damaging. In Oman, overgrazing in watersheds is causing siltation of khawrs and other productive marine environments (Salm and Jensen 1989). Similarly, feral goats, donkeys and introduced cats and rats threaten nesting seabirds and other wildlife in parts of Oman. Wood is harvested for fuel, construction and other purposes. In the Red Sea, mangrove is used in construction of bird traps. In Yemen and Oman, in particular, mist forest and oases form luxuriant vegetation. Heavy loss of trees in these areas could increase run-off and hence erosion and sedimentation. In several parts of the region, inland irrigation schemes and heavy grazing have already resulted in a lowering of the water table and saline intrusion (Brooks and IUCN 1982). Apart from direct consequences to agriculture, such practices may have ecological implications for mangroves and other coastal habitats (Salm and Jensen).

Agricultural practices may have had beneficial ecological effects in some areas. For instance, nutrients from irrigated agriculture in the Qatif oasis of Saudi Arabia appear to have had beneficial effects on the adjacent marine environment of the Gulf (IUCN 1987). On the other hand, agricultural chemicals elsewhere have been harmful to coastal and marine environments, such as the North Sea and the Great Barrier Reef.

(3) Possible longer-term impacts

Among the possible longer-term impacts of concern are global warming and associated effects (e.g. sea level rise), acid deposition which some believe might be exacerbated by the burning oil wells in Kuwait (see section 4b below), and possible destabilization of integrated ecosystems in the region.

A) INCREASING CONCENTRATIONS OF GREENHOUSE GASES

Ecological effects of increasing concentrations of greenhouse gases are multiple and still not clearly understood. Based on studies elsewhere, the following may be of concern in the Arabian region. First, it is known from both field and laboratory studies on plants elsewhere that increased CO_2 levels lead to a degree of stomatal closure. This stomatal response can cause physiological stress to the plant in two main ways: by altering internal water flow; and, in areas of high irradiance and summer temperatures, leaf temperatures may reach their lethal limit. In the Gulf and other highly arid areas of the Arabian region, these factors might assume much significance.

Second, it is now widely held that global temperatures have gradually risen over the past century in accordance with the predicted effects of increased levels of CO_2 and other greenhouse gases. Worldwide, ambient air temperatures are predicted to rise 1–2°C by the middle of next century, despite geographical variation (Houghton *et al.* 1990). In parts of the Arabian region, some intertidal ecosystems and species may well already be approaching their upper and/or lower temperature limits. Additional stress imposed by global warming could therefore potentially disrupt some marine populations, and perhaps even species or ecosystems. For instance, latitudinal displacements of vegetational assemblages have been widely predicted for many regions. Among the other expected effects of global warming are changed weather patterns.

Third, a major consequence of increasing temperatures is sea level rise, due to thermal expansion of sea water and to the melting of polar ice. By the year 2050, many estimates suggest that sea level will have risen about 25 cm. Saline intrusion and coastal flooding already occurring in some islands of the South Pacific and elsewhere is believed to be a manifestation of sea level rise from global warming. In the Gulf, in particular, much of the coastline is characterized by broad intertidal flats of very low elevation (<0.5–1 m). Coastal flooding could affect intertidal and shallow-water ecosystems, but have more immediate effects on coastal villages, desalination plants, oil refineries and other industrial plants. Offshore ecosystems, such as coral reefs and seagrass beds, could also be affected.

Fourth, through habitat loss or change (above) and probably other unforeseen factors, life cycles of biota associated with particular ecosystems might become desynchronized or otherwise disrupted. Included might be birds using coastal vegetation, commercial crustaceans, molluscs, marine reptiles and mammals inhabiting seagrasses, and perhaps endemic and other fish associated with coral reefs. Of much concern is coral "bleaching", which results in reduced chlorophyll of zooxanthellae or even complete expulsion of the symbiotic algae. Causes of bleaching are complex, but appear to involve increasing sea temperatures and/or increased irradiance. While death sometimes results, coral colonies recover in many cases (Jokiel and Coles 1990). Bleaching has been well documented in parts of the Arabian region, such as the Red Sea (Antonius 1988), and probably also occurs in the Gulf (Chapter 12). It is suggested that mass coral bleaching might be a harbinger of climatic change, although firm conclusions would at present be premature (Jokiel and Coles 1990). In the Gulf, low (as well as high) extremes of temperature also stress corals (Chapter 12) and can cause loss of pigment (Coles and Fadlallah 1991).

B) ACID DEPOSITION

Acid deposition (dry and wet) arises principally when water vapour in the atmosphere reacts with certain polluting air emissions, in particular oxides of nitrogen and sulphur. The possible significance in the Arabian region, particularly on marine ecosystems, is not known. While acid deposition may be important intertidally, it will be less so subtidally because of the carbonate sea water buffer mechanism. Gases from the burning oil wells in Kuwait might increase acid deposition, as noted below (section 4b).

C) LARGE-SCALE MARINE ECOSYSTEM INSTABILITY

As discussed, coastal development in the Gulf over the past two to three decades has been unprecedented, adding to the effects of an already naturally stressful marine environment. Because of this and the Gulf's limited water exchange, the possibility arises that continuing human pressures might compromise the resilience of certain ecosystems, or even the Gulf's overall ecological integrity. Impacts in the Gulf from the 1991 war are certainly among the heaviest pressures incurred in recent times. Radically different species and genetic compositions of populations are one manifestation of altered states of an ecosystem, as can occur in fisheries under heavy exploitation (McGlade 1989, Law 1991). Non-linear dynamical models would help determine possible future ecological and socio-economic states of the Gulf under a variety of environmental pressures. Although pressures in the Red Sea (Dicks 1987, Ormond 1987) and elsewhere in the region (Salm and Dobbin 1987) are steadily increasing, they have generally not yet reached levels like that in the Gulf.

(4) The 1991 Gulf war

The 1991 Gulf war was an event involving both short-term and longer-term environmental and geopolitical implications. It was also among the more devastating examples of environmental sabotage (marine and terrestrial) on a massive scale, although some aspects have been exaggerated (Small 1991). For these and other reasons, ecological consequences of the Gulf war are treated separately. War-related impacts clearly did not occur in a pristine environment, and should be viewed in this context (Price and Sheppard 1991, Sheppard and Price 1991). The main environmental impacts originated from events in Kuwait, and included a large oil spill as well as extensive air pollution from the burning of more than 600 oil wells. Aspects of the following are necessarily tentative, and the picture may well change as new information becomes available.

A) MARINE POLLUTION FROM OIL

i) Overall magnitude of oil spill

A large oil spill resulted from destruction of oil tanks and other facilities in Kuwait during early 1991. The size of the spill is not known with certainty. Estimates have ranged from about 1 or 2 up to 11 million barrels (see Price and Sheppard 1991). The present spill therefore appears to be larger than the earlier Nowruz blow-out of 1983 (see section 1a above). Although it now seems that the total amount may not be more than 6 million barrels, the present spill is among the largest known in human history. Part of the Kuwaiti coast and most of the northern half (c. 500 km) of the Saudi Arabian Gulf coast became oiled, in some places very heavily (Figure 14.1). Oiling apparently extended to the Iranian coast. Offshore, the slick occupied several areas and impacted several of the Saudi Arabian coral islands. The actual reefs, however, did not appear to have been

Figure 14.1 *Aerial photograph showing coastal oil pollution on Saudi Arabian Gulf shores from the 1991 Gulf war.*

been oiled. Oil sheen and tar balls reached offshore areas further south, near Dammam and Bahrain.

The extent of the slick using SLAR (Side Looking Airborne Radar) in the western Gulf during March 1991 is shown in Figure 14.2. While the maps show the overall distribution of oil, they do not indicate the abundance or quantity of spilled oil, except in a general sense. Oil spill trajectory models (e.g. "Gulfslik II") have been developed by the Research Institute of King Fahd University of Petroleum and Minerals (KFUPM RI) in Dhahran.

ii) Habitats affected

Intertidally, affected habitats included mud and other tidal flats, saltmarsh and mangrove vegetation and beaches. Approximately 80% of the mangroves on a small island within Dawhat Dafi in Saudi Arabia was heavily oiled, and the chances of recovery are not rated as high. Dawhat Dafi is the most northerly locality for mangroves, at least along the western Gulf. This impact is of concern, because of the few mangroves remaining along the Saudi Gulf coast (Chapter 8). Few assessments have yet been made of subtidal habitats. Observations made by divers on some coral reefs do not suggest evidence of recent oiling. It appears that the oil is now starting to sink in some areas. Difficulties in assessing ecological effects of the spill will be particularly problematic because of the Gulf's highly variable marine environment.

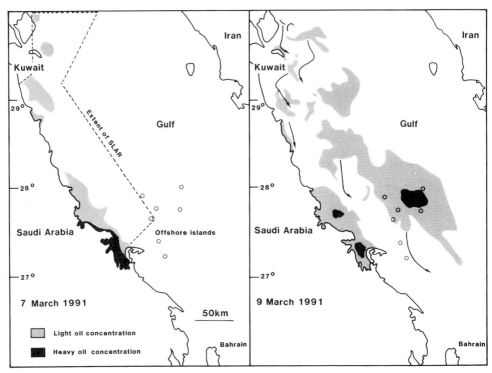

Figure 14.2 *Distribution of surface oil in the Gulf (7 March 1991) and trajectory update for Gulf oil spill (9 March 1991). SLAR = Sideways Looking Airborne Rader. Arrows show expected movements of oil over the following six days. From MEPA (Saudia Arabia)/US Coastguard.*

iii) Fisheries

Some impact to fishing and shrimping areas in the northern sector of the Gulf is likely (Sheppard and Price 1991). Intertidal traps northwest of Abu Ali used by artisanal fishermen were heavily oiled in some areas. Manifa Bay, one of the major shrimping areas of the industrial fleet, was also heavily oiled. Little, if any, fishing/shrimping apparently took place in northern waters during the war. Even if fish are not killed, tainting could result, as occurred during the Nowruz spill. In some instances, fish and shrimp may simply swim away from heavily oiled areas. However, for the reasons explained above (section B1a), shallow-water nursery and spawning areas of penaeid shrimp, other crustaceans and fish might be vulnerable. Major oil spills therefore have the potential to affect recruitment and cause some disruption of the fisheries, particularly in view of previous coastal infilling that has also occurred near some nursery and spawning areas. However, ecological effects on the fisheries and need detailed investigation. Possibly, any adverse effects may have been at least partly offset by the reduced fishing (Chapter 13).

iv) Birds and other fauna

The Gulf is internationally important for wintering waders, passage migrants and breeding seabirds (Gallagher *et al.* 1984, IUCN 1987, Bundy *et al.* 1989, Zwarts *et al.* in press). At least 52 species of dead oiled birds were identified shortly after cessation of hostilities (R. Dennis, pers. comm.). Included were black-necked grebes, greater crested grebes, cormorants, Socotra cormorants and waders such as dunlin. The total number of mortalities is not yet known. Along one stretch of coast 48 dead birds $1500\,\mathrm{m}^{-1}$ of shoreline have been observed (Grainger, pers. comm.). However, these counts cannot be extrapolated and used for estimates for the whole Gulf.

B) ATMOSPHERIC POLLUTION

Atmospheric pollution from burning oil and refinery products was potentially an even greater hazard than the oil slicks. Estimates of quantities of burning oil varied, but based on the 1989 average daily production rate were probably c. 1.6 million barrels per day (Small 1991). As a result, an estimated 16,000 tonnes per day of smoke were produced (Small 1991). Smoke from burning oil fields generally stay below $1\,\mathrm{km}$ altitude, whereas smoke from refineries could have reached 1–$3\,\mathrm{km}$ (Small 1991).

   Apart from the problems on human health, the burning oil wells could have had multiple environmental consequences, which are controversial and by no means fully agreed upon (Pearce 1991, Small 1991). These include:

(1) Possible atmospheric warming as a result of released CO_2 and other greenhouse gases. However, it is now estimated that in one year the additional CO_2 would increase the total CO_2 content of the atmosphere by only 0.00025%, from which no additional global warming would result (Small 1991).

(2) Reduction of incoming solar radiation, including changes in its spectral composition, and depression of temperatures by the atmospheric smoke (Bakan *et al.* 1991, Browning *et al.* 1991). Temperature inversions extended at least as far south as Bahrain. Such effects, if prolonged, could certainly pose a major stress on both terrestrial and marine wildlife. For example, coral reefs are naturally subjected to exceptionally low and high extremes of temperature (Coles and Fadlallah 1991). There is concern that degradation of corals and other photosynthetic communities may result from any reduction in temperatures. On the other hand, Small (1991) predict an actual increase in the daily-average surface temperature, despite a decrease in peak daytime temperatures.

(3) Air pollution from particulate fall-out and increased acid deposition. Marine and terrestrial ecosystems in the vicinity of the burning oil fields were subjected to heavy particulate levels. However, the overall effects are not likely to be known for some time. Increased acid deposition was also anticipated (Browning *et al.* 1991).

C) OTHER IMPACTS

Other possible environmental impacts from war-related activities have also been identified (Price and Sheppard 1991). These included: detonated mines; collisions with warships (several whales washed up on beaches were reported to have been killed in this way); damage to coastal vegetation from tanks and other military vehicles; and sunken ships which leak oil and other potentially toxic substances. The list is not all-encompassing, nor is the ecological significance of these factors known in any detail.

Summary

Coastal and marine environments throughout the Arabian region are becoming subjected to increasing human pressures, most of which appear to have resulted in harmful environmental effects. These are generally more pronounced in the Gulf, but it is clear that even the Red Sea is no longer the pristine sea it largely was only 20 years ago.

Oil, domestic, urban and industrial pollutants are a problem in several areas, notably in the Gulf. Overt effects include local eutrophication and algal blooms. Although some data are available on concentrations of pollutants (e.g. petroleum hydrocarbons, heavy metals) in water, sediments and biota, their effects on ecosystem structure and function are generally not well known. Throughout much of the Arabian region, the coastal zone is fast becoming the repository for solid wastes. In the Red Sea, ecological effects from industrial inputs such as mining are of increasing concern, particularly if extraction of deep sea metalliferous muds becomes extensive.

More acute ecological problems have probably arisen from loss and degradation of productive coastal habitats, caused by coastal landfill, dredging, and sedimentation caused by these and related practices. In some Gulf States (e.g. Saudi Arabia) 40% of the coastline has now been developed, and much of the shoreline of countries such as Kuwait and Bahrain is now artificial. Habitat decline also extends to other parts of the region and to the wider Indian Ocean, where approximately 50% of mangrove forests may have been lost over the last 20 years (IUCN/UNEP, 1985c).

Degradation of coral reefs from heavy collecting and other recreational and touristic uses is becoming more widespread, particularly in the Red Sea. In addition to fishing, hunting of adult turtles and birds (and their eggs) is extensive in some areas. The collection of algae for fish bait in Bahrain is a longstanding practice that appears to be sustainable. Of increasing concern is degradation of the coastal environment caused by agricultural and other activities inland, sometimes far from the coast. In Oman, for instance, overgrazing in watersheds is causing siltation of productive coastal environments.

Among the possible longer-term ecological impacts on marine life of the region are increasing concentrations of greenhouse gases, global warming and sea level rise; acid deposition; and large-scale marine ecosystem instability. The significance of these impacts, in particular, are at present mainly speculative. Environmental consequences of the 1991 Gulf war include marine pollution from oil, atmospheric pollution and other impacts.

CHAPTER 15

Coastal Zone Management

Contents
A. **Challenges and opportunities**   307
    (1) The coastal zone as a complex natural system   307
    (2) The need for integrated approaches   308
B. **National and regional initiatives**   308
    (1) Selected case studies of integrated management   308
    (2) Regional and international initiatives   313
    (3) Environmental responses to the 1991 Gulf war   314
Summary   316

This chapter discusses some marine resource management activities taking place around the Arabian region. The subject is of relevance to marine scientists, just as it is to more direct beneficiaries of the region's coastal and marine resources.

A. Challenges and opportunities

(1) The coastal zone as a complex natural system

Seagrasses, coral reefs and other ecosystems are key renewable resources, associated with diverse exploitation patterns. Like other natural systems, they are complex and influenced not only by natural processes but also by human activities. This concept of interdependence can be extended to larger areas, such as the Red Sea and Gulf, or even to the entire Arabian region. Yet despite increasing concern over the state of the world's coastal environments, knowledge of the connections between the non-living, biological and human domains is still far from complete (see McGlade 1989). Even the workings of just the principal biological elements of a grass bed (e.g. macrophytes, microflora, shrimps and other fauna) are ecologically intricate and only poorly known. But when the interactions of the various coastal users (i.e. the human domain) are super-imposed, the system becomes truly complex.

Both resource use and management entail manipulating individual parts of the system, in the case of management through altering (usually limiting) resource access. However, it has been shown time and time again that altering one part of a system (e.g. mangrove or seagrass removal) can have unforeseen and dramatic effects (e.g. coastal erosion and fishery decline) on another, often distant component. Failure to perceive and analyse the system holistically, then manage it accordingly may be a major reason for widespread resource degradation. The problem is often further compounded by increasing demands on resources for local trade and/or distant international markets. In the Gulf, for instance, high economic returns from oil and the fisheries may, directly or indirectly, create pressures that compromise the resilience of its marine and coastal ecosystems.

(2) The need for integrated approaches

Management clearly cannot proceed effectively through science, economics, socio-politics (or simple intuition) alone. Response to problems and opportunities is seldom straightforward, especially in the coastal zone. But how should management proceed, recognizing the genuine complexities inherent in natural systems? Clearly, an interdisciplinary rather than the more usual sectoral approach to research and management is needed. One promising avenue is to improve understanding of the scientific and human domains through the use of powerful dynamical models. Possible short-term and long-term future states of a system arising from a range of initial conditions (e.g. management strategies or impacts) can thereby be explored (see McGlade 1989). Understanding system complexity and behaviour is clearly important in determining future management needs and reponses. However, management outside the context of the prevailing socio-political customs and background is unlikely to be successful. In many instances, it is inadequate attention to human considerations and economics, as much as shortfalls in scientific knowledge, that has impeded management. The value of extensive public awareness programmes, both in the Arabian region (e.g. MEPA/IUCN 1989) and elswhere (Lemay and Hale 1989) is also becoming increasingly recognized.

   A fundamental objective of countries within the Arabian region, and indeed of most human societies, is sustainable development. This recognizes man's dependence on nature (e.g. the coastal zone) and advocates forms of development that enhance nature's contribution to human welfare, not just anticipating and preventing undesirable side effects. Strategies need to be developed, whereby the needs of conservation and development begin to converge. Experience elsewhere has shown that assessment and management of coastal resources often takes place most effectively using an integrated approach, for instance a coastal zone management programme or plan. Their application in the Arabian region is discussed below.

B. National and regional initiatives

(1) Selected case studies of integrated management

The following are two case studies, illustrating how integrated approaches to management have been adopted by two countries, Saudi Arabia and Oman.

A) SAUDI ARABIA

During the 1980s, Saudi Arabia's Meteorology and Environmental Protection Administration (MEPA) undertook an appraisal of natural resources and management requirements of both its Red Sea and Gulf coasts as part of planning for the region. A major objective was development of a national coastal zone management programme. This provided a framework aimed at balancing the needs of development with those of conservation. Assistance was provided by IUCN (World Conservation Union), through collaboration with coastal planners, ecologists, oceanographers, ornithologists and specialists in Islamic environmental law.

Among the analyses undertaken were identification of principal conflict areas by map analysis. Areas in which concentrated resources overlap with heavy uses/impacts denote principal conflict areas, and hence where management needs may be most pressing. The main areas are shown for the Saudi Arabian Gulf in Figure 15.1. Non-overlapping areas donote resource–use compatibilities, and where there might be opportunities for future sustainable resource use. Bioeconomic assessments, based on earlier research (Basson et al. 1977), revealed that shallow embayments constitute valuable economic as well as ecological resources. It was estimated that Tarut Bay (approx. 400 km^2) might produce annually 2.3×10^8 kg seagrass wet wt. Via a two-step food chain and an overall efficiency of 1%, or of 10% at each step, this might yield annually 2.3×10^6 kg of shrimp (or fish), valued at around US\$10 million. These calculations were approximate and theoretical. Nevertheless, the high value placed on certain marine ecosystems and renewable resources is supported by actual landings by shrimp fisheries in the Gulf (Chapter 13).

The proposed coastal zone management programme identified a number of major tasks, in response to the various problems, opportunities and issues relating to the Gulf and Red Sea coasts. These tasks may be summarized as:

(1) formulation of institutional arrangements for the programme, to determine specific responsibilities;

(2) formulation of management policies for the integrated management of coastal and marine uses and resources;

(3) identification and analysis of human activities that are beneficial to the coastal and marine environment, and to develop initiatives to encourage these activities wherever possible;

(4) formulation of options for legislative and regulatory structures, including an integrating "Umbrella Law" relating to all environmental matters, with specific consideration of the need for coastal zone management;

(5) formulation of options for a national coastal zone management plan which addresses issues at national, provincial and local scales;

(6) continued development of the coastal and marine protected area system for environmentally sensitive areas (ESAs);

(7) development of a comprehensive public awareness programme, using posters, audiovisuals and other materials;

(8) development of a coordinated strategy to ensure effective implementation of the national plan.

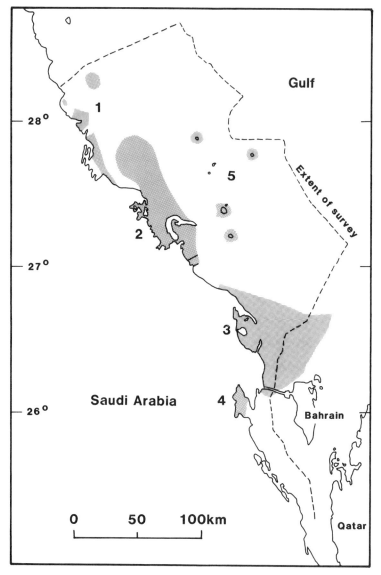

Figure 15.1 *Map showing conflicts between key resources and uses along the Saudi Arabian Gulf coast (from IUCN 1987). 1 = Saffaniyah to Manifa, 2 = Abu Ali area, 3 = Tarut Bay complex, 4 = Dawhat Zalum, and 5 = offshore islands.*

Although the national coastal zone management programme as a whole has yet to be implemented, several of the above tasks are now well developed. For example, as a result of the above and earlier studies 11 ESAs plus seven recreational areas were identified in the Gulf; and 46 ESAs (which include some recreational areas) were identified in the Red Sea (see also Sheppard 1982b). Through subsequent initiatives of another agency, NCWCD (National Commission for Wildlife Conservation & Development), these ESAs are being incorporated into a broader system plan of protected areas for the entire country (Child and Grainger 1990). Protected areas have been classified according to various socio-economic and ecological criteria. It appears that multiple-use areas will assume considerable importance. The dual approach of protected area management (Jungius 1988), in conjunction with geographically broader resource and uses policies, is considered to be effective and is also adopted elsewhere, for instance in Oman (below).

On the other hand, there is a real need for management to keep pace with the continuing human demands placed on the marine environment. For instance, degradation and loss of mangroves is still occurring in some coastal areas of Saudi Arabia, despite their known bioeconomic importance. *Rhizophora* mangroves have been transplanted with some success on the northern Saudi Gulf coast (Chapter 8). On a large-scale, such initiatives should help restoration of heavily degraded ecosystems.

B) OMAN

A particularly successful example of a coastal zone management plan in the Arabian region, both conceptually and operationally, is that of Oman. The plan, which commenced in the early 1980s and is still under development, is cross-sectoral in its approach to wildlife, habitats, human use and management of coastal areas. It is also integrated with management and conservation of extensive terrestrial areas, including inland sites. Indeed, the marine and coastal regions initially were included as a part of a greater terrestrial and land use management project (Clarke *et al.* 1986), which itself appears to have had its roots in the now well known project to rehabilitate the Arabian oryx.

Initially, the goal of this major conservation project was to prepare a system plan for Nature Conservation Areas (NCAs) in Oman. The conservation programme arising from this then formed part of Oman's total land use strategy. Policy, legal aspects and implementation were all considered, and proposals were made for the structure and operation of a Directorate General of Wildlife and Nature Conservation, within the Ministry of Environment and Water Resources. As part of the initial work, Clarke *et al.* (1986) identified and mapped 43 Land Classes and 12 Marine Habitats, which were designed to include all threatened, endemic, economically valuable, biologically important, and simply interesting habitats in Oman. This resulted in the recommendation of 91 NCAs, which fall into three major classes: National Nature Reserves, National Scenic Reserves and National Resource Reserves. The coastal and marine protected areas within this plan encompass all three categories, and were selected from hundreds of sites observed throughout the length of the coastline from Musandam to the Dhofhar region (Anderlini 1985, Sheppard 1986a).

Concurrently, more detailed coastal zone management plans were developed for the Capital Area, the Daymaniyat Islands off the Battinah coast, and of the coastline east of Muscat to the eastern tip of the Arabian peninsula (Phase 1: Salm and Dobbin 1986, 1987). This work continued with similar detailed planning for the use and protection of stretches of coast in the central part of the coastline around Barr al Hikman, opposite Masirah Island (Phase 2: Salm *et al.* 1988), and additional work has continued in the Dhofhar region (Phase 3: Salm and Jensen 1989).

There are three components of the coastal zone management plan (Salm and Dobbin 1986, Salm *et al*. 1988): (i) establishment of planning policies which provide broad-brush development guidance in the coastal zone, and hopefully avert the emergence of new conflicts; (ii) establishment of protected areas to enable intensive management of specific sites of particular value; and (iii) identification of specific remedial actions for the numerous management issues in the coastal zone, and assignment of these to a responsible agency.

Through the approach adopted, the entire coastal zone of Oman essentially functions as a vast multiple use resource (Salm and Jensen 1989). Of special interest are the voluntary conservation measures taken by the Hikmani fishermen. Like other coral reef areas, those around Barr al Hikman are associated with particularly rich fishery resources. However, under normal conditions the Hikmani fishermen do not fish in these waters, but instead deliberately venture further offshore, where fish abundance is lower. It is only during difficult conditions (e.g. bad weather) that they fish the more productive areas further inshore. The inshore areas thereby function as "cold storage areas" (Salm and Jensen 1989). Successful voluntary conservation activities are also known in some other parts of the world (Gubbay, 1988).

The combination of broad-based, nation-wide measures, coupled with the detailed coastal sector plans, constitutes an advanced approach that is very effective. Such forward planning seems timely. Despite low levels of population in most parts of the country, some areas or species are now becoming subjected to moderate exploitation and other less direct pressures. Also of significance is that the programme encompasses aesthetic appeal and recognition of intrinsic wildlife values as well as commercial considerations. The range of marine ecosystems and coastal zone types in Oman is greater than is found in many countries, and the plans encompass all of them. For these and other reasons, many aspects of the Oman coastal zone management plan could probably serve as a useful case study and be applied elsewhere in the Arabian region.

C) OTHER NATIONAL INITIATIVES

In the Red Sea, protected areas have also been established in other countries (e.g. Egypt, Djibouti, Israel and Jordan), often in conjunction with zoning and other resource-use policies. Further information on marine conservation in these and the other Red Sea countries is available in IUCN/UNEP (1985b), Ormond (1987), and Sheppard and Wells (1988). In the Gulf, marine conservation and management has perhaps been most progressive in Iran, although little information is available in the Western scientific and related literature. Included has been development of environmental standards and various wildlife laws and regulations (e.g. for protecting seabirds). Five of the country's protected areas include sections of coast/islands, at least one of which dates back to 1960 (IUCN 1976). The Hara Protected Region contains the largest stand of mangrove (>6500 ha) along the Iranian coast (IUCN 1976, IUCN/UNEP 1985a), and in the Gulf as a whole. Additional information for the Gulf countries is available in IUCN/UNEP (1985a), Halwagy *et al*. (1986), Sheppard and Wells (1988) and Vousden (1988).

(2) Regional and international initiatives

A) THE UNEP REGIONAL SEAS PROGRAMME

Launched in 1974, the UNEP Regional Seas Programme includes some eleven regions and about 140 coastal states. Its action-oriented programme addresses both the consequences and causes of environmental degradation, and adopts integrated approaches to management of marine and coastal areas. Each regional action plan is formulated according to the needs of the region, as perceived by the governments concerned. Legislation and institutional infrastructures in place are broadly aligned with the various Regional Seas divisions. Further, the divisions show general correspondence with recent physical and biogeographical classifications of the world's coastal and marine environments (Hayden *et al.* 1984, Ray and Hayden 1986).

Included in the Regional Seas Programme are the Red Sea and Gulf of Aden region, and the Kuwait Action Plan (KAP) region, i.e. the Gulf (otherwise known as the ROPME Sea Area or Region [ROPME = Regional Organization for Protection of the Marine Environment]).

The Kuwait Action Plan forms part of the broader Kuwait Regional Convention on the Protection of the Marine Environment from Pollution. All Gulf countries are signatories to the convention, whose aims include to prevent and control pollution from ships and other causes, to establish national standards, and to develop national research and monitoring programmes relating to all types of pollution. The KAP operates through close cooperation with international organizations, regional organizations (e.g. ROPME), and also with many national organizations, institutions and focal points. Many of the major conservation and research initiatives in both the Gulf and Red Sea have been part of UNEP's Regional Seas Programme. ROPME's activities, not surprisingly, were hampered during the 1991 Gulf war, but the organization has now been revitalized and moved back to the original secretariat in Kuwait.

The Red Sea and Gulf of Aden Action Plan is based on the Regional Convention for the Conservation of the Red Sea and Gulf of Aden. Actively involved have been regional organizations such as ALECSO (Arab League Educational, Cultural & Scientific Organization), PERSGA (Environmental Programme for the Red Sea and Gulf of Aden) and others, with ALECSO coordinating all activities and providing the interim secretariat for PERSGA.

B) OTHER REGIONAL AND INTERNATIONAL AGREEMENTS

Essential in regional and national management plans is the upholding of any environmental treaties, laws and conventions to which a country is party. In addition to the UNEP Regional Seas Programme, there are other important regional agreements (Johnston 1981, Couper 1983, IUCN/UNEP 1985a,b, IUCN 1987). Included are the African Convention on Conservation of Natural Resources, the Saudi–Sudanese Red Sea Commission (deep sea mining), the Arab Declaration on Environment and Development, the GCC (Gulf Cooperative Council), MEMAC (Marine Emergency Mutual Aid Centre), GAOCMAO (Gulf Area Oil Companies Mutual Aid Organization). These and other agreements relate to environmental management and pollution control.

Important international agreements include parts of UNCLOS (UN Convention on the Law of the Sea), CITES (Convention on International Trade in Endangered Species), RAMSAR (Wetlands Convention), the Bonn Convention on Migratory Species, Indian Ocean Alliance, IBP (International Biological Programme), MAB (Man and the Biosphere Programme), the World Heritage Convention, and others (Johnston 1981, Couper 1983).

Upholding regional and international agreements is particularly important in seas like the Red Sea and Gulf, whose trans-boundary resources consititute a "global commons" shared by many countries.

(3) Environmental responses to the 1991 Gulf war

A separate section is devoted to environmental management following the 1991 Gulf war for the same reasons indicated in the previous chapter. However, existing scientific information and management frameworks (e.g. IUCN 1987) provide a useful basis for present and future activities. International assistance in the Gulf is already considerable and is continuing. Assistance is being provided through task forces, numerous delegations, international commissions and organizations, contractors and companies. Orchestrating outside assistance with national initiatives has proved to be a highly complex task. Details of the main responses to environmental damage from war-related activities to date are summarized below (see also Price and Sheppard, 1991).

A) COMBATING OIL POLLUTION

i) Protection of key areas

Emphasis to date has been given to the protection of desalination plants and key industrial facilities. Several of these areas also contain important living resources. Booms have been deployed to deflect oil from key areas, and also to contain it.

ii) Recovery of floating oil

This is being undertaken principally using skimmers and "vac trucks". Again, emphasis has been in and around the desalination plants and industrial facilities. Operations later included areas such Manifa Bay, Musallimiyah Bay and Dawhat Dafi in Saudi Arabia, all of which contain important biological resources. By April 1991, a total of 500,000 barrels of oil/water were reported to have been recovered by Saudi Aramco (Arabian American Oil Company) and others (NCWCD 1991). If principally oil, this is a significant quantity and operations are continuing. Recovered oil is taken to permanent collection sites, and temporary storage pits dug in the desert for subsequent reprocessing.

iii) Field assessments and guidelines for remediation

Assessments of the extent of damage from the oil spill are underway, and will probably continue for months or years. Summary information was compiled on key coastal areas and their ecological and socio-economic attributes, so that priorities for protection and clean-up could be determined. Among the key coastal areas are the offshore coral islands, tidal flats, mangrove areas and bays.

Field assessments and remediation proposals were prepared for most of the important coastal areas which were oiled along the northern sector of the Saudi Arabian Gulf, in particular Manifa Bay, the offshore coral islands, Musallimiyah and Dawhat Dafi area of Abu Ali. Similar proposals have been developed for Tarut Bay, an area which was not impacted. At each area, habitats were classified in terms of priority for clean-up (e.g. 1. mangroves, 2. saltmarshes, 3. tidal flats, 4. algal flats and rock flats, 5. beaches). Guidelines for clean-up included: removal of oil from saltmarshes using vac trucks, ideally at high tide when oil can be refloated; use of sorbent material and sorbent booms at the periphery of saltmarshes; use of skimmers at high tide; use of drainage ditches, from which oil later can be removed (MEPA 1991). These guidelines will then be passed on to contractors involved in future clean-up operations, so that equipment can be deployed with minimal additional impact. It is understood that actual clean-up activities are likely to be less extensive than oil spill recovery operations (above).

iv) Other approaches to combat oil

Other approaches and products considered at the time of the spill included bioremediation (use of oil digesting bacterial cultures), compounds rendering oil more viscous to facilitate recovery, and sheep's wool devoid of lanolin (wool grease). Sheep's wool has a greater capacity to absorb oil than synthetic sorbent materials commonly used. Also, being a natural product, it is biodegradable which would be useful in instances where removal of oil-coated sorbent is not possible. Several of the above approaches were considered to have potential application in the Gulf, particularly in conjunction with other oil spill clean-up methods (above). Dispersants were tried on test patches during early stages of the oil spill, but were largely ineffective, aside from any ecological considerations. The "do-nothing" approach was probably the best, and only, option in many instances, from both ecological and logistic standpoints.

B) COMBATING AIR POLLUTION

Capping the burning oil wells in Kuwait was clearly of highest priority, to alleviate further human health and environmental problems, and of course to stop the loss of Kuwait's major economic product. All oil burning wells were extinguished by November 1991.

C) ENVIRONMENTAL REHABILITATION AND OTHER INITIATIVES

In Saudi Arabia, a Wildlife Rehabilitation Centre for oiled birds and other fauna, such as turtles, was established at Jubail. Several hundred birds were taken to the Centre, the survival rate from which was about 30% prior to release. Species collected included cormorants, Socotra cormorants, terns and waders. The work was supervised by several veterinarians. Many of the cleaned birds have now been released back into the wild. While ecological benefits on local populations can only be modest or negligible, the Rehabilitation Centre is important in attracting attention to and raising public awareness about environmental problems facing the Gulf and its wildlife.

A major action plan to assist Gulf States to deal with environmental aspects of the Gulf war has been developed through UNEP, in collaboration with ROPME, the International Maritime Organization (IMO), World Conservation Union (IUCN), World Wide Fund for Nature (WWF), Intergovernmental Oceanographic Commission (IOC), World Meteorological Organization (WMO), International Atomic Energy Agency (IAEA) and others.

Creation of transfrontier peace parks may be a useful approach in any future regional development and management plans. It is of significance that there are no officially recognized protected areas in Iraq.

Summary

The coastal zone is a complex natural system, in which integrated resource assessment and management is generally more effective than more sectoral approaches commonly adopted.

The key elements of integrated coastal zone management programmes are described for two countries, Saudi Arabia and Oman, which serve as case studies. A dual approach has been adopted: (i) policies for resources and resource–uses for the entire coastline; and (ii) development of a system of protected areas, in particular in areas containing outstanding marine wildlife, and/or in areas associated with major resource–use conflicts. Habitat restoration has been undertaken on a small scale on the Saudi Gulf coast, and may assume greater importance in the future.

Regional conservation activities include the Kuwait Action Plan and Red Sea and Gulf of Aden Action Plan of the UNEP Regional Seas Programmes. Upholding regional and international agreements is particularly important in seas like the Red Sea and Gulf, whose transboundary resources constitute a "global commons" shared by many countries.

Environmental responses to the 1991 Gulf war include recovery of floating oil, field assessments and guidelines for oil spill clean-up, capping the burning oil wells in Kuwait, and establishment of a wildlife rehabilitation centre. A major action plan to assist Gulf States deal with environmental aspects of the Gulf war has been developed through UNEP, in collaboration with other organizations.

References

ADEY, W.H. (1975) The algal ridges and coral reefs of St Croix, their structure and Holocene development. *Atoll Research Bulletin* **187**: 67 pp.

ADMIRALTY. (1944) *Iraq and the Persian Gulf*. Geographical Handbook Series. Naval Intelligence Division. Oxford and Cambridge University Press, 682 pp.

AL-ATTAR, M.H. (1979) Distribution and abundance of penaeid larvae in Kuwait Bay and adjacent waters. Kuwait Institute of Scientific Research Annual Research Report 1979: 46–48.

ALCALA, A.C. (1981) Fish yield of coral reefs of Sumilon Island, Central Philippines. *Natural Resources Council of the Philippines Research Bulletin* **36**: 1–7.

ALEEM, A.A. (1979) A contribution to the study of seagrasses along the Red Sea coast of Saudi Arabia. *Aquatic Botany* **7**: 71–78.

ALLDREDGE, A.L. and KING, J.M. (1977) Distribution, abundance and substrate preferences of demersal reef zooplankton at Lizard Island lagoon, Great Barrier Reef. *Marine Biology* **41**: 317–333.

ALLEN, G.R. (1972) *Anemonefishes. Their Classification and Biology*. T.F.H. Publications, New Jersey. 288 pp.

ALLEN, G.R. and RANDALL, J.E. (1980) A review of the damselfishes (Teleostei: Pomacentridae) of the Red Sea. *Israel Journal of Zoology* **29**: 1–98.

AL-MUDAFFAR, N., FAWZI, N.O. and AL-EDANEE, J. (1990) Hydrocarbons in surface sediments and bivalves from Shatt al Arab and its rivers, Southern Iraq. *Oil and Chemical Pollution* **7**: 17–28.

AL-SAYARI, S.F., DULLO, C., HOTZL, H., JADO, A.R. and ZOETL, J.G. (1984) The Quaternary along the coast of the Gulf of Aqaba. In: Jado, A.R. & Zoetl, J.G. (eds.) *Quaternary of Saudi Arabia*. Springer Verlag, Berlin. pp. 32–44.

ANDERLINI, V.C. (1985) Protected areas system plan for the Sultanate of Oman. WWF/IUCN Project 9069. IUCN, Gland. 29 pp.

ANDERSON, G.R.V., EHRLICH, A.H., EHRLICH, P.R., ROUGHGARDEN, J.D., RUSSELL, B.C. and TALBOT F.H. (1981) The community structure of coral reef fishes. *American Naturalist* **117**: 476–495.

ANGELUCCI, A., MATTEUCCI, R. and PRATURLON, A. (1981) Outline of geology and sedimentary environments of the Dahlak Islands (Southern Red Sea). *Bollettino della Societa Geologica Italiana* **99**: 405–419.

ANGELUCCI, A., CARBONE, F. and MATTEUCCI, R. (1982a) La scogliera corallina di Ilisi nelle Isole dei Bagiuni (Somalia meridionaleonale) Somalia. *Bollettino della Societa Paleontologica Italiana* **21** (2–3): 201–210.

ANGELUCCI, A. and 13 others. (1982b) Il Ghubbet Entatu Nell Arcipelago Delle Isole Dahlak (Mar Rosso): Un Esempio Di Sedimentazione Carbonatica. *Bollettino della Societa Paleontologica Italiana* **21**: 190–200.

ANGELUCCI, A. and 12 others (1985) L'arcipelago Delle Isole Dahlak Nel Mar Rosso Meridionale: Alcune Caratteristiche Geologiche. *Bollettino della Societa Geografica Italiana Series XI*, Vol. 2. pp. 233–262.

ANONYMOUS. (1988) *Ecological Studies of Southern Oman Kelp Communities*. Summary Report. ROPME, Kuwait, 46 pp.

ANSELL, A.D. (1984) Sandy beaches and shallow subtidal benthos. Abstract in Marine Habitats and Fauna of the Coast of Oman. 289th Scientific Meeting of Challenger Society of Scottish Marine Biological Society (SMBA/Edinburgh University), Edinburgh, 18 December.

ANTONIUS, A. (1988) Distribution and dynamics of coral diseases in the eastern Red Sea. In: Choat, J.H. *et al.* (eds.) *Proceedings of the 6th International Coral Reef Symposium.* Townsville, Australia, Vol. 2. pp. 293–298.

AOKI, S. (1974) Dredging of bottom deposits and benthos, including mineralogy in fine sediments of the Arabian Gulf. In: Kuronuma, K. (ed.) *Arabian Gulf fishery-oceanography survey. Transactions of the Tokyo University of Fisheries* No. 1. pp. 52–60.

BACKMAN, T.W. and BARILOTTI, D.C. (1976) Irradiance reduction: effects on standing crops of the eel grass *Zostera marina* in a coastal lagoon. *Marine Biology* **34**: 33–40.

BAKAN, S., CHLOND, A. and 15 other co-authors. (1991) Climate response to smoke from the burning oil wells in Kuwait. *Nature* **351**: 367–371.

BAKER, J.M. and DICKS, B. (1982) The environmental effects of pollution from the Gulf oil industry. IUCN/MEPA report for the Expert Meeting of the Gulf Co-ordinating Council to review environmental issues.

BAKER, J.M., CLARK, R.B., KINGSTON, P.F. and JENKINS, R.H. (1990) Natural recovery of cold water marine environments after an oil spill. 13th Annual Arctic & Marine Oilspill Program Technical Seminar, 111 pp.

BALLENTINE, D. and HUMM, H.J. (1975) Benthic algae of the Anclote estuary. I. Epiphytes of seagrass leaves. *Florida Science* **38**: 144–149.

BANAIMOON, S.A. (1988) The marine algal flora of Khalf and adjacent regions, Hadramout, PDR Yemen. *Botanica Marina* **31**: 215–221.

BARNES, R.S.K. and HUGHES, R.N. (1982) *An Introduction to Marine Ecology.* Blackwell Scientific Publications, Oxford. 339 pp.

BARRANIA, A. (1981) Socio-economic Aspects of the Saudi Arabian Fisheries in the Red Sea. FAO RAB/77/008/9. 54 pp.

BARRATT, L. (1982) Scientific report of the Joint Services Expedition to the Egyptian Red Sea 1982. 46 pp.

BARRATT, L. and MEDLEY, P. (1990) Managing multi-species ornamental reef fisheries. *Progress in Underwater Science* **15**: 55–72.

BARRATT, L., ORMOND, R.F.G., CAMPBELL, A.C., HISCOCK, S., HOGARTH, P.J. and TAYLOR, J.D. (1984) An ecological study of rocky shores on the south coast of Oman. Report of IUCN to UNEP's Regional Seas Programme.

BARRATT, L., ORMOND, R.F.G. and WRATHALL, T. (1986) Ecology and productivity of the sublittoral Algae *Ecklonia radiata* and *Sargassopsis zanardini*. Part 1. *Ecological Studies of Southern Oman Kelp Communities.* Council for the Conservation of the Environment and Water Resources, Muscat, Oman, and Regional Organisation for the Protection of the Marine Environment, Kuwait, pp. 2.1–2.20.

BARRATT, L., DAWSON SHEPHERD, A.R., ORMOND, R.F.G. and McDOWALL, R. (1987) Yemen Arab Republic marine conservation survey. Vol. I. Distribution of habitats and species along the YAR coastline. IUCN Red Sea and Gulf of Aden Environment Programme/TMRU York, UK. 110 pp.

BARRATT, L., ORMOND, R.F.G. and WRATHALL, T.J. (1988) Ecology and productivity of the sublittoral algae *Ecklonia radiata* and *Sargassopsis zanardinii. Ecological Studies of Southern Oman Kelp Communities. Summary Report.* ROPME/GC-6/001. pp. 22–48.

BASSON, P.W. (1979) Marine algae of the Arabian Gulf coast of Saudi Arabia. *Botanica Marina* **22**: 47–64.

BASSON P.W. (1989) Notes on economic plants: fish bait algae. *Economic Botany* **42**: 271–278.

BASSON, P.W. and HARDY, J.T. unpublished. Report on diatom abundance and diversity in Tarut Bay, Arabian Gulf. Aramco, Dhahram.

BASSON, P.W., BURCHARD, J.E., HARDY, J.T. and PRICE, A.R.G. (1977) *Biotopes of the Western Arabian Gulf.* Aramco, Dhahran. 284 pp.

BEER, S. and WAISEL, Y. (1982) Effects of light and pressure on photosynthesis in two seagrasses. *Aquatic Botany* **13**: 331–337.

BEHAIRY, A.K.A. (1983) Marine transgressions in the west coast of Saudi Arabia (Red Sea) between mid-Pleistocene and Present. *Marine Geology* **52**: M25–M31.

BEHAIRY, A.K.A. and EL-SAYED, M.M. (1984) Dissolved organic matter in coastal waters at Jeddah, Saudi Arabia. *Marine Pollution Bulletin* **15**: 113–116.

BEHAIRY, A.K.A., EL-SAYED, M.K. and DURGAPRASDA RAO, N.V.N. (1985) Eolian dust in the coastal area north of Jeddah, Saudi Arabia. *Journal of Arid Environments* **8**: 89–98.

BEINSSEN, K.H. (1979) Fishing Power of Divers in the Abalone Fishery, Victoria, Australia. International Council for the Exploration of the Sea. Special Meeting on Population Assessment of Shellfish Stocks (47). 9 pp.

BELLWOOD, D.R. (1988) Seasonal changes in the size and composition of the fish yield from reefs around Apo Island, Central Philippines, with notes on methods of yield estimation. *Journal of Fish Biology* **32**: 881–893.

BEMERT. G. and ORMOND, R. (1981) *Red Sea Coral Reefs.* Kegan Paul International, London. 192 pp.

BENAYAHU, Y. (1985) Faunistic composition and patterns in the distribution of soft corals (Octocorallia Alcyonacea) along the coral reefs of Sinai Peninsula. *Proceedings 5th International Coral Reef Congress,* Tahiti. **6**: 255–260.

BENAYAHU, Y. and LOYA, Y. (1977a) Seasonal occurrence of benthic algae communities and grazing regulation by sea urchins at the coral reefs of Eilat, Red Sea. *Proceedings 3rd International Coral Reef Symposium*, Miami, pp. 383–389.

BENAYAHU, Y. and LOYA, Y. (1977b) Space partitioning by stony corals, soft corals and benthic algae on the coral reefs of the northern Gulf of Eilat (Red Sea). *Helgolander wissenschaftliche Meeresuntersuchungen* **30**: 362–382.

BENAYAHU, Y. and LOYA, Y. (1981) Competition for space among coral-reef sessile organisms at Eilat, Red Sea. *Bulletin of Marine Science* **31**: 514–522.

BENAYAHU, Y. and LOYA, Y. (1983) Surface brooding in the Red Sea soft coral *Parerythropodium fulvum fulvum* (Forskal 1775). *Biological Bulletin* **165**: 353–369.

BENAYAHU, Y. and LOYA, Y. (1984a) Life history studies on the Red Sea soft coral *Xenia macrospiculata* Gohar, 1940. II. Planulae shedding and post larval development. *Biological Bulletin* **166**: 44–53.

BENAYAHU, Y. and LOYA, Y. (1984b) Substratum preferences and planulae settling of two Red Sea Alcyonaceans: *Xenia macrospiculata* Gohar and *Parerythropodium fulvum fulvum* (Forskal). *Journal of Experimental Marine Biology and Ecology* **83**: 249–261.

BENAYAHU, Y. and LOYA, Y. (1985) Settlement and recruitment of a soft coral: Why is *Xenia macrospiculata* a successful colonizer? *Bulletin of Marine Science* **36**: 177–188.

BENAYAHU, Y. and LOYA, Y. (1986) Sexual reproduction of a soft coral: synchronous and brief annual spawning of *Sarcophyton glaucum* (Quoy and Gaimard, 1833). *Biological Bulletin* **170**: 32–42.

BERMAN, D. (1981) *Assault on the Largest Unknown. The International Indian Ocean Expedition.* Unesco Press, Paris. 96 pp.

BIRKELAND, C. (1977) The importance of rate of biomass accumulation in early successional stages of benthic communities to the survival of coral recruits. *Proceedings of 3rd International Coral Reef Symposium*, Miami **1**: 15–21.

BLAXTER, J.H.S. (1986) Development of sense organs and behavior of teleost larvae with special reference to feeding and predator avoidance. *Transactions of the American Fisheries Society* **115**: 98–114.

BOROWITZKA, M.A. (1981) Algae and grazing in coral reef ecosystems. *Endeavour, New Series* **5**: 99–106.

BOSCH, D. and BOSCH, E. (1982) *Seashells of Oman*. Longman, London, 206 pp.

BOUCHON-NAVARO, Y. and BOUCHON, C. (1989) Correlations between chaetodontid fishes and coral communities of the Gulf of Aqaba (Red Sea). *Environmental Biology of Fishes* **25**: 47–60.

BOUCHON-NAVARO, Y. and HARMELIN-VIVIEN, M.L. (1981) Quantitative distribution of the herbivorous reef fishes in the Gulf of Aqaba (Red Sea). *Marine Biology* **63**: 76–86.

BRAITHWAITE, C.J.R. (1982) Patterns of accretion of reefs in the Sudanese Red Sea. *Marine Geology* **46**: 297–325.

BRAITHWAITE, C.J.R. (1987) Geology and palaeogeography of the Red Sea region. In: Edwards, A.J. & Head, S.M. (eds.) *Red Sea*. Pergamon Press, Oxford. pp. 22–44.

BRAMKAMP, R.A. and Powers, R.W. (1955) Two Persian Gulf lagoons. *Journal of Sedimentary Petrology* **25**: 139–140.

BROOKS, W.H. and IUCN. (1982) Conservation and sustainable use of natural resources: Part II terrestrial. IUCN/MEPA report for the Expert Meeting of the Gulf Co-ordinating Council to review environmental issues.

BROWNING, K.A., ALLAM, R.J. and 10 other co-authors. (1991) Environmental effects from burning oil wells in Kuwait. *Nature* **351**: 363–367.

BRIGGS, J.C. (1974) *Marine Zoogeography*. McGraw-Hill, New York. 475 pp.

BUNDY, G., CONNOR, R.J. and HARRISON, J.O. (1989) *Birds of the Eastern Province of Saudi Arabia*. H.F.&G. Witherly Ltd/Aramco. 224 pp.

BURCHARD, J.E. (1979) *Coral fauna of the Arabian Gulf*. Aramco, Dhahran. 36 pp.

BURCHARD, J.E. (1983) Synopsis of Arabian Gulf Corals. Manuscript. 7 pp.

BURCHARD, J.E. unpublished. Report to Aramco on benthic faunal abundance and diversity in Tarut Bay, Arabian Gulf.

BURNS, K.A., VILLENEUVE, J.P., ANDERLINI, V.C. and FOWLER, S.W. (1982) Survey of tar, hydrocarbon and metal pollution in the coastal waters of Oman. *Marine Pollution Bulletin* **7**: 240–247.

BÜTTIKER, W. (1988) Trapping of turtle doves (*Streptopelia turtur* Linnaeus, 1758) in Saudi Arabia. *Fauna of Saudi Arabia* **9**: 12–18.

CAMPBELL, A.C. (1987) Echinoderms of the Red Sea. In: Edwards, A.J. & Head, S.M. (eds.) *Red Sea*. Pergamon Press, Oxford. pp. 215–232.

CAMPBELL, A.C. and MORRISON, M. (1988) The echinoderm fauna of Dhofar (southern Oman) excluding holothuroids. In: Burke, R.D. *et al.* (eds.) *Echinoderm Biology. Proceedings of the 6th International Echinoderm conference*. A.A. Balkema, Rotterdam. pp. 369–378.

CARPENTER, K.E., MICLAT, R.I., ALBALADEJO, V.D. and CORPUZ, V.T. (1981) The influence of substrate structure on the local abundance and diversity of Philippine reef fishes. *Proceedings of the 4th International Coral Reef symposium*, Manila **2**: 497–502.

CARTER, R.N. and PRINCE, S.P. (1981) Epidemic models used to explain biogeographic distribution limits. *Nature (London)* **293**: 644–645.

CHAPMAN, V.J. (1984) Mangrove biogeography. In: Por, F.D. and Dor, I. (eds.) *Hydrobiology of the Mangal*. Dr W. Junk Publishers, The Hague. pp. 15–24.

CHAPPELL, J. (1983) Evidence for smoothly falling sea level relative to North Queensland, Australia, during the past 6,000 yr. *Nature* **302**: 406–408.

CHILD, G. and GRAINGER, J. (1990) A plan to protect areas in Saudi Arabia. National Commission for Wildlife Conservation and Development, Saudi Arabia.

CHRISTENSEN, B. (1978) Primary production of mangrove forests. In: Proceedings of International Workshop for Mangrove and Estuarine Area Development for the Indo-Pacific region. Los Banos, Laguna. Philippine Council for Agriculture and Resources Research, pp. 131–136.

CHUANG, S.H. (1977) Ecology of Singapore and Malayan coral reefs — preliminary classification. *Proceedings of 3rd International Symposium on Coral Reefs*, Miami **1**: 55–61.

CIVITELLI, G. and MATTEUCCI, R. (1980) La scogliera a frangia di Tanam (Isole Dahlak, Mar Rosso). *Bollettino della Societa Geologica Italiana* **99**: 517–530.

CLARK, R.B. (1989) *Marine Pollution* (2nd edn). Clarendon Press, Oxford. 220 pp.

CLARKE, J.E., AL-LUMKI F., ANDERLINI, V.C. and SHEPPARD, C.R.C. (1986) Sultanate of Oman. Proposals for a system of Nature Conservation Areas. IUCN, Gland. 477 pp.

CLARKE, R.D. (1977) Habitat distribution and species diversity of chaetodontid and pomacentrid fishes near Bimini, Bahamas. *Marine Biology* **40**: 277–289.

COHEN, D.M. (1973) Zoogeography of the fishes of the Indian Ocean. In: Zeitzschel, B. (ed.) *The Biology of the Indian Ocean*. Springer Verlag, New York. pp. 451–463.

COLES, S.L. (1988) Limitations on reef coral development in the Arabian Gulf: Temperature or algal competition? *Proceedings of 6th International Coral Reef Symposium*, Townsville, Australia. **3**: 211–216.

COLES, S.L. and FADLALLAH, Y.H. (1991) Reef coral survival and mortality at low temperatures in the Arabian Gulf: new species-specific lower temperature limits. *Coral Reefs* **9**: 231–237.

COLES, S.L. and GUNAY, N. (1989) Tar pollution on Saudi Arabian Gulf beaches. *Marine Pollution Bulletin* **18**: 214–218.

COLES, S.L. and McCAIN, J.C. (1990) Environmental factors affecting benthic communities of the western Arabian Gulf. *Marine Environmental Research* **29**: 289–315.

COLES, S.L. and TARR, A.B. (1990) Reef fish assemblages in the western Arabian Gulf: A geographically isolated population in an extreme environment. *Bulletin of Marine Science* **47**: 696–720.

COLL, J.C., LA BARRE, S., SAMMARCO, P.W., WILLIAMS, W.T. and BAKUS, G.J. (1982) Chemical defenses in soft corals (Coelenterata: Octocorallia) of the Great Barrier Reef: a study of comparative toxicities. *Marine Ecology Progress Series* **8**: 271–278.

COLLENETTE, S. (1985) *An Illustrated Guide to the Flowers of Saudi Arabia*. Scorpion, London. 514 pp.

CORNELIUS, P.F.S., FALCON, N.L., SOUTH, D. and VITA-FINZI, C. (1973) The Musandam expedition 1971–2. Scientific Results: Part 1, Biological aspects. *Geographical Journal* **139**: 400–403.

COUPER, A. (ed.) (1983) *The Times Atlas of the Oceans*. Times Books, London. 272 pp.

COWELL, E.B. (1978) Pollution of coastal zones by hydrocarbons. Oral submission to the European Parliament. European Parliamentary Hearings, 4 July, pp. 14–15.

CRANE, J. (1975) *Fiddler Crabs of The World. (Ocypodidae: genus* Uca*)*. Princeton University Press, Princeton. 660 pp.

CROSSLAND, C. (1938) The coral reefs at Ghardaqa, Red Sea. *Proceedings of the Zoological Society, London* **108**: 513–523.

CROSSLAND, C.J. (1983) Seasonal growth of *Acropora cf. formosa* and *Pocillopora damicornis* on a high latitude reef (Houtman Abrolhos, western Australia), *Proceedings of 4th International Symposium on coral reefs*, Manila **1**: 663–667.

CROSSLAND, C.J., DAWSON SHEPHERD, A., STAFFORD SMITH, M. and MARSHALL, J.I. (1987) Saudi Arabia: An Analysis of Coastal and Marine Habitats of the Red Sea. Saudi Arabia Marine Conservation Programme. Synoptic Report. International Union for Conservation of Nature: Geneva.

CURRIE, R.I., FISHER, A.E. and HARGREAVES, P.M. (1973) Arabian Sea upwelling. In: Zeitzschel, B. (ed.) *The Biology of the Indian Ocean*. Springer Verlag, New York. pp. 37–52.

CUSHING, D.H. (1973) Production in the Indian Ocean and the transfer from the primary to the secondary level. In: Zeitzschel, B. (ed.) *The Biology of the Indian Ocean*. Springer Verlag, New York. pp. 475–486.

DANA, J.D. (1843) On the temperature limiting the distribution of corals. *American Journal of Science* **45**: 130–131.

DART, J.K.G. (1972) Echinoids, algal lawn and coral recolonisation. *Nature* **239**: 50–51.

DARWIN, C. (1842) *The Structure and Distribution of Coral Reefs*. Smith Elder and Co, London.

DEN HARTOG, C. (1970) *The Seagrasses of the World*. North Holland Publishing Company, Amsterdam. 275 pp.

DEN HARTOG, C. (1979) Seagrasses and ecosystems, an appraisal of the research approach. *Aquatic Botany* **7**: 105–117.

DEN HARTOG, C. (1980) In: Phillips, R.C. and McRoy, C.P. (eds.) *Handbook of Seagrass Biology: An Ecosystem Perspective*. Garland STPM Press, New York. pp. x–xiii.

DEXTER, P.E. (1973) A shallow water design wave procedure applicable to small cays and submerged reefs. Engineering dynamics of the coastal Zone. 1st Australian Conference on Coastal Engineering, pp. 74–81.

DIAB, A. and AL-GHONAIME, M. (1985) An ecological study of the oil degradation bacteria in the Arabian Gulf waters. *Progress Report*, submitted to Environmental Protection Council, Kuwait.

DIAMOND, J.M. (1984) Evolution of ecological segregation in the New Guinea montane avifauna. In: Diamond, J. and Case, T.J. (eds.) *Community Ecology*. Harper and Row, New York. pp. 98–125.

DICKS, B. (1986) Oil and the black mangrove, *Avicennia marina* in the northern Red Sea. *Marine Pollution Bulletin* **17**: 500–503.

DICKS, B. (1987) Pollution. In: Edwards, A. & Head, S.M. (eds.) *Key Environments: The Red Sea*. Pergamon Press, Oxford. pp. 383–404.

DIETRICH, G. (1973) The unique situation in the environment of the Indian Ocean. In: Zeitzschel, B. (ed.) *The Biology of the Indian Ocean*. Springer Verlag, New York. pp. 1–6.

DIPPER, F. and WOODWARD, T. (1989) *The Living Seas. Marine Life of the Southern Gulf*. Motivate Publishing, Dubai. 95 pp.

DOHERTY, P.J. and SALE, P.F. (1986) Predation on juvenile coral reef fishes: an exclusion experiment. *Coral Reefs* **4**: 225–234.

DOHERTY, P.J. and WILLIAMS, D.McB. (1988) The replenishment of coral reef fish populations. *Oceanography and Marine Biology Annual Review* **26**: 487–551.

DONE, T.J. (1983) Coral zonation: its nature and significance. In: Barnes, D.J. (ed.) *Perspectives on Coral Reefs*. Brian Cloustan Publishing, Manuka, Australia. pp. 107–147.

DOR, I. (1975) The blue-green algae of the mangrove forests of Sinai. *Rapp. P.-v. Réun Commn int. Explor. scient. Mer Méditerr.* **23**: 109–110.

DOR, I. (1984) Epiphytic blue-green algae (Cyanobacteria) of the Sinai mangal: Considerations on vertical zonation and morphological adaptations. In: Por, F.D. and Dor, I. (eds.) *Hydrobiology of the Mangal*. Dr W. Junk Publishers, The Hague. pp. 35–54.

DOR, I. and LEVY, I. (1984) Primary productivity of the benthic algae in the hard-bottom mangal of Sinai. In: Por, F.D. and Dor, I. (eds.) *Hydrobiology of the Mangal.* Dr W. Junk Publishers, The Hague. pp. 179–191.

DOR, M. (1984) *Checklist of Fishes of the Red Sea.* Israel Academy of Science and Humanities. 437 pp.

DOTY, M.S. (1974) Coral Reef Roles Played by Free-Living Algae. *Proceedings of 2nd International Coral Reef Symposium* **1**: 27–33.

DOWIDAR, N.M. (1983) Primary production in the central Red Sea off Jiddah. In: Proceedings of International Conference on Marine Science in the Red Sea. Eds. Latif, A.F.A., Bajoumi, A.R. and Thompson, M.F. *Bulletin of Institute of Oceanography and Fisheries* **9**: 160–170.

DOWNING, N. (1985) Coral Reef communities in an extreme environment: The northwest Arabian Gulf. *Proceedings of 5th International Coral Reef Congress,* Tahiti. Vol. 6. pp. 343–348.

DREW, E.A. (1983) Light. In: Earll, R. & Erwin, D. (eds.) *Sublittoral Ecology. The Ecology of the Shallow Sublittoral Benthos.* Clarendon Press, Oxford. pp. 10–57.

DUGAN, P.J. (ed.) (1990) *Wetland conservation: A review of current issues and required action.* IUCN, Gland. 96 pp.

DULLO, W. (1984) Progressive diagenetic sequence of aragonite structures: Pleistocene coral reefs and their modern counterparts on the eastern Red Sea coast, Saudi Arabia. *Palaeontographica Americana* **54**: 254–260.

DUMONT, J.P.C. (1981) A report on the cheilostome Bryozoa of the Sudanese Red Sea. *Journal of Natural History* **15**: 623–637.

EAKIN, C.M. (1988) Avoidance of damselfish lawns by the sea urchin *Diadema mexicanum* at Uva Island, Panama. *Proceedings 6th International Coral Reef Symposium, Townsville* **2**: 21–26.

EDWARDS, A.J. and HEAD, S.M. (eds.) (1987) *Red Sea.* Pergamon Press, Oxford. 441 pp.

EDWARDS, F.J. (1987) Climate and oceanography. In: Edwards, A.J. & Head, S.M. (eds.). *Red Sea.* Pergamon Press, Oxford. pp. 45–69.

EDWARDS, S., HIND, S. and ROSENTHAL, D. (1981) Red Sea reef study 1981. Report of Cambridge University Underwater exploration Group. 44 pp.

EHRLICH, A. and DOR, I. (1985) Photosynthetic microorganisms of the Gavish Sabkha. In: Friedman, G.M. & Krumbein, W.E. (eds.) *Hypersaline Ecosystems. The Gavish Sabkha.* Springer Verlag, Berlin. pp. 296–321.

EHRLICH, P.R. (1975) The population biology of coral reef fishes. *Annual Reviews of Ecology and Systematics* **6**: 211–247.

EISLER, R. (1973) Latent effects of Iranian crude oil and a chemical oil dispersant on Red Sea molluscs. *Israel Journal of Zoology* **22**: 97–105.

EISLER, R. (1975) Toxic, sublethal and latent effects of petroleum on Red Sea macrofauna. Proceedings of the 1975 Conference on Prevention and Control of Oil Pollution, San Francisco. pp. 535–540.

EL-SAYED, S.Z. and JITTS, H.R. (1973) Phytoplankton production in the southeastern Indian Ocean. In: Zeitzschel, B. (ed.) *The Biology of the Indian Ocean.* Springer Verlag, New York. pp. 131–142.

EMARA, H.I. (1990) Oil pollution in the southern Arabian Gulf and Gulf of Oman. *Marine Pollution Bulletin* **21**: 399–401.

EMERY, K.O. (1956) Sediments and water of the Persian Gulf. *Bulletin of American Association of Petroleum Geologists* **40**: 2354–2383.

ENOMOTO, Y. (1971) Oceanographic survey and biological study of shrimps in the waters adjacent to the eastern coast of the state of Kuwait. *Bulletin of the Tokai Regional Fisheries Research Laboratory* **65**: 1–74.

EVANS, G. (1964) The Recent sedimentary facies of the Persian Gulf region. *Philosophical Transactions of the Royal Society, London. A.* **259**: 291–298.

EVANS, G., KENDAL, C.G.St.C and SKIPWITH, P.A.d'E. (1964) Origin of the coastal flats, the sabkha, of the Trucial Coast, Persian Gulf. *Nature, London* **202**: 759–761.

EVANS, G., SCHMIDT, V., BUSH, P. and NELSON, H. (1969) Strategy and geologic history of the sabkha, Abu Dhabi, Persian Gulf. *Sedimentology* **12**: 145–159.

EVENARI, M., GUTTERMAN, Y. and GAVISH, E. (1985) Botanical studies on coastal salinas and sabkhas of the Sinai. In: Friedman, G.M. & Krumbein, W.E. (eds.) *Hypersaline Ecosystems. The Gavish Sabkha.* Springer Verlag, Berlin. pp. 145–184.

FALKOWSKI, P.G. and DUBINSKY, Z. (1981) Light-shade adaptation of *Stylophora pistillata*, a hermatypic coral from the Gulf of Eilat. *Nature, London* **289**: 172–174.

FAO. (1980) Shrimp Resources Evaluation and Management. FAO UTFN/KUW/06/KUW. 10 pp.

FAO. (1989a) Review of the Shrimp and Fish Resources of the Gulfs. IOFC: DMG/89/Inf.5. 29 pp.

FAO. (1989b) Review of the Large Pelagic Resources in the Arabian Sea and Gulfs Area. IOFC: DMG/89/Inf.6. 14 pp.

FARMER, A.S.D. (1981) Prospects for peneid shrimp culture in arid lands. In: *Advances in Food Producing Systems for Arid and Semi Arid Lands.* Academic Press, New York. pp. 859–897.

FISHELSON, L. (1971) Ecology and distribution of the benthic fauna in the shallow water of the Red Sea. *Marine Biology* **10**: 113–133.

FISHELSON, L. (1973a) Ecology of coral reefs in the Gulf of Aqaba (Red Sea) influenced by pollution. *Oecologia* **12**: 55–67.

FISHELSON, L. (1973b) Ecological and biological phenomena influencing coral-species composition on the reef tables at Eilat (Gulf of Aqaba, Red Sea). *Marine Biology* **19**: 183–196.

FISHELSON, L. (1980) Marine reserves along the Sinai peninsula (northern Red Sea). *Helgolander wissenschaftliche Meeresuntersuchungen* **33**: 624–640.

FOWLER, S.W. (1985) Coastal baseline studies of pollutants in Bahrain, UAE and Oman. *Proceedings of ROPME Symposium on Regional Marine Pollution Monitoring and Research Programmes.* (ROPME/GC-4/2). pp. 155–180.

FRAZIER, J.G., BERTRAM, G.C. and EVANS, P.G.H. (1987) Turtles and marine mammals. In: Edwards, A.J. & Head, S.M. (eds.) *Red Sea.* Pergamon Press, Oxford. pp. 288–314.

FRICKE, H.W. and SCHUHMACHER, H. (1983) The depth limits of Red Sea stony corals, an ecophysiological problem. (A deep diving survey by submersible). P.S.Z.N.I. *Marine Ecology* **4**: 163–194.

FRIEDMAN, G.M. (1968) Geology and geochemistry of reefs, carbonate sediments and waters, Gulf of Aqaba (Elat), Red Sea. *Journal of Sedimentary Petrology* **38**: 895–919.

FRIEDMAN, G.M. (1972) Significance of Red Sea in problems of evaporites and basinal limestones. *American Association of Petroleum Geologists Bulletin* **56**: 1072–1086.

FRIEDMAN, G.M. (1985) Chapter 3. Gulf of Elat (Aqaba). Geological and sedimentological framework. In: Friedman, G.M. & Krumbein, W.E. (eds.) *Hypersaline Ecosystems. The Gavish Sabkha.* Springer Verlag, Berlin. pp. 39–71.

FRIEDMAN, G.M. and KRUMBEIN, W.E. (eds.) *Hypersaline Ecosystems. The Gavish Sabkha.* Springer Verlag, Berlin. 484 pp.

FRIEDMAN, G.M., SNEH, A. and OWEN, R.W. (1985) The Ras Mohammed Pool. Implications for the Gavish Sabkah. In: Friedman, G.M. & Krumbein, W.E. (eds.) *Hypersaline Ecosystems. The Gavish Sabkha.* Springer Verlag, Berlin. pp. 218–237.

FRYDL, P. (1979) The effect of parrotfish (Scaridae) on coral in Barbados, W.I. *International*

Revue Gesampten Hydrobiologie **64**: 737–748.

FRYDL, P. and STEARN, C.W. (1978) Rate of bioerosion by parrotfish in Barbados reef environments. *Journal of Sedimentary Petrology* **48**: 1149–1158.

FUGRO INC. (1977) Analyses and recommendations, offshore soil investigation, Yanbu complex, Saudi Arabia. *Report 76–106*, Vol. 1. Royal Commission for Jubail and Yanbu, Kingdom of Saudi Arabia. Yanbu.

GAB-ALLA, A.A.-F.A., HARTNOLL, R.G., GHOBASHY, A-F. and MOHAMMED, S. (1990) Biology of peneid prawns in the Suez Canal lakes. *Marine Biology* **107**: 417–426.

GALLAGHER, M.D., SCOTT, D.A., ORMOND, R.F.G., CONNOR, R.J. and JENNINGS, M.C. (1984) The distribution and conservation of seabirds breeding on the coasts and islands of Iran and Arabia. *ICBP Technical Publication* no. 2. pp. 421–456.

GASPERETTI, J. (1988) Snakes of Arabia. *Fauna of Saudi Arabia* **9**: 169–400.

GATTUSO, J.P. (1985) Features of depth effects on *Stylophora pistillata*, an hermatypic coral in the Gulf of Aqaba. Proceedings of 5th International Coral Reef Congress, Tahiti, Vol. 6: 95–100.

GEOREDA LTD. (1982) *Oceanography, final report.* Royal Commission for Jubail and Yanbu Contract PID-0203 Kingdom of Saudi Arabia.

GINSBERG, R.N. and SCHROEDER, J.H. (1973) Growth and fossilization of algal cup reefs, Bermuda. *Sedimentology* **20**: 575–614.

GJØSAETER, J. (1981) Abundance and production of lanternfish (Myctophidae) in the western and northern Arabian Sea. *Fiskeridirektoratets skrifter serie Havundersokelser* **17**: 215–257.

GJØSAETER, J. (1984) Mesopelagic fish, a large potential resource in the Arabian Sea. *Deep-Sea Research* **31**: 1019–1035.

GLADFELTER, W.B. and JOHNSON, W.S. (1983) Feeding niche separation in a guild of tropical reef fishes (Holocentridae). *Ecology* **64**: 552–563.

GLYNN, P.W. (1983) Final report on the effects of the sea star *Acanthaster* on Omani coral reefs, with some recommendations for further study. Report to Ministry of Agriculture and Fisheries and the Omani-American Joint Commission for Economic and Technical Cooperation, Muscat, Oman, 50 pp.

GOERING, J.J. and PARKER, P.L. (1972) Nitrogen fixation by epiphytes on sea grasses. *Limnology and Oceanography* **17**: 320–323.

GOHAR, H.A.F. (1957) The Red Sea dugong, *Dugong dugon* (Erxlb.), subspecies *tabernaculi* (Ruppelli). *Publications of the Marine Biological Station, Ghardaqa* **9**: 3–49.

GOLDMAN, B. and TALBOT, F.H. (1976) Aspects of the ecology of coral reef fishes. In: Jones, O.A. & Endean, R. (eds.) *The Biology and Geology of Coral Reefs*. 3, Biology 2. Academic Press, New York. pp. 125–154.

GOLOB, R. (1980) Statistical analysis of oil pollution in the Kuwait Action Plan region, and the implications of selected oil spills worldwide to the region. In: Proceedings of International Workshop on Combating Marine Pollution from Oil Exploration, Exploitation and Transportation in the Kuwait Action Plan region. December, Manama, Bahrain. IMCO/UNEP.

GOODMAN, S.M. (1985) Natural resources and management considerations: Gebel Elba Conservation Area, Egypt/Sudan. WWF/IUCN project No. 3612.

GORDEYEVA, K.T. (1970) Quantitative distribution of zooplankton in the Red Sea. *Oceanology (Washington)* **10**: 867–871.

GOREAU, T.F. (1969) Post-Pleistocene urban renewal in coral reefs. *Micronesica* **5**: 323–326.

GRAVIER, C. (1910a) Sur quelques particularités biologique des récifs madréporiques de la Baie de Tadjourah: Gulf of Aden. *C.r. Ass. Avanc. Sci.* **39**: 167–169.

GRAVIER, C. (1910b) Sur quelques formes nouvelles de Madréporaires de la Baie de

Tadjourah: Gulf of Aden. *Bull. Mus. natn. Hist nat. Paris* **16**: 273–276.

GRAVIER, C. (1910c) Sur les récifs coralliens de la Baie de Tadjourah et leurs Madréporaires: Gulf of Aden. *C.r. hebd. Seanc. Acad. Sci. Paris* **151**: 650–652.

GRAVIER, C. (1911) Les récifs de coraux et les madréporaires de la Baie de Tadjourah: Gulf of Aden. *Ann. Inst. Oceanogr. Paris.* **2**: 99.

GREEN, F.W. (1983) Comparison of present-day coral communities off the Oman coast with mid-tertiary corals from the Mam reef, near Seeb, Oman. Paper read to International Society for Reef Studies 8–9 December, Nice. 6 pp.

GREEN, F.W. (1984) Oman's coral. *PDO News*. Petroleum Development Organisation, Muscat, **3**: 6–11.

GREEN, F. and KEECH, R. (1986) *The Coral Seas of Muscat*. MEED, London. 106 pp.

GROOMBRIDGE, B. (1982) *The IUCN Amphibia-Reptilia Red Data Book, 1. (Testudines, Crocodylia, Rhynchocephalia)*. IUCN, Gland.

GROOMBRIDGE, B. (1984) India's sea turtles in world perspective. Manuscript. IUCN, Cambridge. 15 pp.

GUBBAY, S. (1988) *A Coastal Directory for Marine Nature Conservation*. Marine Conservation Society, UK, 319 pp.

GUBBAY, S. and ROSENTHAL, D. (1982) Reefwatch Egypt 1982. Report of the expedition. Cambridge University Underwater Exploration Group.

GUILCHER, A. (1988) A heretofore neglected type of coral reef: the ridge reef. Morphology and origin. *Proceedings of 6th International Coral Reef Symposium, Australia*, **3**: 399–402.

GULLAND, J.A. (ed.) (1971) *Fish Resources of the Ocean*. Fishing News Books, Surrey. 225 pp.

GUPTA, R.S. and KUREISHI, T.W. (1981) Present state of oil pollution in the Northern Indian Ocean. *Marine Pollution Bulletin* **12**: 295–301.

GVIRTZMAN, G., BUCHBINDER, B., SNEH, A., NIR, Y. and FRIEDMAN, G.M. (1977) Morphology of the Red Sea fringing reefs: a result of the erosional pattern of the last glacial low-stand sea level and the following Holocene recolonisation. *Mem. B.R.G.M.* **89**: 480–491.

GYGI, R.A. (1975) *Sparisoma viride* (Bonnaterre), the stoplight parrotfish, a major sediment producer on coral reefs of Bermuda. *Ecologae Geologicae Helvetiae* **68**: 327–359.

HALWAGY, R. (1986) Subkingdom Angiospermae. In: Jones, D.A (ed.) *A Field Guide to the Shores of Kuwait*. Blandford, Poole and University of Kuwait Press. pp. 36–39.

HALWAGY, R., CLAYTON, D. and BEHBEHANI, M. (eds.) (1986) *Marine Environment and Pollution. Proceedings of First Arabian Gulf Conference on Environment and Pollution*. University of Kuwait, 7–9 February 1982. 348 pp.

HAMNER, W.M. and CARLTON, J.H. (1979) Copepod swarms: attributes and role in coral reef ecosystems. *Limnology and Oceanography* **24**: 1–14.

HAMNER, W.M., JONES, M.S., CARLETON, J., HAURI, I.R. and WILLIAMS, D.McB. (1988) Zooplankton, planktivorous fish, and water currents on a windward reef face: Great Barrier Reef, Australia. *Bulletin of Marine Science* **42**: 459–479.

HARLIN, M.M. (1973) Transfer of products between epiphytic marine algae and host plants. *Journal of Phycology* **9**: 243–248.

HARRINGTON, F.A. (1976) Iran: surveys of the southern Iranian coastline with recommendations for additional marine reserves. In: *Promotion of the Establishment of marine parks and Reserves in the Northern Indian Ocean including the Red Sea and Persian Gulf*. IUCN Publ. New Series, no. 35: 5075.

HATCHER, B.G. (1985) Ecological research at the Houtman's Abrolhos: High latitude reefs of Western Australia. *Proceedings of 5th International Coral Reef Congress, Tahiti* **6**: 291–297.

HATCHER, B.G. and RIMMER, D.W. (1985) The role of grazing in controlling benthic community structure on a high latitude coral reef: measurements of grazing intensity. *Proceedings of 5th International Coral Reef Congress, Tahiti* **6**: 229–236.

HATCHER, B.G., KIRKMAN, H. and WOOD, W.F. (1987) Growth of the kelp *Ecklonia radiata* near the northern limit of its range in Western Australia. *Marine Biology* **95**: 63–73.

HAY, M.E. (1984) Patterns of fish and urchin grazing on Caribbean coral reefs: are previous results typical? *Ecology* **65**: 446–454.

HAYDEN, B.P., RAY, G.C. and DOLAN, R. (1984) Classification of coastal and marine environments. *Environmental Conservation* **11**: 199–207.

HAYWARD, A.B. (1982) Coral reefs in a clastic sedimentary environment: Fossil (Miocene, SW Turkey) and Modern (Recent, Red Sea) analogues. *Coral Reefs* **1**: 109–114.

HEAD, S.M. (1987a) Corals and coral reefs of the Red Sea. In: Edwards, A.J. & Head, S.M. (eds.) *Red Sea*. Pergamon Press, Oxford. pp. 128–151.

HEAD, S.M. (1987b) Red Sea fisheries. In: Edwards, A.J. & Head, S.M. (eds.) *Red Sea*. Pergamon Press, Oxford. pp. 363–382.

HEAD, S.M. (1987c) Minor invertebrate groups. In: Edwards, A.J. & Head, S.M. (eds.) *Red Sea*. Pergamon Press, Oxford. pp. 233–250.

HEYDORN, A. (1972) Tongaland's coral reefs — an endangered heritage. *African Wildlife* **26**: 20–23.

HIRTH, H.F., KLIKOFF, L.G. and HARPER, K.T. (1973) Sea grasses at Khor Umaira, People's Democratic Republic of Yemen with reference to their role in the diet of the green turtle *Chelonia mydas*. *Fishery Bulletin* **71**: 1093–1097.

HOGARTH, P.J. (1986) Occurrence of *Uca (Deltuca) urvillei* (Milne Edwards, 1852) in the Saudi Red Sea (Brachyura: Ocyopodidae). *Crustaceana* **51**: 221–223.

HOLTHUIS, L.B. (1973) Caridean shrimps found in land-locked saltwater pools of four Indopacific localities (Sinai peninsula, Funafuti Atoll, Maui and Hawaii Islands) with a description of one new genus and four new species. *Zoologische Verhandelingen* **128**: 1–48.

HOPLEY, D. (1982) *The geomorphology of the Great Barrier Reef. Quaternary development of coral reefs*. Wiley Interscience, New York.

HOPLEY, D., SLOCOMBE, A.M., MUIR, F. and GRANT, C. (1983) Nearshore fringing reefs in North Queensland. *Coral Reefs* **1**: 151–160.

HOUGHTON, J.T., JENKINS, G.T. and EPHRAUMS, J.J. (eds.) (1990) *Climate Change: The IPCC Scientific Assessment*. Cambridge University Press, Cambridge. 364 pp.

HULINGS, N.C. (1979) The ecology, biometry and biomass of the seagrass *Halophila stipulacea* along the Jordanian coast of the Gulf of Aqaba. *Botanica Marina* **22**: 425–430.

HULINGS, N.C. and KIRKMAN, H. (1982) Further observations and data on seagrasses along the Jordanian and Saudi Arabian coasts of the Gulf of Aqaba. *Tethys* **10**: 218–220.

HUMM, H.J. (1964) Epiphytes of the seagrass *Thalassia testudinum* in Florida. *Bulletin of Marine Science of the Gulf and Caribbean* **14**: 306–341.

HUMPHRIES, C.J. and PARENTI, L.R. (1986) *Cladistic Biogeography*. Clarendon Press, Oxford. 98 pp.

HUNTER, J.R. (1986) The physical oceanography of the Arabian Gulf: a review and theoretical interpretation of previous observations. In: Halwagy, R. *et al.* (eds.) *First Gulf Conference on Environment and Pollution*, Kuwait, February 7–9th, 1982. University of Kuwait. pp. 1–23.

HUSTON, M.A. (1985) Patterns of species diversity on coral reefs. *Annual Review of Ecology and Systematics* **16**: 149–177.

HUTCHINGS, P. and SAENGER, P. (1987) *Ecology of Mangroves*. University of Queensland Press, Queensland. 388 pp.

HUTCHINGS, P.A. (1986) Biological destruction of coral reefs. *Coral Reefs* **4**: 239–52.

ISH-SALOM-GORDON, N. and DUBINSKY, Z. (1990) Possible modes of salt secretion in *Avicennia marina* in the Sinai Egypt. *Plant Cell Physiology* **31**: 27–32.

IUCN. (1976) Papers and proceedings of the regional meeting at Tehran, Iran, 6–10 March. IUCN New Series 35. IUCN, Gland.

IUCN. (1982) Management requirements for natural habitats and biological resources on the Arabian Gulf coast of Saudi Arabia. IUCN/University of York report to MEPA.

IUCN. (1983) *Global status of mangrove ecosystems*. IUCN Commission on Ecology Papers No. 3, 88 pp.

IUCN. (1987) Arabian Gulf. Saudi Arabia: An Assessment of Biotopes and Coastal Zone Management Requirements for the Arabian Gulf. MEPA Coastal and Marine Management Series, Report No. 5. IUCN, Gland.

IUCN/UNEP. (1985a) The management and conservation of renewable marine resources in the Indian Ocean Region in the Kuwait Action Plan region. UNEP Regional Seas Reports & Studies. No. 63, 63 pp.

IUCN/UNEP. (1985b) The management and conservation of renewable marine resources in the Indian Ocean Region in the Red Sea and Gulf of Aden region. UNEP Regional Seas Reports & Studies. No. 64.

IUCN/UNEP. (1985c) Management and conservation of renewable marine resources in the Indian Ocean region: Overview. UNEP Regional Seas Reports & Studies. No. 60. 78 pp.

JACOBS, R.P.W.M. and DICKS, B. (1985) Seagrasses in the Zeit Bay area and at Ras Gharib (Egyptian Red Sea coast). *Aquatic Botany* **23**: 137–147.

JADO, A.R. and ZOETL, J.G. (eds.) (1984) *Sedimentological, hydrogeological, hydrochemical, geomorphological, geochronological and climatological investigations in Western Saudi Arabia. Quaternary Period in Saudi Arabia*. Vol. 2. Springer Verlag, Berlin.

JOHANNES, R.E., COLES, S.L. and KUENZEL, N.T. (1970) The role of zooplankton in the nutrition of some scleractinian corals. *Limnology and Oceanography* **15**: 579–586.

JOHANNES, R.E., WIEBE, W.J., CROSSLAND, C.J., RIMMER, D.W. and SMITH, S.V. (1983) Latitudinal limits of coral reef growth. *Marine Ecology Progress Series* **11**: 105–111.

JOHNSTON, D.J. (ed.) (1981) *The Environmental Law of The Sea*. IUCN Environmental Policy and Law Paper No. 18, 419.

JOKIEL, P.L. and COLES, S.L. (1990) Response of Hawaiian and other Indo-Pacific reef corals to elevated temperature. *Coral Reefs* **8**: 155–162.

JONES, D.A. (1984) Crabs of the mangal ecosystem. In: Por, F.D. and Dor, I. (eds.) *Hydrobiology of the Mangal*. Dr W. Junk Publishers, The Hague. pp. 89–109.

JONES, D.A. (1985) The biological characteristics of the marine habitats found within the ROPME Sea Area. *Proceedings of ROPME Symposium on Regional Marine Pollution Monitoring and Research Programmes*. (ROPME/GC-4/2). pp. 71–89.

JONES, D.A. (1986) *A Field Guide to the Sea shores of Kuwait and the Arabian Gulf*. University of Kuwait and Blandford Press, Poole. 192 pp.

JONES, D.A., PRICE, A.R.G. and HUGHES, R.N. (1978) Ecology of a high saline lagoon, Dawhat as Sayh, Arabian Gulf, Saudi Arabia. *Estuarine and Coastal Marine Science* **6**: 253–262.

JONES, D.A., GHAMRAWY, M. and WAHBEH, M.I. (1987) Littoral and shallow subtidal environments. In: Edwards, A. and Head S.M. (eds.) *Red Sea*. Pergamon Press, Oxford. pp. 169–193.

JUNGIUS, H. (1988) The national parks and protected area, concept and its application to the Arabian Peninsula. *Fauna of Saudi Arabia* **9**: 3–11.

KARBE, L. (1987) Hot brines and the deep sea environment. In: Edwards, A.J. & Head, S.M. (eds.) *Red Sea*. Pergamon Press, Oxford. pp. 70–89.

KEDIDI, S.M. (1984a) Description of the Artisanal Fishery at Tuwwal, Saudi Arabia. Catches, Efforts and Catches per Unit Effort Survey Conducted During 1981–1982. UNDP/FAO RAB/81/002/16. 17 pp.

KEDIDI, S.M. (1984b) The Red Sea Reef Associated Fishery of the Sudan. Catches, Efforts and Catches per Fishing Effort Survey Conducted During 1982–1984. UNDP/FAO RAB/83/023/06. 18 pp.

KENDALL, C.G.St.C. and SKIPWITH, P.A.d'E. (1968) Recent algal mats of a Persian Gulf lagoon. *Journal of Sedimentary Petrology* **38**: 1040–1059.

KENDALL, C.G.St.C. and SKIPWITH, P.A.d'E. (1969a) Holocene shallow water carbonate and evaporite sediments of Khor al Bazam, Abu Dhabi, southwest Persian Gulf. *Bulletin of American Association of Petroleum Geologists* **53**: 841–869.

KENDALL, C.G.St.C. and SKIPWITH, P.A.d'E. (1969b) Geomorphology of a recent shallow water carbonate province: Khor al Bazam, Trucial Coast, southwest Persian Gulf. *Bulletin of Geological Society of America* **80**: 865–892.

KIMOR, B. (1973) Plankton relations of the Red Sea, Persian Gulf and Arabian Sea. In: Zeitzschel, B. (ed.) *The Biology of the Indian Ocean*. Springer Verlag, New York. pp. 221–232.

KINSMAN, D.J.J. (1964) Reef coral tolerance of high temperatures and salinities. *Nature, London* **202**: 1280–1282.

KLEIN, R., LOYA, Y., GVIRTZMAN, G., ISDALE, P.J. and SUSIC, M. (1990) Seasonal rainfall in the Sinai desert during the late Quaterny inferred from flourescent bands in fossil corals. *Nature, London*, **345**: 145–147.

KLUMPP, D.W. and McKINNON, D. (1989) Temporal and spatial patterns in primary production of a coral-reef epilithic algal community. *Journal of Experimental Marine Biology and Ecology* **131**: 1–22.

KLUMPP, D.W., McKINNON, D. and DANIEL, P. (1987) Damselfish territories: zones of high productivity on coral reefs. *Marine Ecology Progress Series* **40**: 41–51.

KLUNZINGER, C.B. (1878) *Upper Egypt: Its People and its Products*. Blackie and Sons, Glasgow. 408 pp.

KOEMAN, J.H. (1982) Environmental effects of pesticides and other chemicals. IUCN/MEPA report for the Expert Meeting of the Gulf Co-ordinating Council to review environmental issues.

KOGO, M. (1986) Introduction: Our activities on mangroves in Arabia. In: *A report of mangrove research in Sultanate of Oman. The Fourth Research on Mangroves in the Middle East*. Japan Cooperation Center for the Middle East, No. 191, pp. 9–16.

KREY, J. (1973) Primary production in the Indian Ocean. In: Zeitzschel, B. (ed.) *The Biology of the Indian Ocean*. Springer Verlag, New York. pp. 115–126.

KRUMBEIN, W.E. (1985) Introduction and definitions. In: Friedman, G.M. and Krumbein, W.E. (eds.) *Hypersaline Ecosystems. The Gavish Sabkha*. Springer Verlag, Berlin. pp. 13–17.

KURONUMA, K. and ABE, Y. (1972) *Fishes of Kuwait*. Kuwait Institute for Scientific Research. 123 pp.

LA BARRE, S. and COLL, J.C. (1983) Movement in soft corals. The growth interaction between *Nephthea brasica* (Coelenterata: Octocorallia) and *Acropora hyacinthus* (Coelenterata: Scleractinia). *Marine Biology* **72**: 119–124.

LA BARRE, S., COLL, J.C. and SAMMARCO, P.W. (1986) Competitive strategies of soft corals (Coelenterata: Octocorallia): III. Spacing and aggressive interactions between alcyonarians. *Marine Ecology Progress Series* **28**: 147–156.

LASKER, R. (1981) The role of a stable ocean in larval fish survival and subsequent recruitment. In: Lasker, R. (ed.) *Marine Fish Larvae. Morphology, Ecology and Relation to Fisheries.* University of Washington Press, Seattle. pp. 80–88.

LAW, R. (1991) Fishing in evolutionary waters. *New Scientist* No. 1758, pp. 35–37.

LEGORE, R.S., MARSZALEK, D.S., HOFMANN, J.E. and CUDDEBACK, J.E. (1983) A field experiment to assess impact of chemically dispersed oil on Arabian Gulf corals. SPE 11444, Middle East Technical Conference, Bahrain, 14–17 March, pp. 51–57.

LEIS, J.M. (1986) Ecological requirements of Indo-Pacific larval fishes: a neglected zoogeographic factor. In: Uyeno, T., Arai, R., Taniuchi, T. and Matsuura, K. (eds.) *Indo-Pacific Fish Biology: Proceedings of the 2nd International Conference on Indo-Pacific Fishes. Ichthyological Society of Japan,* Tokyo. pp. 759–766.

LEMAY, M.H. and HALE, L.Z. (1989) *Coastal Resources Management: A Guide to Public Education Programs and Materials.* Kumarian Press, Connecticut. 58 pp.

LENZ, J. (1973) Zooplankton biomass and its relation to particulate matter in the upper 200 m of the Arabian Sea during the NE monsoon. In: Zeitschel, B. (ed.) *The Biology of the Indian Ocean.* Springer Verlag, New York. pp. 239–243.

LEWIS, A.H., JONES, D.A., GHAMRAWI, M. and KHOSHAM, S. (1973) An Analysis of the Arabian Gulf Shrimp Resources Landed in Saudi Arabia — 1965–71. Bulletin of the Marine Research Centre, Saudi Arabia, No. 4. 8 pp.

LEWIS, J.B. (1977) Processes of organic production on coral reefs. *Biological Reviews* 305–346.

LEWIS, J.B. (1982) Coral reef ecosystems. In: Longhurst, A.R. (ed.) *Analysis of marine ecosystems.* Academic Press, London. 99, 127–158.

LIGHTY, R.G. (1983) Fleshy-algal domination of a modern Bahamian barrier reef: example of an alternate climax reef community. Abstract. *Proceedings of 4th International Symposium on coral reefs,* Manila 2: 722.

LINDEN *et al.* (1990) State of the marine environment in the ROPME Sea Area. UNEP Regional Seas Reports and Studies. No. 112, Rev. 1. UNEP, Nairobi.

LIPKIN, Y. (1975) Food of the Red Sea *Dugong* (Mammalia: Sirenia) from Sinai. *Israel Journal of Zoology* 24: 81–98.

LIPKIN, Y. (1977) Seagrass vegetation of Sinai and Israel. In: McRoy, C.P. and Helfferich, C. (eds.) *Seagrass Ecosystems: A Scientific Perspective.* Marcel Dekker, New York. pp. 263–293.

LIPKIN, Y. (1979) Quantitative aspects of seagrass communities, particularly of those dominated by *Halophila stipulacea,* in Sinai (northern Red Sea). *Aquatic Botany* 7: 119–128.

LITTLER, M.M. and DOTY, M.S. (1975) Ecological components structuring the seaward edges of tropical Pacific Reefs: The distribution, communities and productivity of *Porolithon. Journal of Ecology* 63: 117–129.

LONGHURST, A.R. and PAULY, D. (1987) *Ecology of Tropical Oceans.* Academic Press, London. 407 pp.

LOYA, Y. (1972) Community structure and species diversity of hermatypic corals at Eilat, Red Sea. *Marine Biology* 13: 100–123.

LOYA, Y. (1975) Possible effects of water pollution on the community structure of Red Sea corals. *Marine Biology* 29: 177–185.

LOYA, Y. (1976a) Recolonisation of Red Sea corals affected by natural catastrophes and man-made perturbations. *Ecology* 57: 278–289.

LOYA, Y. (1976b) Settlement, mortality and recruitment of a Red Sea coral population. In: Mackie, G.O. (ed.) *Coelenterate Ecology and Behavior.* Plenum Press, New York. pp. 89–100.

LOYA, Y. and RINKEVICH, B. (1980) Effects of oil pollution on coral reef communities. *Marine Ecology Progress Series* **3**: 167–180.

LUCKHURST, B.E. and LUCKHURST, K. (1978) Analysis of the influence of substrate variables on coral reef fish communities. *Marine Biology* **49**: 317–323.

LUGO, A.E., EVINK, G., BRINSON, M., BROCE, A. and SNEDAKER, S.C. (1975) Diurnal rates of photosynthesis, respiration and transpiration in mangrove forests of south Florida. In: Golley, F.B. and Medina, E. (eds.) *Tropical Ecological Systems.* Springer-Verlag, New York.

MacARTHUR, R.H. and WILSON, E.O. (1967) *The Theory of Island Biogeography.* Princeton University Press, Princeton. 203 pp.

McCAIN, J.C. (1984a) Marine ecology of Saudi Arabia. The intertidal infauna of the sand beaches in the northern area, Arabian Gulf, Saudi Arabia. *Fauna of Saudi Arabia* **6**: 53–78.

McCAIN, J.C. (1984b) Marine ecology of Saudi Arabia. The nearshore, soft bottom benthic communities of the northern area, Arabian Gulf, Saudi Arabia. *Fauna of Saudi Arabia* **6**: 79–97.

McCAIN, J.C., TARR, A.B., CARPENTER, K.E. and COLES, S.L. (1984) Marine Ecology of Saudi Arabia. A survey of Coral Reefs and Reef fishes in the Northern Area, Arabian Gulf, Saudi Arabia. *Fauna of Saudi Arabia* **6**: 102–120.

McCLANAHAN, T.R. (1988) Coexistence in a sea urchin guild and its implications to coral reef diversity and degradation. *Oecologia* **77**: 210–218.

McDOWALL, R.J. unpublished. *Aspects of zoogeography, diversity and ecology of molluscs in the Red Sea.* D.Phil. University of York (in prep.).

McGILL, D.A. (1973) Light and nutrients in the Indian Ocean. In: Zeitzschel, B. (ed.) *The Biology of the Indian Ocean.* Springer Verlag, New York. pp. 53–102.

McGLADE, J.M. (1989) Integrated fisheries management models: Understanding the limits to marine resource exploitation. *American Fisheries Society Symposium* **6**: 139–165.

MACINTYRE, I.G. and PILKEY, O.H. (1969) Tropical reef corals: tolerance of low temperatures on the North Carolina continental shelf. *Science* **166**: 374–375.

MACNAE, W. (1968) A general account of the fauna and flora of mangrove swamps and forests in the Indo-Pacific region. *Advances in Marine Biology* **6**: 73–270.

McROY, C.P. and McMILLAN, C. (1977) Production ecology and physiology of seagrasses. In: McRoy, C.P. and Helfferich, C. (eds.) *Seagrass Ecosystems: A Scientific Perspective.* Dekker, New York. pp. 53–88.

MADANY, I.M., ALI, S.M. and AKHTER, M.S. (1987) The impact of dredging and reclamation in Bahrain. *Journal of Shoreline Management* **3**: 255–268.

MANDURA, A.S., SAIFULLAH, S.M. and KHAFATI, A.K. (1987) Mangrove ecosystem of Southern Red Sea Coast of Saudi Arabia. *Proceedings of Saudi Biological Society* **10**: 165–193.

MANN, K.H. (1982) *Ecology of Coastal Waters: A Systems Approach.* Studies in Ecology, Vol. 8. Blackwell Scientific Publications, Oxford. 322 pp.

MARINI, L. (1985) Study of a locality in Iran suitable for a marine biological station. In: I Parchi Costieri Mediterranei. *Proceedings of International Conference, Castellabate,* June (1973. Regione Campania Assesorato per il Turismo. pp. 685–706.

MASTALLER, M. (1978) The marine molluscan assemblages of Port Sudan, Red Sea. *Zoologische Mededelingen* **53**: 117–144.

MASTALLER, M. (1987) Molluscs of the Red Sea. In: Edwards, A.J. & Head, S.M. (eds.) *Red Sea.* Pergamon Press, Oxford. pp. 194–214.

MEPA/IUCN. (1989) The development and conservation of Saudi Arabia's coastal resources: The need for a coastal zone management programme. Audiovisual presentation (English and Arabic).

MEPA. (1991) Report on Dawhat and Daji field assessment and remediation proposal.

MERGNER, H. and SCHUHMACHER, H. (1981) Quantitative Analyse der Korallenbesied-lung eines Vorriffareals bei Aqaba (Rotes Meer). *Helgoländer Wissenschaftliche Meeresuntersuchungen* **34**: 337–354.

MERGNER, H. and SCHUHMACHER, H. (1985) Quantitative Analyse von Korallen-gemeinschaften des Sanganeb Atolls (mittleres Rotes Meer) I. Die Besiedlungs-struktur hydrodynamisch unterschiedlich exponierter Auben- und Innenriffe. *Helgoländer Wissenschaftliche Meeresuntersuchungen* **26**: 238–358, or **39**: 375–417.

MERGNER, H.H. and SVOBODA, A. (1977) Productivity and seasonal changes in selected reef areas in the Gulf of Aqaba (Red Sea). *Helgolander wissenschaftliche Meeresunter-suchungen* **30**: 383–399.

MESHAL, A.H. (1987) Hydrography of a hypersaline coastal lagoon in the Red Sea. *Estuarine and Coastal Marine Science* **24**: 167–175.

MESSIHA-HANNA, R. and ORMOND, R.F.G. (1982) Oil pollution, urchin erosion and coral reef deterioration in the Egyptian Red Sea. *Iraqi Journal of Marine Science* **1**: 35–57.

MEYER, J.L. and SCHULTZ, E.T. (1985) Migrating haemulid fishes as a source of nutrients and organic matter on coral reefs. *Limnology and Oceanography* **30**: 146–156.

MEYER, J.L., SCHULTZ, E.T. and HELFMAN, G.S. (1983) Fish schools: an asset to corals. *Science* **220**: 1047–1049.

MILLER, J. (1989) Marine turtles. Vol 1. An assessment of the conservation status of marine turtles in Saudi Arabia. MEPA, Coastal and Marine Management Series Report, No 9. MEPA, Jeddah. 209 pp.

MILLER, T.J., CROWDER, L.B., RICE, R.A. and MARSCHALL, E.A. (1988) Larval size and recruitment mechanisms in fishes: Toward a conceptual framework. *Canadian Journal of Fisheries and Aquatic Science* **45**: 1657–1670.

MILLIMAN, J.D. and EMERY, K.O. (1968) Sea levels during the past 35,000 years. *Science* **162**: 1121–1123.

MORAN, P.J. (1988) The *Acanthaster* phenomenon. Australian Institute of Marine Science Monograph Series, Volume 7. Australian Institute of Marine Science, Townsville, Queensland. 178 pp.

MORCOS, S.E. (1970) Physical and chemical oceanography of the Red Sea. *Oceanographic and Marine Biology Annual Reviews* **8**: 73–202.

MORGAN, G.R. (1985) Status of the Shrimp and Fish Resources of the Gulf. FAO Fisheries Circular, No. 792. 49 pp.

MORLEY, N.J.F. (1975) The coastal waters of the Red Sea. Bulletin of Marine Research Centre, Saudi Arabia, No 5. 19 pp.

MORTENSEN, Th. and GISLEN, T. (1941) Echinoderms from the Iranian Gulf. Asteroidea, Ophiuroidea and Echinoidea. In: Jessen, K. and Spärck, R. (eds.) *Danish Scientific Investigations in Iran*. Part II. Einar Munksgaard, Copenhagen. pp. 55–112.

MUNK, W.H. and SARGENT, M.C. (1948) Adjustment of Bikini Atoll to Ocean Waves. *American Geophysical Union* **29**: 855–860.

MUNRO, J.L. and WILLIAMS, D.McB. (1985) Assessment and management of coral reef fisheries: biological, environmental and socio-economic aspects. *Proceedings of the 5th International Coral Reef Congress*, Tahiti **4**: 545–572.

MUSTAFA, Z., NAWAB, Z., HRON, R. and LE LANN, F. (1980) Role of physical oceanography and environmental studies. In: *Proceedings of Symposium on the Coastal and Marine Environment of the Red Sea, Gulf of Aden and Tropical Western Indian Ocean*. Khartoum 9–14 January. Vol. III. University of Khartoum. pp. 7–32.

NCWCD. (1991) Gulf spill. A periodic brief newsletter. Issue 9. National Commission for

Wildlife Conservation and Development (NCWCD), Riyadh, Saudi Arabia.

NELSON SMITH, A. (1984) Effects of oil-industry related pollution on marine resources of the Kuwait Action Plan region. In: UNEP Regional Seas Reports & Studies. No. 44, pp. 35–52.

NEUMANN, L.D. (1979) The protection and development of the marine environment and coastal areas of the Kuwait Conference region: The programme of the United Nations System. In: *Proceedings of 1979 Oil Spill Conference (Prevention, Behavior, Control, Cleanup)*. 19–22 March, Los Angeles, USA. EPA/API/ISCG, pp. 287–291.

NEVE, P. and AL-AIIDY, H. (1973) The Red Sea Fisheries of Saudi Arabia. Bulletin of the Marine Research Centre, Jeddah, Saudi Arabia, No. 3. 32 pp.

NEYMAN, A.A., SOKOLOVA, M.N., VINOGRADOVA, N.G. and PASTERNAK, F.A. (1973) Some patterns of the distribution of bottom fauna in the Indian Ocea. In: Zeitzschel, B. (ed.) *The Biology of the Indian Ocean*. Springer-Verlag, New York. pp. 467–473.

NIEBUHR, C. (1792) *Travels through Arabia, and other countries in the Far East, performed by M. Niebuhr, now a Captain of Engineers in the service of the King of Denmark*. (Translated into English by Robert Heron. With notes by the translator; and illustrated with engravings and maps. Vol. I, xx + 424 pp. Vol. II, xii + 437 pp. Librairie du Liban, Beirut, reprint (no date).)

NIZAMUDDIN, M., HISCOCK, S., BARRATT, L. and ORMOND, R.F.G. (1986) The occurrence and morphology of *Sargassopsis* gen nov. (Phaeophyta, Fucales). Chapter 7. In: Ecology and productivity of the sublittoral Algae *Ecklonia radiata* and *Sargassopsis zanardini*. Part 1. *Ecological Studies of Southern Oman Kelp Communities*. Council for the Conservation of the Environment and Water Resources, Muscat, Oman, and Regional Organisation for the Protection of the Marine Environment, Kuwait, pp. 7.1–7.13.

NORTE, A.G.C. del, NANOLA, C.L., McMANUS, J.W., REYES, R.B., CAMPOS, W.L. and CABANSAG, J.B.P. (1989) Overfishing on a Philippine Coral Reef: A glimpse into the future. In: Magoon, O.T., Converse, H., Miner, D., Tobin, L.T. and Clark, D. (eds.) *Coastal Zone '89: Proceedings of the Sixth Symposium on Coastal and Ocean Management*. American Society of Civil Engineers, New York. pp. 3087–3097.

NOVACZEK, I. (1984) Response of *Ecklonia radiata* (Laminariales) to light at 15°C with reference to the field light budget at Goat Island Bay, New Zealand. *Marine Biology* **80**: 263–272.

OAKLEY, S.G. (1984) The effects of spearfishing pressure on grouper (Serranidae) populations in the eastern Red Sea. In: *Proceedings of Symposium on the Coral Reef Environment of the Red Sea*. Jeddah, Saudi Arabia, January. King Abdul Aziz University Press, Jeddah. pp. 341–359.

OAKLEY, S.G. (in press a) Ecological factors affecting the management of marine resources. In: *Proceedings of Workshop on the Ecological Imperatives for Sustainable Resource Use*. June 1989, National Commission for Wildlife Conservation and Development, Riyadh, Saudi Arabia.

OAKLEY, S.G. (in press b) Illuminating the lack of knowledge on the status of the Saudi Arabia shrimp fisheries. In: *Proceedings of Workshop on the Ecological Imperatives for Sustainable Resource Use*. June 1989, National Commission for Wildlife Conservation and Development, Riyadh, Saudi Arabia.

OAKLEY, S.G. and BAKHSH, A.A. (in press) The trawl fishery for shrimp and fin fish in the Jizan region of the Red Sea (1985–87). In: *Proceedings of Saudi Biological Society*, 1988.

ODUM, H.T. and ODUM, E.P. (1955) Trophic structure and productivity of a windward coral reef community on Eniwetok Atoll. *Ecological Monographs* **25**: 291–320.

OGDEN, J. (1988) The influence of adjacent systems on the structure and function of coral reefs. *Proceedings of the 6th International Coral Reef Symposium*, Townsville **1**: 123–130.

OGDEN, J.C. and GLADFELTER, E.H. (1983) Coral reefs, seagrass beds and mangroves: Their interaction in the coastal zones of the Caribbean. UNESCO 23: 133 pp.

OGDEN, J.C. and LOBEL, P.S. (1978) The role of herbivorous fishes and urchins in coral reef communities. *Environmental Biology of Fishes* **3**: 49–63.

OGDEN, J.C. and QUINN, T.P. (1984) Migration in coral reef fishes: ecological significance and orientation mechanisms. In: Mcleave, J.D., Arnold, G.P., Dodson, J.J. & Neill, W.H. (eds.) *Mechanisms of Migration in Fishes*. Plenum Press, New York. pp. 293–308.

OREN, O.H. (1962) The Israel south Red Sea Expedition. *Nature, London* **194**: 1134–1137.

OREN, D.H. (1964) Hydrography of Dahlak Archipelago (Red Sea). *Sea Fisheries Research Bulletin* **35**: 3–22.

ORME, A.R. (1982) Africa: Coastal Ecology. In: Schwartz, M.L. (ed.) *Encyclopedia of Beaches and Coastal Environments*. Hutchinson Ross, Stroudsburg. pp. 3–16.

ORMOND, R.F.G. (1987) Conservation and management. In: Edwards, A. and Head, S.M. (eds.) *Red Sea*. Pergamon Press, Oxford. pp. 405–423.

ORMOND, R.F.G. and CAMPBELL, A.C. (1974) Formation and breakdown of *Acanthaster planci* aggregations in the Red Sea. *Proceedings of 2nd International Coral Reef Symposium* **1**: 596–619.

ORMOND, R.G.F. and EDWARDS, A.J. (1987) Red Sea Fishes. In: Edwards, A.J. & Head, S.M. (eds.) *Red Sea*. Pergamon Press, Oxford. pp. 251–287.

ORMOND, R.F.G., CAMPBELL, A.C., HEAD, S.M., MOORE, R.J., RAINBOW, P.S. and SANDERS, A.P. (1973) Formation and breakdown of aggregations of the crown of thorns starfish *Acanthaster planci* in the Red Sea (L). *Nature, London* **246**: 167–169.

ORMOND, R.F.G., DAWSON SHEPHERD, A.R., PRICE, A.R.G. and PITTS, J.R. (1984a) Report on the distribution of habitats and species in the Saudi Arabian Red Sea. 1. IUCN/MEPA, Kingdom of Saudi Arabia. 123 pp.

ORMOND, R.F.G., DAWSON SHEPHERD, A.R., PRICE, A.R.G. and PITTS, J.R. (1984b) Report on the distribution of habitats and species in the Saudi Arabian Red Sea. 2. IUCN/MEPA, Kingdom of Saudi Arabia. 151 pp.

ORMOND, R.F.G., DAWSON SHEPHERD, A.R., PRICE, A.R.G. and PITTS, J.R. (1984c) Management of Red Sea coastal resources: recommendations for protected areas. IUCN/MEPA, Kingdom of Saudi Arabia. 113 pp.

ORMOND, R.F.G., DAWSON SHEPHERD, A.R., PRICE, A.R.G. and PITTS, R.J. (1986a) Saudi Arabian Marine Conservation Programme. Data from Detailed Study Sites. IUCN/MEPA Report, No. 10. Kingdom of Saudi Arabia.

ORMOND, R.F.G., DAWSON SHEPHERD, A.R., PRICE, A.R.G. and PITTS, R.J. (1986b) Distribution of Habitats and Species Along the Southern Red Sea Coast of Saudi Arabia. IUCN/MEPA, Report, No. 11. Kingdom of Saudi Arabia. 61 pp.

ORMOND, R.F.G., PRICE, A.R.G. and DAWSON SHEPHERD, A.R. (1988) The distribution and character of mangroves in the Red Sea, Arabian Gulf and southern Arabia. In: Field, C.D and Vannucci, M. (eds.) *Symposium on New Perspectives in Research and Management of Mangrove Ecosystems*. Proceedings. November 11–14, 1986, Colombo, Sri Lanka. UNDP/UNESCO. pp. 125–130.

PAPENFUSS, G.E. (1968) A history, catalogue and bibliography of benthic Red Sea algae. *Israel Journal of Botany* **17**: 1–118.

PATZERT, W.C. (1974) Wind-induced reversal in Red Sea circulation. *Deep-Sea Research* **21**: 109–121.

PAULY, D. and MATHEWS, C.P. (1986) Kuwait's finfish catch three times more shrimp than its trawlers. *Naga, ICLARM Quarterly, Philippines* **9**: 11–12.

PEACOCK, N.A. (1979) Final Report: the Fishery Resource Survey of the Saudi Arabian Red Sea, February 1977–October 1979. Field Report, No. 40. Fisheries Development Project, Kingdom of Saudi Arabia, Ministry of Agriculture and Water. 28 pp.

PEARCE, F. (1991) Desert fires cast a shadow over Asia. *New Scientist* **1751**: 30–31.

PENHALE, P.A. (1977) Macrophyte-epiphyte biomass and productivity in an eelgrass (*Zostera marina* L.) community. *Journal of Experimental Marine Biology & Ecology* **26**: 211–224.

PENN, J.W. (1975) The influence of tidal cycles on the distributional pathway of *Penaeus latisulcatus* Kushinouye in Shark Bay, Western Australia. *Australian Journal of Marine & Freshwater Research* **26**: 93–102.

PETERSEN, C.G.J. (1913) Valuation of the sea, II: The animal communities of the sea bottom and their importance for marine zoogeography. *Report of the Danish Biological Station* **21**: 1–44.

PETERSEN, C.G.J. (1918) The sea bottom and its production of fish food; a survey of the work done in connection with valuation of the Danish waters from 1883–1917. *Report of the Danish Biological Station* **23**: 1–82.

PHILLIPS, R.C. and McROY, C.P. (1980) *Handbook of Seagrass Biology: An Ecosystem Perspective*. Garland STPM Press, New York. pp. 353.

PHILLIPS, R.C. and MENEZ, E.G. (1988) *Seagrasses*. Smithsonian Contributions to the Marine Sciences, No. 34. 104 pp.

PICHON, M. (1971) Comparative study of the main features of some coral reefs of Madagascar, La Reunion and Mauritius. In: Stoddart, D.R. and Yonge, C.M. (eds.) *Regional variation in Indian Ocean reefs. Symposium of the Zoological Society of London, 28.* Academic Press, London. pp. 185–216.

PICHON, M. (1978) Recherches sur les peuplements a dominance d'anthozoaires dans les recifs coralliens de Tulear (Madagascar). *Atoll Research Bulletin* **222**: XXXV 477.

PIELOU, E.C. (1975) *Ecological Diversity*. Wiley-Interscience, New York.

PIELOU, E.C. (1979) *Biogeography*. John Wiley, New York. 351 pp.

PILCHER, N., NAWWAB, A-R., OAKLEY, S. and AL-MANSI, A. (in press) An updated summary on the artisanal fishery off the Saudi Arabian Gulf coast. In: *Proceedings of Workshop on the Ecological Imperatives for Sustainable Resource Use.* June 1989, National Commission for Wildlife Conservation and Development, Riyadh, Saudi Arabia.

POLUNIN, N.V.C. and KOIKE, I. (1987) Temporal focusing of nitrogen release by a periodically feeding herbivorous reef fish. *Journal of Experimental Marine Biology and Ecology* **111**: 285–96.

PONOMAREVA, L.A. (1968) Quantitative distribution of zooplankton in the Red Sea as observed in the period May to June 1966. *Oceanology Washington.* (Translation of Okeanologiya) **8**: 240–242.

POR, F.D. (1968) Copepods of some land-locked basins on the Islands of Entebeir and Nocra (Dahlak Archipelago, Red Sea). *Sea Fisheries Research Station, Haifa Bulletin* **49**: 32–50.

POR, F.D. (1972) Hydrobiological notes on the high salinity waters of the Sinai Peninsula. *Marine Biology* **14**: 111–119.

POR, F.D. (1975) The Coleoptera-dominated fauna of the hypersaline Solar Lake (Gulf of Elat, Red Sea). *10th European Symposium of Marine Biology*, Osttend **2**: 563–573.

POR, F.D. (1978) *Lessepsian Migration*. Ecological Studies 23. Springer Verlag, Berlin. 228 pp.

POR, F.D. (1984) The ecosystem of the mangal: general considerations. In: Por, F.D. &

Dor, I. (eds.) *Hydrobiology of the Mangal*. Dr W. Junk Publishers, The Hague. pp. 1–14.

POR, F.D. (1985) Anchialine pools — comparative hydrobiology. In: Friedman, G.M. and Krumbein, W.E. (eds.) *Hypersaline Ecosystems. The Gavish Sabkha*. Springer Verlag, Berlin. pp. 136–144.

POR, F.D. and DOR, I. (eds.) (1984) *Hydrobiology of the Mangal*. Dr W. Junk Publishers, The Hague. 260 pp.

POR, F.D. and TSURNAMAL, M. (1973) Ecology of the Ras Mohammed crack in Sinai. *Nature, London* **241**: 43–44.

POR, F.D., DOR, I. and AMIR, A. (1977) The mangal of Sinai: Limits of an ecosytem. *Helgoländer Meeresuntersuchungen* **30**: 295–314.

PORTER, J.W., PORTER, K.G. and BATAC-CATALAN, Z. (1977) Quantitative sampling of Indo-Pacific demersal reef plankton. *Proceedings of 3rd International Symposium on Coral Reefs. Miami* **1**: 105–112.

POTTS, D.C. (1983) Evolutionary disequilibrium among Indo-Pacific corals. *Bulletin of Marine Science* **33**: 619–632.

PREEN, A. (1989) Dugongs. Vol. 1. The status and conservation of Dugongs in the Arabian Region. MEPA Coastal and Marine Management Series, report No 10. MEPA, Jeddah, 200 pp.

PRELL, W.L. (1984) Variation of monsoonal upwelling: a response to changing solar radiation. Climate Processes and Climate Sensitivity. *Geophysical Monographs* No. 29. pp. 48–57.

PRELL, W.L. and KUTZBACH, J.E. (1987) Monsoon variability over the past 150,000 years. *Journal of Geophysical Research* **92**: 8411–8425.

PRELL, W.L. and VAN CAMPO, E. (1986) Coherent response of Arabian Sea upwelling and pollen transport to late Quaternary monsoonal winds. *Nature, London* **323**: 526–528.

PRICE, A.R.G. (1976) The Peneid Shrimp Fishery of the NW Arabian Gulf including the biology of *Peneus semisulcatus*. MSc Thesis, Univ. of Wales.

PRICE, A.R.G. (1979) Temporal variations in abundance of penaeid shrimp larvae and oceanographic conditions off Ras Tanura, western Arabian Gulf. *Estuarine & Coastal Marine Science* **9**: 451–465.

PRICE, A.R.G. (1982a) Echinoderms of Saudi Arabia. Comparison between echinoderm faunas of Arabian Gulf, SE Arabia, Red Sea and Gulfs of Aqaba and Suez. *Fauna of Saudi Arabia* **4**: 3–21.

PRICE, A.R.G. (1982b) Distribution of penaeid shrimp larvae along the Arabian Gulf coast of Saudi Arabia. *Journal of Natural History* **16**: 745–757.

PRICE, A.R.G. (1982c) Western Arabian Gulf echinoderms in high salinity waters and the occurrence of dwarfism. *Journal of Natural History* **16**: 519–527.

PRICE, A.R.G. (1983) Echinoderms of Saudi Arabia. Echinoderms of the Arabian Gulf coast of Saudi Arabia. *Fauna of Saudi Arabia* **5**: 28–108.

PRICE, A.R.G. (1990) Rapid assessment of coastal zone management requirements: case study in the Arabian Gulf. *Ocean & Shoreline Management* **13**: 1–19.

PRICE, A.R.G. and COLES, S.L. (in press) Aspects of seagrass ecology along the Western Arabian Gulf. *Hydrobiologia*.

PRICE, A.R.G. and JONES, D.A. (1975) Commercial and biological aspects of the Saudi Arabian Gulf shrimp fishery. *Bulletin of the Marine Research Centre, Jeddah, Saudi Arabia*, No. 6. 24 pp.

PRICE, A.R.G. and NELSON-SMITH, A. (1986) Observations on surface pollution in the Indian Ocean during the Sindbad Voyage (1980–81). *Marine Pollution Bulletin* **17**: 60–62.

PRICE, A.R.G. and SHEPPARD, C.R.C. (1991) The Gulf: Past, present and possible future states. *Marine Pollution Bulletin* **22**: 222–227.

PRICE, A.R.G., VOUSDEN, D.H.P. and ORMOND, R.F.G. (1983) Ecological study of sites on the coast of Bahrain, with special reference to the shrimp fishery and possible impact from the Saudi-Bahrain Causeway under construction. IUCN report to the UNEP Regional Seas Programme, Geneva.

PRICE, A.R.G., CHIFFINGS, T.W., ATKINSON, M.J. and WRATHALL, T.J. (1987a) Appraisal of Resources in the Saudi Arabian Gulf. In: Magoon, O.T., Converse, H., Miner, D., Tobin, L.T., Clark, D. and Domurat, G. (eds.) *5th Symposium on Coastal and Ocean Management*. Vol. 1. American Society of Coastal Engineers, New York. pp. 1031–1045.

PRICE, A.R.G., MEDLEY, P.A.H., McDOWALL, R.J., DAWSON SHEPHERD, A.R., HOGARTH, P.J. and ORMOND, R.F.G. (1987b) Aspects of mangal ecology along the Red Sea coast of Saudi Arabia. *Journal of Natural History* **21**: 449–464.

PRICE, A.R.G., WRATHALL, T.J. and BERNARD, S.M. (1987c) Occurrence of tar and other pollution along the Saudi Arabian shores of the Gulf. *Marine Pollution Bulletin* **18**: 650–651.

PRICE, A.R.G., CROSSLAND, C.J., DAWSON SHEPHERD, A.R., McDOWALL, R.J., MEDLEY, P.A.H., ORMOND, R.F.G., STAFFORD SMITH, M.G. and WRATHALL, T.J. (1988) Aspects of seagrass ecology along the eastern coast of the Red Sea. *Botanica Marina* **31**: 83–92.

PROBYN, T.A and McQUAID, C.D. (1985) In-situ measurements of nitrogenous uptake by kelp (*Ecklonia maxima*) and phytoplankton in a nitrate-rich upwelling environment. *Marine Biology* **88**: 149–154.

PURSER, B.H. (1973) *The Persian Gulf*. Springer Verlag, New York.

PURSER, B.H. (1985) Coastal evaporite systems. In: Friedman, G.M. and Krumbein, W.E. (eds.) *Hypersaline Ecosystems. The Gavish Sabkha*. Springer Verlag, Berlin. pp. 72–102.

PURSER, B.H. and SEIBOLD, E. (1973) The principal environmental factors influencing Holocene sedimentation and diagenesis in the Persian Gulf. In: Purser, B.H. *The Persian Gulf*. Springer Verlag, New York. pp. 1–9.

QUINN, N.J. and KOJIS, B.J. (1985) Does the presence of coral reefs in proximity to a tropical estuary affect the estuarine fish assemblage? Proceedings of 5th International Coral Reef Congress, Tahiti **5**: 445–450.

RABANAL, H.R. and BEUSCHEL, G.K. (1978) The mangroves and related coastal fishery resources in the United Arab Emirates. FAO report W/L6740.

RAO, T.S.S. (1973) Zooplankton studies in the Indian Ocean. In: Zeitzschel, B. (ed.) *The Biology of the Indian Ocean*. Springer Verlag, New York. pp. 243–256.

RAY, G.C. (1976) Critical marine habitats. In: IUCN. An international conference on marine parks and reserves. IUCN Publication New Series No. 37, pp. 15–60.

RAY, G.C. and HAYDEN, B.P. (1986) Classification of coastal and marine environments for identification of biosphere reserves. IUCN Project No. 9148 (Final Report).

REICE, S.R., SPIRA, Y. and POR, F.D. (1984) Decomposition in the mangal of Sinai: The effect of spatial heterogeneity. In: Por, F.D. and Dor, I. (eds.) *Hydrobiology of the Mangal*. Dr W. Junk Publishers, The Hague. pp. 193–199.

REISS, Z. (1979) Foraminiferal research in the Gulf of Aqaba: a review. *Utrecht Micropaleontology Bulletin* **15**: 7–25.

REISS, Z. and HOTTINGER, L. (1984) *The Gulf of Aqaba*. Ecological Studies 50, Springer Verlag, Berlin.

REISS, Z., LUZ, B., ALMOGI-LABIN, A., HALICZ, E., WINTER, A. and WULF, M. (1980) Late Quaternary paleooceanography of the Gulf of Aqaba (Eliat) Red Sea.

Quaternary Research **14**: 294–308.

RHEINHEIMER, G. (1973) Bacteriological investigations in the Arabian Sea. In: Zeitzschel, B. (ed.) *The Biology of the Indian Ocean.* Springer Verlag, New York. pp. 187–188.

RINKEVICH, B. and LOYA, Y. (1979) Laboratory experiments on the effects of crude oil on the Red Sea coral *Stylophora pistillata. Marine Pollution Bulletin* **10**: 328–330.

RINKEVICH, B. and LOYA, Y. (1984a) Does light enhance calcification in hermatypic corals? *Marine Biology* **80**: 1–6.

RINKEVICH, B. and LOYA, Y. (1984b) Coral illumination through an optic glass-fiber: incorporation of 14C photosynthates. *Marine Biology* **80**: 7–15.

RISK, M.J. (1972) Fish diversity on a coral reef in the Virgin Islands. *Atoll Research Bulletin* **193**: 1–6.

ROBERTS, C.M. (1985) Resource sharing in territorial herbivorous reef fishes. *Proceedings of the 5th International Coral Reef Congress,* Tahiti **4**: 17–22.

ROBERTS, C.M. (1986) Aspects of Coral Reef Fish Community Structure in the Saudi Arabian Red Sea and on the Great Barrier Reef. D.Phil. Thesis, University of York, UK.

ROBERTS, C.M. (1991) Larval mortality and the composition of coral reef fish communities. *Trends in Ecology and Evolution* **6**: 83–87.

ROBERTS, C.M. and ORMOND, R.F.G. (1987) Habitat complexity and coral reef fish diversity and abundance on Red Sea fringing reefs. *Marine Ecology Progress Series* **41**: 1–8.

ROBERTS, C.M., ORMOND, R.F.G and DAWSON SHEPHERD, A.R. (1988) The usefulness of butterflyfishes as environmental indicators on coral reefs. *Proceedings of the 6th International Coral Reef Symposium, Townsville* **2**: 331–336.

ROBERTS, C.M., DAWSON SHEPHERD, A.R. and ORMOND, R.F.G. Large scale variation in assemblage structure of Red Sea butterflyfishes and angelfishes. Unpublished manuscript.

ROCHFORD, D.J. (1966) Source regions of oxygen maxima in intermediate depths of the Arabian Sea. *Australian Journal of Marine and Freshwater Research* **17**: 1–30.

ROCHFORD, D.J. (1967) The phosphate levels of major surface currents of the Indian Ocean. *Australian Journal of Marine and Freshwater Research* **18**: 1–22.

ROGERS, C.S. (1990) Responses of coral reefs and reef organisms to sedimentation. *Marine Ecology Progress Series* **62**: 185–202.

ROPER, C.F.E., SWEENEY, M.J. and NAUEN, C.E. (1984) FAO Species Catalogue. Vol. 3. Cephalopods of the World. An Annotated and Illustrated Catalogue of Species of Interest to Fisheries. FAO Fish. Synop., (125) Vol. 3. 277 pp.

ROPME. (1987) Dust fallout in the northern part of the ROPME Sea Area. Regional Organisation for the Protection of the Marine Environment. GC-5/005. Kuwait.

ROPME/UNEP. (1988) Proceedings of the ROPME Workshop on coastal area development. UNEP Regional Seas Reports and Studies, No. 90.

ROSEN, B.R. (1971) The distribution of reef coral genera in the Indian Ocean. *Symposium of the Zoological Society, London,* **28**: 263–299.

ROSEN, B.R. (1981) The tropical high diversity enigma — the corals' eye view. In: Forey, P.L. & Greenwood, G.P.M. (eds.) *Chance, Change and Challenge: The Evolving Biosphere.* British Museum (Natural History) and Cambridge University Press, pp. 103–129.

ROSEN, B.R. (1984) Reef coral biogeography and climate through the late Cainozoic: just islands in the sun or a critical pattern of islands? In: Brenchley, P.J. (ed.) *Fossils and Climate.* pp. 201–262.

ROSEN, B.R. (1988) Progress, problems and patterns in the biogeography of reef corals and other tropical marine organisms. *Helgoländer Wissenschaftliche Meeresunter-suchungen* **42**: 269–301.

ROSS, J.P. (1979) Sea turtles in the Sultanate of Oman. Report of World Wildlife Fund Project 1320.

ROSS, J.P. and BARWANI, M.A. (1979) Review of sea turtles in the Arabian Area. In: Bjorndal, K.A. (ed.) *Biology and Conservation of Sea Turtles.* Smithsonian Institution Press, Washington DC. pp. 373–383.

RUSS, G. (1987) Is rate of removal of algae by grazers reduced inside territories of tropical damselfishes? *Journal of Experimental Marine Biology and Ecology* **110**: 1–17.

RYTHER, J.H. and MENZEL, D.W. (1965) On the production, composition and distribution of organic matter in the western Arabian Sea. *Deep Sea Research* **12**: 199–209.

RYTHER, J.H., HALL, J.R., PEASE, A.K., BAKUM, A. and SONES, M.M. (1966) Primary organic production in relation to the chemistry and hydrology of the western Indian Ocean. *Limnology and Oceanography* **11**: 371–380.

SAAD, M.A.H. (1978) Seasonal variations of some physiochemical conditions of Shatt al Arab estuary, Iraq. *Estuarine and Coastal Marine Science* **6**: 503–513.

SAAD, M.A.H. and ARLT, G. (1977) Studies on the bottom deposits and the meiofauna of Shatt al Arab and the Arabian Gulf. *Cah. Biol. Mar.* **18**: 71–84.

SALE, P.F. (1980) The ecology of fishes on coral reefs. *Oceanography and Marine Biology Annual Review* **18**: 367–421.

SALM, R. and DOBBIN, J.A. (1986) Oman Coastal Zone Management Plan. Greater Capital Area. IUCN, Gland. 78 pp.

SALM, R.V. and DOBBIN, J.A. (1987) A coastal zone management strategy for the Sultanate of Oman. In: Magoon, O.T., Converse, H., Miner, D., Tobin, L.T., Clark, D. & Domurat, G. (eds.) *5th Symposium on Coastal and Ocean Management.* Vol. 1. American Society of Coastal Engineers, New York. pp. 97–106.

SALM, R.V. and JENSEN, R.A.C. (1989) Oman. Coastal zone management plan. Dhofar. Volumes 1 and 2. IUCN, Gland.

SALM, R., DOBBIN, J.A. and PAPASTAVROU, V.A. (1988) Oman Coastal Zone Management Plan. Quiryat to Ras Al Hadd. IUCN, Gland. 56 pp.

SAMMARCO, P.W. (1985) The Great Barrier Reef vs the Caribbean: comparisons of grazers, coral recruitment patterns and reef recovery. *Proceedings of 5th International Coral Reef Congress,* Tahiti **4**: 391–397.

SAMMARCO, P.W. and CARLETON, J.H. (1981) Damselfish territoriality and coral community structure: reduced grazing, coral recruitment, and effects on coral spat. *Proceedings of the 4th International Coral Reef Symposium,* Manila **2**: 525–535.

SAMMARCO, P.W., COLL, J.C., LA BARRE, S. and WILLIS, B. (1983) Competitive strategies of soft corals (Coelenterata: Octocorallia): Allelopathic effects on selected scleractinian corals. *Coral Reefs* **1**: 173–178.

SAMMARCO, P.W., COLL, J.C. and LA BARRE. (1985) Competitive strategies of soft corals (Coelenterata: Octocorallia) II. Variable defensive response and susceptibility to scleractinian corals. *Journal of Experimental Marine Biology and Ecology* **91**: 199–215.

SAMRA EL. M.I., EMARA, H.I. and SHUNBO, F. (1986) Dissolved petroleum hydrocarbons in the Northwest Arabian Gulf. *Marine Pollution Bulletin* **17**: 65–68.

SANDERS, M.J. (1981a) Preliminary Stock Assessment for the Deep Sea Lobster *Puerulus sewelli* taken off the coast of the People's Democratic Republic of Yemen. UNDP/FAO RAB/77/008/18. 48 pp.

SANDERS, M.J. (1981b) Revised Stock Assessment for the Cuttlefish *Sepia pharaonis* taken off the coast of the People's Democratic Republic of Yemen. UNDP/FAO RAB/77/008/13. 44 pp.

SANDERS, M.J. 1982) Preliminary Stock Assessment for the Abalone Taken off the South East Coast of Oman. FAO/UNDP FI:DP/RAB/80/015/3. 48 pp.

SANDERS, M.J. (1984) Development of Fisheries in Areas of the Red Sea and Gulf of Aden. Final Report. UNDP/FAO RAB/83/023/INT/02. 24 pp.

SANDERS, M.J. and BOUHLEL, M. (1982) Prediction of the Annual Catch of Cuttlefish from the People's Democratic Republic of Yemen During 1982. UNDP/FAO RAB/81/002/3. 20 pp.

SANDERS, M.J. and KEDIDI, S.M. (1981) Summary Review of Red Sea Commercial Fisheries. Catches and Stock Assessments Including Maps of Actual and Potential Fishing Grounds. UNDP/FAO RAB/77/008/19. 49 pp.

SANDERS, M.J. and KEDIDI, S.M. (1984a) Stock Assessment for the Spangled Emperor *Lethrinus nebulosus* Caught by Small Scale Fishermen Along the Egyptian Red Sea Coast. UNDP/FAO RAB/83/023/01. 41 pp.

SANDERS, M.J. and KEDIDI, S.M. (1984b) Catches, Fishing Efforts, Catches per Fishing Effort, and Fishing Locations for the Gulf of Suez and Egyptian Red Sea Fishery for Reef Associated Fish During 1979 to 1982. UNDP/FAO RAB/83/023/02. 65 pp.

SANDERS, M.J. and KEDIDI, S.M. (1984c) Stock Assessment for the Brushtooth Lizardfish (*Saurida undosquamis*) Caught by Trawl in the Gulf of Suez. UNDP/FAO RAB/83/023/05. 72 pp.

SANDERS, M.J. and KEDIDI, S.M. (1984d) Stock Assessment for the Round Scad (*Decapterus maruadsi*) Caught by the Purse Seine and Trawl from the Gulf of Suez and More Southern Red Sea Waters. UNDP/FAO RAB/81/002/26. 41 pp.

SANDERS, M.J. and KEDIDI, S.M (1984e) Stock Assessment for the Horse Mackerel *Trachurus indicus* Caught by Purse Seine and Trawl in the Gulf of Suez. UNDP/FAO RAB/81/002/20.

SANDERS, M.J. and KEDIDI, S.M. (1984f) Catches, Fishing Efforts, Catches per Fishing Effort and Fishing Locations for the Gulf of Suez and Egyptian Red Sea Coast Trawl Fishery During 1979 to 1982. UNDP/FAO RAB/81/002/22. 56 pp.

SANDERS, M.J. and KEDIDI, S.M. (1984g) Summary Report of Stock Assessments, with Management Implications, Concerning the Egyptian Red Sea Fisheries. UNDP/FAO RAB/83/023/09. 54 pp.

SANDERS, M.J. and KEDIDI, S.M. (1984h) Stock Assessment for the Spotted Sardinella (*Sardinella sirm*) Caught by Purse Seine Adjacent to the Border Between Egypt and Sudan. UNDP/FAO RAB/83/023/04. 28 pp.

SANLAVILLE, P. (1982) Asia, Middle East, Coastal Morphology. In: Schwartz, M.L. (ed.) *Encyclopedia of Beaches and Coastal Environments*. Hutchinson Ross, Stroudsburg. pp. 98–102.

SAVIDGE, G., LENNON, H.J. and MATTHEWS, A.D. (1988) A shore based survey of oceanographic variables in the Dhofar region of southern Oman, August–October 1985. *Ecological Studies of Southern Oman Kelp Communities. Summary Report. ROPME/GC-6/001.* pp. 4–21.

SCHEER, G. (1971) Coral reefs and coral genera in the Red Sea and Indian Ocean. In: Stoddart, D.R. and Yonge, C.M. (eds.) *Symposium of the Zoological Society, London,* No. 28. Academic Press, London. pp. 329–367.

SCHEER, G. (1984) The distribution of reef corals in the Indian Ocean with a historical review of its investigation. *Deep Sea Research. A* **31**: 885–900.

SCHLICHTER, D., FRICKE, H. and WEBER, W. (1986) Light harvesting by wavelength transformation in a symbiotic coral of the Red Sea twilight zone. *Marine Biology* **91**: 403–407.

SCHLICHTER, D., WEBER, W. and FRICKE, H. (1985) A chromatophore system in the hermatypic, deep water coral *Leptoseris fragilis* (Anthozoa: Hexacorallia). *Marine Biology* **89**: 143–147.

SCHROEDER, J.H. (1981) Man versus the reef in Sudan: threats destruction, protection. *Proceedings of the Fourth International Coral Reef Symposium*, Manila **1**: 253–257.

SCHUHMACHER, H. and MERGNER, H. (1985) Quantitative Analyse von Korallengemeinschaften des Sanganeb-Atolls (mittleres Rotes Meer) II. Vergleich mit einem Riffareal bei Aqaba (nordliches Rotes Meer) am nordrande des indopazifischen Riffgurtels. *Helgolander Wissenschaftliche Meeresuntersuchungen* **39**: 419–440.

SCHUHMACHER, H. and ZIBROWIUS, H. (1985) What is hermatypic? A redefinition of ecological groups in corals and other organisms. *Coral Reefs* **4**: 1–10.

SHAIKH, E.A., ROFF, J.C. and DOWIDAR, N.M. (1986) Phytoplankton ecology and production in the Red Sea off Jiddah, Saudi Arabia. *Marine Biology* **92**: 405–416.

SHARABATI, D. (1984) *Red Sea Shells*. Routledge & Kegan Paul, London. 128 pp.

SHEPPARD, C.R.C. (1981) The reef and soft-substrate coral fauna of Chagos, Indian Ocean. *Journal of Natural History* **15**: 607–621.

SHEPPARD, C.R.C. (1982a) Coral populations on reef slopes and their major controls. *Marine Ecology Progress Series* **7**: 83–115.

SHEPPARD, C.R.C. (1982b) Overview and status report of the mangroves at Madinat Yanbu al Sinaiyah. Report for Royal Commission for Jubail and Yanbu. E&E Buffalo.

SHEPPARD, C.R.C. (1983a) *A Natural History of the Coral Reef*. Blandford Press, Poole. 168 pp.

SHEPPARD, C.R.C. (1983b) Overview and status report on coral reefs at Madinat Yanbu al Sinaiyah. Report for Royal Commission for Jubail and Yanbu. E&E Buffalo.

SHEPPARD, C.R.C. (1985a) Reefs and coral assemblages of Saudi Arabia. 2. Fringing reefs in the southern region, Jeddah to Jizan. *Fauna of Saudi Arabia* **7**: 37–58.

SHEPPARD, C.R.C. (1985b) The unspoiled Little Barrier Reef of Saudi Arabia. *Sea Frontiers* **31**: 94–103.

SHEPPARD, C.R.C. (1985c) Corals of Oman. Project 9070, IUCN/Sultanate of Oman, Ministry of Commerce and Industry. 11 pp.

SHEPPARD, C.R.C. (1985d) Corals, coral reefs and other hard substrate biota of Bahrain. ROPME Marine Habitat Survey Environmental Protection Unit Bahrain. 25 pp.

SHEPPARD, C.R.C. (1986a) Marine Habitats and Species in Oman. IUCN Project 9069 for Sultanate of Oman. 50 pp.

SHEPPARD, C.R.C. (1986b) Preliminary observations on the status of *Panulirus homarus* in southern Oman. Report for Office of Adviser for Conservation, Muscat, Sultanate of Oman, 7 pp.

SHEPPARD, C.R.C. (1987) Coral species of the Indian Ocean and adjacent seas: a synonymised compilation and some regional distributional patterns. *Atoll Research Bulletin* **307**: 1–32.

SHEPPARD, C.R.C. (1988) Similar trends, different causes: Responses of corals to stressed environments in Arabian seas. *Proceedings of 6th International Coral Reef Symposium, Townsville Australia.* **3**: 297–302.

SHEPPARD, C.R.C. and PRICE, A.R.G. (1991) Will marine life survive in the Gulf? *New Scientist* **1759**: 36–40.

SHEPPARD, C.R.C. and SALM, R.V. (1988) Reef and coral communities of Oman, with a description of a new coral species (Order Scleractinia, genus *Acanthastrea*). *Journal of Natural History* **22**: 263–279.

SHEPPARD, C.R.C. and SHEPPARD, A.L.S. (1985) Reefs and coral assemblages of Saudi Arabia 1. The central Red Sea at Madinat al Sinaiyah. *Fauna of Saudi Arabia* **7**: 17–36.

SHEPPARD, C.R.C. and SHEPPARD, A.L.S. (1991) Corals and coral communities of Arabia. *Fauna of Saudi Arabia* **12**: 3–10.

SHEPPARD, C.R.C. and WELLS, S. (1988) *Directory of Coral Reefs of International Importance. Vol. 2. Indian Ocean Region.* IUCN Gland, and UNEP, Nairobi. 389 pp.

SHINN, E.A. (1976) Coral reef recovery in Florida and the Persian Gulf. *Environmental Geology* **1**: 241–254.

SHLESINGER, Y. and LOYA, Y. (1985) Coral community reproductive patterns: Red Sea versus Great Barrier Reef. *Science* **228**: 1333–1335.

SHULMAN, M.J. and OGDEN, J.C. (1987) What controls tropical reef fish populations: recruitment or benthic mortality? An example in the Caribbean reef fish *Haemulon flavolineatum. Marine Ecology Progress Series* **39**: 233–242.

SIDDEEK, M.S.M., BISHOP, J.M., EL-MUSA, M., ABDUL-GHAFFAR, A.F., LEE, J.U., AL-YAMANI, F., JOSEPH, P.S., ALMATAR, S. and ABDULLAH, M.S. (1990) Reduction in effort and favourable environment helped to increase shrimp catch in Kuwait. *Fishbyte* **8**: 13–15.

SIEBOLD, E. (1973) Biogenic sedimentation of the Persian Gulf. In: Zeitzschel, B. (ed.) *The Biology of the Indian Ocean.* Springer Verlag, New York. pp. 103–114.

SMALL, R.D. (1991) Environmental impact of fires in Kuwait. *Nature* **350**: 11–12.

SMITH, G.B., SALEH, M. and SANGOOR, K. (1987) The reef ichthyofauna of Bahrain (Arabian Gulf) with comments on its zoogeographic affinities. *Arabian Gulf Journal of Scientific Research* **B5**: 127–146.

SMITH, P.J., FRANCIS, R.I.C.C. and McVEAGH, M. (1991) Loss of genetic diversity due to fishing pressure. *Fisheries Research* **10**: 309–316.

SNEH, A. and FRIEDMAN, G.M. (1980) Spur and groove patterns on the reefs of the northern gulfs of the Red Sea. *Journal of Sedimentary Petrology* **50**: 981–986.

SPOONER, M.F. (1970a) Oil spill in Tarut Bay, Saudi Arabia. *Marine Pollution Bulletin* **1**: 166–167.

SPOONER, M.F. (1970b) Oil spill in Tarut Bay: follow-up observations (unpublished). Plymouth Marine Biological Association.

STEHLI, F.G. and WELLS, J.W. (1971) Diversity and age patterns in hermatypic corals. *Systematic Zoology* **20**: 115–126.

STENECK, R.S. (1988) Herbivory on coral reefs: a synthesis. *Proceedings of 6th International Coral Reef Symposium. Townsville, Australia,* Vol. 1: 37–49.

STRÖMME, T. (1983) Survey of the abundance and distribution of the fish resources off Oman from Muscat to Salalah, 1–19th March. Reports on surveys with the R/V "Dr Fridtjof Nansen", Institute of Marine Research Bergen, 12 pp.

SULLIVAN, W. (1974) *Continents in Motion: The New Earth Debate.* McGraw-Hill Book Company, New York. 399 pp.

SWEATMAN, H.P.A. (1983) Influence of conspecifics on choice of settlement sites by larvae of two pomacentrid fishes (*Dascyllus aruanus* and *D. reticulatus*) on coral reefs. *Marine Biology* **75**: 225–229.

SWEATMAN, H.P.A. (1984) A field study of the predatory behaviour and feeding rate of a piscivorous coral reef fish, the lizardfish *Synodus englemani. Copeia* 1984: 187–194.

SZMANT-FROEHLICH, A., JOHNSON, V., HOEHN, T., BATTEY, J., SMITH, G.J., FLEISCHMAN, E., PORTER, J. and DALLMEYER, D. (1981) The physiological effects of oil drilling muds on the Caribbean coral *Montastrea annularis.* In: *Proceedings of 4th International Coral Reef Symposium,* Manila **1**: 163–168.

TETRA TECH LTD. (1982) Section V. A survey of the coral reefs in the vicinity of the northern area development project. Prepared for Environmental Unit, ARAMCO, Dhahram, Saudi Arabia by Saudi Arabian Tetra Tech Ltd. 19 pp.

THIEL, H. (1980) Community structure and biomass in the central Red Sea. In: *Proceedings of a Symposium on the Coastal and Marine Environment of the Red Sea, Gulf of Aden and Tropical Western Indian Ocean.* Khartoum 9–14th January. Vol. III.

University of Khartoum. pp. 127–134.

THIEL, H. (1987) Benthos of the deep Red Sea. In: Edwards, A. and Head, S.M. (eds.) *The Red Sea*. Pergamon Press, Oxford. pp. 112–127.

THIEL, H., WEIKERT, H. and KARBE, L. (1986) Risk assessment for mining metalliferous muds in the deep Red Sea. *Ambio* **15**: 34–41.

THORSON, G. (1957) Bottom communities. In: Hedgpeth, J.W. (ed.) *Treatises on Marine Ecology and Paleoecology, 1*. The Geological Society of America 17: 501 pp.

TURKAY, M. (1986) Cristacea, Decapoda der Tiefsee des Roten Meeres. *Senckberg Marit*. **18**: 123–185.

TYLER, J.C. (1971) Habitat preferences of the fishes that dwell in shrub corals on the Great Barrier Reef. *Proceedings of National Academy of Sciences of Philadelphia* **123**: 1–25.

UNDERWOOD, A.J. and FAIRWEATHER, P.G. (1989) Supply-side ecology and benthic marine assemblages. *Trends in Ecology and Evolution* **4**: 16–20.

UNEP. (1985) Ecological interactions between tropical coastal ecosystems. UNEP Regional Seas Reports & Studies No. 37. 71 pp.

UNEP. (1987) State of the marine environment in the Red Sea region. (Draft). UNEP Regional Seas Reports and Studies.

VANDERMEULEN, J.H. (1982) Some conclusions regarding long-term biological effects of some major oil spills. *Philosophical Transactions of the Royal Society of London* **R297**: 335–351.

VAN ZALINGE, N.P. (1984) The shrimp fisheries of the Gulf between Iran and the Arabian Peninsula. In: Gulland, J.A. and Rothschild, B.J. (eds.) *Penaeid Shrimps — Their Biology and Management*. Fishing News Books Ltd, Farnham.

VAUGELAS, J. DE and SAINT-LAURENT, M. DE (1984) Premières données sur l'écologie de *Callichirus Laurae* de Saint-Laurent sp. nov (crustacé décapode Callianassidae): son action bioturbatrice sur les formations sédimentaires du Golfe d'Aqaba (Mer Rouge). *Comptes Rendues des Sennees de L'Academie des Sciences Serie III. Sciences de la Vie* **6**: 147–152.

VAUGHAN, T.W. (1916) The results of investigations of the ecology of the Floridan and Bahamian shoal-water corals. *Proceedings of National Academy of Science, USA* **2**: 95–100.

VENEMA, S.C. (1984) Fishery resources in the north Arabian Sea and adjacent waters. *Deep Sea Research* **31**: 1001–1018.

VERON, J.E.N. (1985) Aspects of the Biogeography of hermatypic corals. *Proceedings of 5th International Coral Reef Congress*, Tahiti **4**: 83–88.

VERON, J.E.N. (1986) *Corals of Australia and the Indo-Pacific*. Angus Robertson, North Ryde, Australia. 644 pp.

VERON, J.E.N and KELLEY, R. (1988) Species stability in reef corals of the Papua New Guinea and the Indo Pacific. Association of Australasian Palaeontologists, Sydney, Memoir 6. 86 pp.

VINE, P.J. (1974) Effects of algal grazing and aggressive behaviour of the fishes *Pomacentrus lividus* and *Acanthurus sohal* on coral reef ecology. *Marine Biology* **24**: 131–136.

VINE, P.J. (1986) *Pearls in Arabian Waters*. Immel Publishing, London. 59 pp.

VINYARD, G.L. and O'BRIAN, W.J. (1976) Effects of light and turbidity on the reactive distance of bluegill (*Lepomis macrochirus*). *Journal of the Fisheries Research Board of Canada* **33**: 2845–2849.

VITA-FRINZI, C. and CORNELIUS, P.F.S. (1973) Cliff sapping by molluscs in Oman. *Journal of Sedimentary Petrology* **43**: 31–32.

VITA-FRINZI, C. and PHETHEAN, S.J. (1980) Recent inshore sediments in Musandam, Oman. *Marine Geology* **36**: 241–251.

VOUSDEN, D.H.P. (1988) The Bahrain marine habitat survey. Vol. 1. The Technical Report. ROPME, 103 pp.

VOUSDEN, D.H.P. and PRICE, A.R.G. (1985) Bridge over fragile waters. *New Scientist* No. 1451: 33–35.

WAHBEH, M.I. (1980) Studies on the ecology and productivity of the seagrass *Halophila stipulacea*, and some associated organisms in the Gulf of Aqaba (Jordan). D.Phil. Thesis, University of York, UK.

WAHBEH, M.I. (1981) Distribution, biomass, biometry and some associated fauna of the seagrass community in the Jordan Gulf of Aqaba. *Proceedings of 4th International Coral Reef Symposium* **2**: 453–459.

WAHBEH, M.I. and ORMOND, R.F.G. (1980) Distribution and productivity of a Red Sea seagrass community. Symposium on the Coastal and Marine Environment of the Red Sea, Gulf of Aden and Tropical Western Indian Ocean, Khartoum, 9–14 January. Manuscript.

WAINWRIGHT, S.A. (1965) Reef communities visited by the South Red Sea Expedition 1962. Bulletin of Sea Fishsheries Research Station, Israel 38: 40–53.

WALCZAK, P. (1977) The Yemen Arab Republic. A study of the marine resources of the Yemen Arab republic. A report prepared for the Fisheries Development Project, FAO, Rome. FAO-FL-YEM-74/003/5. 67 pp.

WALKER, D.I. (1984) Algal lawns in the Gulf of Aqaba, Red Sea. D.Phil. Thesis. University of York, UK.

WALKER, D.I. (1985) Correlations between salinity and growth of the seagrass *Amphibolis antarctica* (Labill.) Sonder & Aschers., in Shark Bay, Western Australia, using a new method for measuring production rate. *Aquatic Botany* **23**: 13–26.

WALKER, D.I. (1987) Benthic algae. In: Edwards, A.J. and Head, S.M. (eds.) *Red Sea*. Pergamon Press, Oxford. pp. 152–168.

WALKER, D.I. and McCOMB, A.J. (1990) Salinity response of the seagrass *Amphibolis antarctica*: An experimental validation of field results. *Aquatic Botany* **36**: 359–366.

WALKER, D.I. and ORMOND, R.F.G. (1982) Coral death from sewage and phosphate pollution at Aqaba, Red Sea. *Marine Pollution Bulletin* **13**: 21–25.

WALKER, D.I. and WOELKERLING, Wm.J. (1988) Quantitative study of sediment contribution by epiphytic coralline red algae in seagrass meadows in Shark Bay, Western Australia. *Marine Ecology Progress Series* **43**: 71–77.

WARD, J.H. (1963) Hierarchical groupings to optimize an objective function. *Journal of the American Statistical Association* **58**: 236–246.

WASS, R.C. (1982) The shoreline fishery of American Samoa — past and present. In: Munro, J.L. (ed.) Ecological Aspects of Coastal Zone Management. Proceedings of a Seminar on Marine and Coastal Processes in the Pacific, Motupore Island Research Centre, July 1980. pp. 51–83. Jakarta: UNESCO-ROSTSEA.

WEGENER, A. (1929) *The Origin of Continents and Oceans*. English translation, 4th revised edn. J. Biram, Dover, New York.

WEIKERT, H. (1987) Plankton and the pelagic environment. In: Edwards, A.J. and Head, S.M. (eds.) *Red Sea*. Pergamon Press, Oxford. pp. 90–111.

WELLINGTON, G.M. (1982) Depth zonation of corals in the Gulf of Panama: control and facilitation by resident reef fishes. *Ecological Monographs* **52**: 223–241.

WELLS, J.W. (1954) Recent corals of the Marshall Islands, Bikini and nearby atolls. *US Geological Survey papers* **260**: 385–486.

WELLS, J.W. (1969) Aspects of Pacific coral reefs. *Micronesica* **5**: 317–322.

WELLSTED, J.R. (1838) *Travels in Arabia*. Vols 1 and 2. John Murray, London.

WELLSTED, J.R. (1840) *Travels to the City of the Caliphs, along the Shores of the Persian Gulf and Mediterranean. Including a Tour on the Island of Socotra*. Vols I and II. Henry Colburn, London.

WENNINK, C.J. and NELSON SMITH, A. (1977) Coastal oil pollution study for the Kingdom of Saudi Arabia. Vol. 1 Red Sea coast; Vol. 2 Gulf coast. IMCO, London.

WENNINK, C.J. and NELSON SMITH, A. (1979) Coastal oil pollution study for the Gulf of Suez and Red Sea coast of the Republic of Egypt. IMCO, London.

WILKINSON, C.R. (1980) Red Sea sponges of Sudan. In: Final Report, A description of studies conducted in the Sudanese Red Sea. Saudi–Sudanese Joint Red Sea Commission, pp. 20–27.

WILKINSON, C.R. (1987) Productivity and abundance of large sponge populations on Flinders Reef flats, Coral Sea. *Coral Reefs* **5**: 183–188.

WILKINSON, C.R. and SAMMARCO, P.W. (1983) Effects of fish grazing and damselfish territoriality on coral reef algae. II. Nitrogen fixation. *Marine Ecology Progress Series* **13**: 15–19.

WILLIAMS, A.H. (1979) Interference behavior and ecology of threespot damselfish (*Eupomacentrus planifrons*). *Oecologia* **38**: 223–230.

WILSON, H., unpublished. PhD Thesis in prep. University of Warwick.

WOOD, E.M. and WELLS, S.M. (1989) *The Marine Curio Trade. Conservation Issues*. Marine Conservation Society, Ross-on-Wye, UK.

WOOD, W.F. (1987) Effect of solar ultra-violet radiation on the kelp *Ecklonia radiata*. *Marine Biology* **96**: 143–150.

WRIGHT, L. (1982) Deltas. In: Schwartz, M.L. (ed.) *Encyclopedia of Beaches and Coastal Environments*. Hutchinson Ross, Stroudsburg. pp. 358–369.

WYRTKI, K. (1973) Physical oceanography of the Indian Ocean. In: Zeitzschel, B. (ed.) *The Biology of the Indian Ocean*. Springer Verlag, New York. pp. 18–36.

ZEITZSCHEL, B. (ed.) (1973) *The Biology of the Indian Ocean*. Springer Verlag, New York. 549 pp.

ZWARTS, L., FELEMBAN, H. and PRICE, A.R.G. (in press) Wader counts along the Saudi Arabian Gulf coast suggest that the Gulf harbours millions of waders. Wader Study Group Bulletin.

ZYADAH, M.A. (1989) Community studies on coral reef fish (Family: Serranidae) of South Sinai. M.Sc. thesis, Suez Canal University, Ismailia, Egypt.

Index

abalone 281–2
Abd el Kuri 31, 33, 68
abundance of fishes 98–9
Acacia 177
Acanthaster 114, 298
Acanthopleura 115, 191
Acanthurus 104–5, 113
acid deposition 289, 301, 305
Acropora 64, 69, 70, 76, 79–80, 82–5, 86, 115, 230,
    246–8, 251–4
  and stress 246–7, 248, 250–2, 253, 256
Aden (name) 6
Aden, Gulf of *see* Gulf of Aden
Afar Triangle 15, 16, 33
agriculture and agrochemicals 289, 294, 299–300
air pollution 297–8, 305–6, 315
Alcyonarians 114, 116–19, 134
Alcyonium 116
algae/algal 111–13, 121–40, 242–5
  Arabian Sea 134–7
  biogeography 236
  blooms 294
  blue-green *see* Cyanophyta
  browns 90, 127–37, 248–55
  coralline 70–3, 111–12, 116, 131–3, 254
  cover 129–30, 250–5
  diatoms *see* diatoms
  diversity
    in Gulf 127
    in Oman 191
  dominance 243–6
  Dunaliella 182, 185, 187
  Ecklonia 40, 43, 51, 135–6, 137, 254
  endemism 236
  and environmental pressures 242–5, 294, 295,
      299, 306
  and fishes 90, 101, 104–7
  grazing 111–16, 124, 126, 129–30, 134, 136–7
  greens 124–5, 127–31, 134–5, 186, 251, 253–4
  Gulf 127–8, 133
  lawns *see* turf algae and algal lawns
  and mangroves 166–8
  mats 167, 182–7, 191
  Oman 131–7
  plankton *see* plankton
  productivity 127, 129–30, 133, 136–40
  Ras Mohammed fissures 178, 193–4
  red 111, 125–31, 166, 168

Red Sea 128–34
  reefs 33, 65, 73, 131–3, 243
  ridge 72
  sabkha species 185–6
  Sargassopsis 135, 136, 137
  Sargassum see Sargassum
  and seagrasses 150, 153–4, 157
  and seasonality 121–40
    distribution of free-living seaweeds 127–37
    microscopic forms and productivity 137–40
  Turbinaria see Turbinaria
  turf algae 111, 113, 121, 127–31
alluvial
  discharges 177
  fans 20–1, 24, 33, 39, 67
  plains 66
Alphanius 233
Alveopora 230
Amphioplus 157
Amphiroa 136
anemonefish 98
angelfish 88, 93
Anomastrea 225
Aphanius 168
Aqaba, Gulf of *see* Gulf of Aqaba
aquarium fishes 284
Arabian Gulf *see* Gulf
Arabian plate 14
Arabian Sea
  algae 134–7
  barrier 55–8, 232
  climate 43, 50–2, 57–9
  corals 76, 84–5
  currents 43, 50–2
  kelp 40, 43, 51, 135–7, 254
  map 8
  plankton 138–9, 197–201
  productivity 135–7, 139, 197–201
  reefs and coral 68, 84, 86
  sea snakes 211–12, 235
  soft corals 117–18
  turtles 213–14
  upwelling 40, 51, 57–9, 68, 117, 134–9, 238–9
archipelagoes
  structure and formations 31–4
  see also islands
Arius 278
Arothron 298

Arthrocnemum 177, 193
artisanal fisheries 264, 266, 281
Ashrafi Islands 32, 64
Aspidosiphon 159
Astraeosmilia 225
Atlantis Deep 46
atmospheric cycles *see under* seasonality
atmospheric pollution 297–8, 305–6, 315
atolls 66, 73–5, 86
Avicennia 162–9, 170, 173, 175, 255, 293

Bab el Mandeb
    biogeographic barrier 55–8, 232
    closure 18, 33
    fisheries 262
    map 8
    plankton transport 201–3
    water flows 47–8, 56
bacteria 166, 185–7, 197–8
    productivity 197–8
Bahrain
    corals 80–4, 243–4
    environmental pressures 288, 292, 296–7, 301,
        305–6
    fisheries 282
    intertidal biota 186–91
    mangroves 164, 169–71
    map 9
    reefs 70, 83–4
    seagrasses 145, 153, 155
Balanus 190–1
barnacles 169
barracuda 155, 278
barriers
    dispersal 55–8, 232, 237–40
    reefs and coral 66, 70–1, 73–5, 79
Battinah coast 9, 31, 188, 239
beaches 67–9, 188–90
beachrock 188
benthic community structure
    and grazing and bioerosion on reefs 114–15
    and seagrasses 159–60
Benthosoma 272–3
billfish 272
bioerosion 109–16, 191
biogeography
    algae 236
    barriers 55–8, 232, 237–40
    corals 222–30
    echinoderms 234
    fishes 88–98
    mangroves 236–7, 258
    molluscs 234
    sea snakes 235–6
    seagrasses 236
biomass *see* productivity

birds
    and environmental pressures 292, 295, 297–9,
        304–6
    and mangroves 162, 168, 169, 173
Bitter Lakes 45, 233
bleaching 244–6, 301
blenny 95, 105
blue-green algae *see* Cyanophyta
Boleophthalmus 171
Botryocladia 193
bream 97, 267–8, 2643
brine pools 46–7, 198
Brissopsis 157
brown algae 90, 127–37, 248–55
Bruguiera 163, 237
Bryozoans 109–10
buildings 289, 294, 296, 297, 306
butterflyfish 88–90, 93–7, 100

calcification 109–11, 113
Calliasmata 193
Carangoides 271
carbonate platforms 26–8, 32–3, 69, 73–5
Cardisoma 170
Cardium 157
Caretta 213
Carlsberg Ridge 14
Cassiopeia 168, 184
catfish 278
Caulastrea 230
Caulerpa 157, 166
cays 34, 65, 69–70, 84
Cellana 115
cementation 109–11, 113, 248
cephalopods 283
Ceriops 163, 237
Cerithidea 167, 168, 169, 171, 189
Cerithium 151, 168
Chaetodon 90, 93, 94, 95
Chaetomorpha 187
Chain Deep 46
Chelonia 137, 151, 153, 155, 213
chemicals and environmental pressures 289, 294,
        299–300, 306
chlorophyll 54, 200–2, 301
Chromis 88, 91, 93, 94, 95
Cirithidea 189
Cladiella 116
Cleistostoma 169, 170
cliffs 26–7, 30, 115–16, 134, 190–1
climate 36–60
    Holocene 55–9, 238–9
    see also hydrographical influences; seasonality
Cliona 116
Clymene 168
Clypeomorus 168

coastal plants *see* plants
Codium 193
coelenterates 168, 293
Coenbita 189
Colpomenia 133
competition 91, 133–5
conservation *see* management and conservation
controls *see* environmental controls
Copepods 203–7
coral
  Arabian Sea 76, 84–5
  Bahrain 80–4, 243–4
  biogeography 222–30
  bleaching 244–6, 301
  cays 69–70
  communities 77–84, 229–30
  cover 82–4
  diseases 299
  dispersal 239–40
  distributions 77–84, 222–30
  diversity 69–70, 77–84, 108, 222–30, 247
  ecological patterns 70–4, 86
  endemism and speciation 226–9
  and environmental pressures 295–6, 298, 301
  framework 76, 110, 248–54
  growth and extreme natural stress 243–52
  low temperature control 51, 69, 133, 243–55
  major species *see Acropora; Porites*
  non-reefal communities 75–6
  Oman 76, 84, 247, 251–2
  ornamental 284
  recruitment 113, 115
  sands 159
  and seagrasses 159, 160
  sediment, control by 64, 67
  soft 114, 116–19, 134
  zonation 78, 80–4
  see also fishes, coral reef; reefs and coral
coralline algae 70–3, 111–12, 116, 131–3, 254
Coscinaraea 246–7
crab 157, 168–71, 273
Craterastrea 225
Crenidens 233
Creseis 204
croaker 268
Crown-of-Thorns 114, 298
crustaceans
  and environmental pressures 293, 304
  fisheries 263, 273–80, 286
  and mangroves 167–8
  and seagrasses 151–3, 157, 159
  see also shrimps
Cryptocentrus 101
Ctenella 225
Ctenochaetus 151
currents, ocean 42–52

cuttiefish 283, 285, 287
Cyanohydnum 168
Cyanophyta and nitrogen fixation 299
  and coral reefs 104–6, 107, 110
  and intertidal areas 182–7, 191
  and mangroves 166–9, 173
  and pelagic system 198, 203
  and seagrasses 150, 156
  and seasonality 128, 134. 139
Cycloseris 230
Cymodocea 144, 145, 149
Cynarina 230
Cyphastrea 243–4, 246–7, 251
Cystoseira 131, 157

Dahlak Archipelago 8, 32–3, 67, 193–4
damselfish 88–9, 91, 93–6, 100–2, 104–5
Dasyatis 168
Daymaniyat Islands 76, 213
Decapterus 271
deep currents 43
deep temperatures 44, 46–7
demersal fishes 156, 266–7, 271–2, 285
  fisheries 264–9, 281, 286
demersal plankton 207–8
Dendronephthya 117, 118
density currents
  Gulf 48–50
  Red Sea 43–6
depth and seagrasses 142, 146–7
Dermochelys 213
desalination 289, 294
dessication of plants 178–9
Dhofhar 8, 31, 51, 312
Diadema 112, 113, 130, 134, 159, 298
Diaseris 230
diatoms 150, 156, 185–96, 198–200, 294
Digenea 168
dinoflagellates 198
Diodon 298
Diplodus 97, 137
Discovery Deep 46
diseases, coral 299
dispersal
  barriers 55–8, 232, 237–40
  geographic 239–40
diversity
  algal 127, 191
  Arabian 108, 221–41
  coral 69–70, 77–84, 108, 222–30, 247
  fishes, coral reef 98–9
  turf algae and algal lawns 127–9
Djibouti 8, 67, 95, 312
  fisheries 96, 279, 284
dolphins 216
domes, salt 19, 28, 33, 73, 122, 248

Dotilla 168, 189
dottyback 95
dredging 289, 296–7, 306
Drupa 191
dugong 151, 153, 214–15
Dunaliella 182, 185, 187
dunes 68, 177, 188
dust fallout 42
dynamic system, seagrass beds as 146

echinoderms
  and algae 130, 136
  biogeography 234
  and environmental pressures 293
  grazing effect 112–14
  and mangroves 168, 169
  and seagrasses 151, 157, 159
  and seaweeds 130, 136
Echinodiscus 189
Echinometra 113, 130, 136, 159, 169
Echinostrephus 113
Echinothrix 113
Ecklonia 40, 43, 51, 135–6, 137, 254
ecosystem *see* environmental controls
effluent 180–2, 202, 289, 294–5
Egypt 7, 165, 312
  fisheries 262–5, 268, 284
emperor fish 88, 264–5, 267
endemism 221–33
  algae 236
  corals 226–9
  fishes 230–2
energy *see* productivity
Enhalus 144, 146
Enteromorpha 187, 299
environmental controls 76, 84, 117–18, 128,
    242–57
  competition 133–5
  and fishes 101
  light 83, 123–5
  salinity 70–1, 133, 178, 243–7, 251–5
  sedimentation 119
  shading 133–5
  temperature 41–2, 49–50, 51–2, 118, 129–31,
    176, 243–55
  turbidity 252–5
  water movement 126
Epinephelus 95, 265, 267
epiphytes 150
Eretmochelys 213
erosion 20, 24, 26, 33, 34, 72, 116, 291, 307
  bioerosion 109–16, 191
  intertidal 166, 190
Erythrastrea 225, 227
Escenius 95
Ethiopia 8, 67, 170–1

fisheries 266–7, 269, 273, 278, 285–6
Etrumeus 233, 271
Eucalanus 204
Eucidaris 113
Euphausiids 203–7
Euphrates river 29, 180–2
Euretaster 159
Eurycarcinus 169, 170–1
eutrophication 289, 293–4, 299
evaporation 18, 19, 23, 28, 33
  Gulf 49, 180
  Red Sea 47
exotic intertidal habitats 192–4

Farasan Islands 6, 8, 32–3, 67
faulting 14–18, 73–5
Favia 244, 246–7
Favites 244, 246–7
fisheries 261–87
  aquarium and ornamental 284
  crustacean 263, 273–80, 286
  and environmental pressures 289, 293, 298–9,
    304
  future prospects and management 285–7
  interdependence of systems 263
  management and conservation 285–7, 307,
    309–10
  mollusc 280–3
  production bases 261–3
  reefs and coral 264–6, 272
  see also fishes *and under* demersal; pelagic
fishes
  biogeography 88–98, 230–3
  coral reef 87–107, 264–7, 272
    grazing effect 111–13
    regional patterns in structure 88–98, 106–7
    role of fish in reef processes 104–6, 107
    vertical distribution 98–104, 107
  and environmental pressures 293, 294, 295,
    297–8, 306
  fishing *see* fisheries
  larvae 92–3, 106–7
  and mangroves 162, 168–9, 173
  and seagrasses 151, 153–5, 156, 157, 159
fissures, coastal 178, 193–4
flash floods 39
flashlight fish 93
flats, reef 111, 116, 119, 130–2
flooding 301
food *see* grazing; nutrients
Foraminifera 18–19, 58, 109–10, 157, 168
fossil reefs 11, 24–8, 33, 65–7, 166, 228, 248
framework 76, 110, 248–52
fresh water *see* page 192
fringing reefs 20–4, 64–84
Fungia 230

fusilier 264, 266

Galaxea 82
gastropods
    and mangroves 167, 168–9
    and seagrasses 151, 152, 156, 159
Gebel Elba 66
Gena 159
Genicanthus 94
Gerres 263
Globigerina 58
goatfish 98, 267, 278
goby 101
Goniopora 79, 82
Gonodactylus 191
Grapsus 191
grasses, supratidal 177
    see also seagrasses
grazing
    algal 111–16, 124, 126, 129–30, 134, 136–7
        see also Cyanophyta
    fishes 105–6
    overgrazing by livestock 300, 306
    in seagrasses 151
    see also nutrients
green algae *see under* algae
greenhouse gases, concentration of 289, 300–1
Green turtle *see Chelonia*
ground water 177
grouper 94–5, 99–100, 102–3
    fisheries 264–267–8, 278
grunt 104, 106
Gulf
    algae 127–8, 133
    corals 80–6, 243
    currents and climate 43, 48–50
    environmental pressures 288–97, 299–306
    evaporation 49, 180
    fisheries 154, 262–5, 267–77, 281–3, 285–7
    fishes 97, 106–7
    islands 9, 33, 54, 69–70, 254
    management and conservation 307–9, 312–16
    mangroves 161–5, 166–73
    map 9
    name 5
    plankton 138–9, 201–4
    pollution 285–6, 287
    productivity 54, 138, 202, 203
    reef fishes 97, 106–7
    reefs distribution 69–70
    sabkha 182–7
    salinity 49–50, 83–4
    Sargassum community 131–4
    seagrasses 143–9, 152–3, 155–9
    sediments 18–19, 25, 29
    soft coral 118

    structure and formation 28–9
    temperature 41–2, 48–50
    tides 54
    turnover 50
    turtles 212–14
    water balance 49–50
    winds 40–2
Gulf of Aden
    corals 77, 85
    currents and climate 47–8
    evaporation 18
    fisheries 262–4, 267–8, 270, 272, 279, 283, 285
    fishes 96, 107
    management and conservation 313, 316
    map 8
    productivity 201–3
    reefs 67–8, 77, 85, 96, 107
    sea snakes 212
    seagrasses 144
    soft corals 117–18
Gulf of Aqaba
    algae 128–9
    corals 64, 78–83
        endemism 230
    environmental pressures 293, 295
    fisheries 263, 266, 279
    fishes 93–5, 107
    formation 17
    mangroves 163
    map 7
    name 6
    productivity 203
    reefs 64
        fishes 93–5, 107
    seagrasses 143–4, 146–50, 152–3, 155–6, 159
    soft corals 116–17
Gulf of Elat 6
Gulf of Oman
    environmental pressures 291
    fisheries 264, 268, 272–3
    fishes 96, 107
    map 9
    reefs and coral 68, 86, 96, 107
    structure and formation 30–1
Gulf of Salwah 9, 29, 50, 53, 69, 250–2
Gulf of Suez
    corals 64, 78–9
    currents and climate 44–6
    environmental pressures 291, 299
    fisheries 262–6, 268–70, 273–4, 278, 283–4, 286
    fishes 95, 107
    mangroves 163
    map 7
    productivity 203
    reefs 63–4, 95, 107, 249
    seagrasses 143–4, 146–8

Gulf of Suez (*continued*)
  shore vegetation 178–80, 192–3
  structure and formation 17, 18, 28–9
Gulf war 285, 287, 289, 290, 298, 302–6, 314–16
Gyrosmilia 225

habitat
  conservation *see* management and conservation
  controls *see* environmental controls
  damaged *see* environmental stresses
Haemulon 104
Halcurias 157
Haliclona 168
Haliotis 136, 281
Halocnemon 170, 178–81
Halodule 142, 144–9, 151, 159, 168, 236
Halophila 142, 144–9, 151, 155, 159, 166, 236
Halophytes 176–82
Haloptilus 207
Halopyrum 177
Hawar Archipelago 9, 33, 54, 69–70, 254
Hawksbill turtle *see Eretmochelys*
Helice 170
herring 271
Heterocentrotus 113
Heterocyathus 159
Heteronema 168
Heteropsammia 159
Heteroxenia 116
Hippa 157, 189
Hipposcarus 264
Holocanthus 93
Holocene
  climate 55–9, 238–9
    monsoons 57–9, 238–9
  salinity 56–7
  transgression 22–3, 56–8, 59
  upwelling 57–9, 238–9
Holothuria 168
Homotrema 109–10, 119
Horastrea 225
horizontal zonation of mangroves 167–9
Hormophysa 131, 133, 148, 153, 275
Hormuz, Strait of 9, 15, 30, 97, 165, 290
hot brine pools 46–7
hot spring vegetation 192–3
human impact 288–9
  see also environmental controls; pollution;
    management and conservation
hydrographical influences on climate 42–55
  Arabian seas currents 43–52
  Indian Ocean currents 42–3
  oxygen and dissolved nutrients 54–5
  wave energy and tidal patterns 52–4
  see also water

hypersalinity
  Red Sea 56–7
  sabkha pools 183–7

Ilyograpsus 170
Ilyoplax 169, 170
Indian Northeast Monsoon and Indian
    Southwest Monsoon currents 42–3
Indian Ocean 172, 306
  and climate
    currents 42–3
    cycles 37–40
  coral zoogeography 222–6
  fisheries 271
industry and environmental pressures 289,
    294–5, 297–8, 306, 313
  see also oil
Inner Gulf 5
international organizations and Conventions
    313–14, 316
intertidal areas 115–16, 174–94, 301
  beaches 188–90
  exotic habitats 192–4
  marshes 176–82
  rocky shores 190
  sabkha, pools and salt flats 182–7
  see also mangroves
Inter-Tropical Convergence Zone 37
ionic composition of soils 176
Iran 290, 302, 312
  fisheries 268, 272, 278, 282, 285
  mangroves 163–5
  seagrasses 145
Iraq 282, 290
islands 31–4
  Abd el Kuri 31, 33, 68
  Ashrafi 32, 64
  cays 34, 65, 69–70, 84
  Dahlak 8, 32–3, 67, 193–4
  Daymaniyat 76, 213, 311
  Farasans 8, 32–3, 67
  Hawar 9, 33, 54, 69–70, 254
  Jana 84
  Karan 84
  Kuria Muria 8, 33, 51, 68, 213, 293, 299
  management and conservation 311, 315
  Masirah 8, 31, 33, 96, 134, 213–14, 299, 311
  Qeshim 9, 69
  Shotur 69
  structure and formations 31–4
  Suakin Archipelago 67
  Tiran 32
  turtle nesting 213–14
  volcanic 16, 31
  Wedj 67

jack 267, 271–2, 278
Jana 84
Juncus 170, 181, 193

Karan 84
Karun river 180–2
Khor al Bazam sabkha 182
Kuria Muria Islands 8, 33, 51, 68, 213, 293, 299
Kuwait
  algae 128
  corals 70
  environmental pressures 296, 300–2, 304, 306
  fisheries 268, 273–9, 282
  Gulf war *see* Gulf war
  intertidal fauna 191
  management and conservation 313, 315, 316
  mangroves 169–72
  marshes 180–2
  rain 180
  reefs 70, 83
  seagrasses 145, 153–4

lagoons 66, 74–5, 119, 183–4, 188
Lambis 281
larvae
  fish 92–3, 106–7
  dispersal 93
  effects of turbidity on 92, 106–7
  mortality 92, 106–7
Lepidochelys 213
Lepomis 92
Leptastrea 244, 246–7
Leptoseris 83
Lessepian migration 233
Lethrinus 265, 267
light 83, 123–6, 202, 204
  and seagrasses 142, 146–7, 156
Limacina 204, 207
limestone domes 65, 69, 122, 243, 248
Limonium 177–80
Linga 151
Lithophaga 116, 191
Lithophyllum 70
'Little Barrier Reef' 66
Littorina 168, 169, 170, 191
lizardfish 103, 167–8
Lobophytum 116, 119
lobster 279–80
long-lining 263, 267, 272
Lucicutia 207
Lutjanus 168, 268
Lyngbya 150

mackerel 267, 271–2, 285
macroalgae *see* algae
Macrophthalmus 168–9, 170–1, 189

Mactra 157
mammals 212–16
  and environmental pressures 292, 298, 300, 306
management and conservation 307–16
  complexity of system 307–8
  fisheries 285–7, 307, 309–10
  integrated approaches 308
    case studies 308–12
  oil 308, 314–15, 316
  regional and international
    initiatives 313–14
  responses to Gulf war 314–16
mangroves
  associated ecosystems 161–74
    Arabian 162–6
    ecological role of 162
    Red Sea and Gulf 166–73
  biogeography 236–7
  environmental pressures 293, 296, 300, 303,
    306
  and fisheries 263
  management and conservation 307, 311, 312
  temperature control 41
Manifa 244–6, 304, 315
maps 7–9
marshes 176–82
  supratidal vegetation 177–8
Masirah Island 8, 31, 33, 96, 134, 213–14, 299, 311
Mediterranean immigrants 233
Melobesia 111
mesopelagic fishes 272–3, 286
metals, heavy 306, 289. 294
Metapenaeus 168, 274–5, 278
Metaplax 171
Metopograpsus 169, 170
Miemna 168
Millepora 79, 98
Minilabrus 95
mining 289, 295–6, 306, 313
Mitra 151
Mitrella 151
mojarra 263, 267, 278
molluscs 116, 192
  biogeography 234
  and environmental pressures 293, 298
  fisheries 278, 280–3, 286
    see also cephalopods
  and mangroves 167–8, 170–1
  and seagrasses 151–3, 157, 159
monsoons 37, 39, 50, 55, 57–8
  in Holocene 57–9, 238–9
Montipora 76, 84, 85, 251
mudskipper 169
mullet 168, 264, 268
Murex 157, 281
Murray Ridge 16

Musandam 311
  erosion 116, 190
  fishes 97
  formation 15, 30
  map 9
  photograph 219

Neopomacentrus 88
Nephthea 116
Nephthys 168
netting, gill, drift and trammel 263, 264, 267, 271, 272
Nitraria 177, 179–81
nitrates 51, 54–5, 126, 133, 135–6, 138–9, 177
nitrogen fixation 110, 139, 185, 203
Nitzschia 186
Nodolittorina 191
non-reef
  green algal flora 134–5
  substrates 75–6, 122, 248–52
Northeast Monsoon 37
nutrients 51, 126, 133, 135–6, 138–9, 177
  dissolved in waters 54–5
  enrichment
    sewage 180–2, 202
    upwelling 40, 51, 68, 117, 134–9
  and environmental pressures 295, 300
  fishes 93, 96, 262, 263
  and mangroves 165, 173
  and seagrasses 143, 151–3, 155–6, 159–60
  transfer and fishes, coral reef 106
  see also grazing

Ocypode 170, 189
oil pollution 285–7, 289–93, 295, 297, 300, 302–6
  management and conservation 308, 314–15, 316
Oman
  algae 131–7
  corals 76, 84, 247, 251–2
  environmental pressures 293–6, 299–300, 306
  fisheries 262, 268, 270–2, 279, 281–2
  Gulf of *see* Gulf of Oman
  islands 8, 31, 33, 51, 68, 96, 134, 213–14, 293, 299, 311
  management and conservation 311–12, 316
  mangroves 163–5, 170–1, 173
  maps 8, 9
  reef disappearance 251–2
  seagrasses 144, 157
  seasonal kelp communities 135–7
  turtles 213–14
  upwelling 40, 51, 68, 117, 134–9
Onchidium 191
Operculina 157
Ophiocoma 168
origins, geography and substrates 14–35

  see also structure and formations
ornamental shells and corals 284, 286, 298, 306
Oscillatoria 139, 150, 198, 201, 203, 205, 217
Otolithes 268
Outer Gulf *see* Gulf of Oman
oxygen 54–5, 206, 294
  oxygen minimum layers 55, 206
oysters, pearl 152, 153, 282–3

Padina 157, 166
Palythoa 188, 191
Panulirus 279–80
Paracalanus 204
Paracleistostoma 170
Parasalenia 113
Parasimplastrea 227–8
Parerythropodium 116
Parexocoetus 233
parrotfish 88, 89, 94, 97, 101, 104–6, 111–13, 264
Parupeneus 98
Parvilucina 151
Pavona 115, 247
pelagic fishes 92, 269–73, 285
  mesopelagic 272–3, 286
  production levels 272
  reef associated 272
  see also netting; purse seining
pelagic system 197–217
  mammals 214–16
  reptiles 211–14
  see also pelagic fishes; plankton
Pelamis 212, 235
Penaeus 151, 152, 153–5, 159, 274–5, 278, 293
Peneroplis 168
Pentaceraster 159
Periclimenes 193
Perielectrona 152, 293
Perinereis 168
Periophthalmus 168, 171
Perisesarma 168
Persian Gulf 5
Phasianella 151
Phormidium 150
phosphates 51, 54–5, 150, 177, 293, 295
  and seaweeds 126, 133, 135–6, 138–9
Photobleraphon 93
Phragmites 180, 181
phytogenic hillocks 178
phytoplankton *see* plankton
Pinctada 153, 282
Pinna 169
Pirinella 167, 168, 171, 189
Planaxis 170
plankton 123, 197–208
  Arabian Sea 138–9, 197–201
  biomass 138–9, 201

demersal 208
diatoms 185–96, 198–200
dinoflagellates 198
distribution 198–9
and fisheries 270
and fishes 102
Gulf 138–9, 201–4
and mangroves 166–7
pelagic bacteria 197–8
productivity and biomass 43, 138–9, 198–204,
    208–10
Red Sea and Gulf phytoplankton 201–4
and seagrasses 157
seasonality 137–9, 199–203
vertical migration 204, 205–7, 208
zooplankton 138–9, 198–9, 203–8, 210
Platygyra 244, 246–7
Plectroglyphidodon 91, 104
Pleuromamma 207
Pocillopora 76, 84, 85, 86, 115, 251, 252
pollution 152, 285–306
control 308, 314–15, 316
polychaetes 151, 152, 156, 168, 293
Pomacentrus 91
pony fish 278
pools
    brine 46–7, 198
    rock 193–4
    sabkha 183–5
porgies 168
Porites 64, 76, 79, 82–5, 86, 109, 117
    and stress 243–4, 246–7, 249, 250, 251–2, 253–4,
        256
Porolithon 70–2, 111–12
Portunus 169, 171, 273
Posidonia 144
Pristipomoides 267
productivity
    algae 127, 129–30, 133, 136–40
    Arabian Sea 135–7, 139, 197–201
    bacteria 197–8
    Cyanophyta 110, 198
    fishes 104–6, 261–3, 272
    Gulf 54, 138, 202, 203
    Gulf of Aden 201–3
    Gulf of Aqaba 203
    Gulf of Suez 203
    mangroves 162, 166–7
    plankton 43, 138–9, 198–204, 208–10
    primary 54, 110, 197–203, 208–11
    Red Sea 54–5, 129–30, 197–208
    seagrasses 143, 148–51, 152, 155–6, 158, 159
    supratidal 177
    turf algae and algal lawns 129–30
prokaryotic biota of sabkha mats 185–7
protection *see* management and conservation

Psammocora 247
Pseudochromis 93, 95
Pseudotriacanthus 278
Puerulus 279, 280
purse seining 263, 267, 269–71
Pyrene 151

Qatar
    mangroves 145, 164–169
    marshes 182
    reefs and coral 69, 83
Qeshim Island 9, 69
Quaternary sea level change 19–24

rabbitfish 151, 264, 266
radiolaria 202
rainfall 39, 56, 180
    acid 289, 301, 305
raised reefs 11, 24–8, 33, 65–7, 228, 248
Ras al Hadd 68, 134, 213
Ras Mohammed 7, 64, 265
    fissures, pool 178, 193–4
Rastrelliger 271
ray 278
reclamation 289, 296
recreation and tourism 289, 295, 298, 310
red algae 111, 125–31, 166, 168
Red Sea
    algae 128–34
    currents and climate 43, 44–8
    environmental pressures 288, 290–302, 306
    evaporation 47
    fisheries 262–74, 279–86
    fishes 87–93, 98–101, 104, 106–7
    Gulfs *see* Gulf of Aqaba; Gulf of Suez
    hypersalinity 56–7
    islands 6, 8, 32–3, 67, 193–4
    isolated in Holocene 56–7
    management and conservation 307, 309,
        312–13, 316
    mangroves 161–73
    map 8
    nutrients 54–5, 138
    plankton 201–7
    pollution 285–6
    productivity 54–5, 129–30, 197–208
    reefs and coral
        distribution 65–7, 85, 86
        fishes 87–93, 98–101, 104, 106–7
        zonation 78–83, 229–30
    salinity 44–8, 56
    Sargassum community 131–4
    seagrasses 143–52, 156–60
    soft corals 116–19
    temperatures 44–8
    tides 52–4

Red Sea (*continued*)
    turtles 212–14
    water balance, exchange and level 45–8
red tides 185, 203–4, 294
reeds 170, 177–82, 193
reefs 63–86
    algal 33, 65, 73, 131–3, 243
    atolls 66, 74–5, 86
    barrier 66, 70–1, 73–5, 79
    biogeography 222–30
    cays 65, 70, 84
    cementation 109–11, 248
    contour 64
    corals
        Arabian Sea 84
        Gulf 83–4
        Red Sea zonation 78–83
    disappearance 250–2
    distribution and development 64–76, 85–6
        geographical 64–70
        non-reef coral populations 75–6
        southern algal constructions 73
    fisheries 264–6, 272, 286
    fishes on *see* fishes, coral reef
    flats 111, 116, 119, 130–2
    fossil 11, 24–8, 33, 65–7, 166, 228, 248
    fringing 20–4, 64–84
    grazing and bioerosion 111–16
    growth 243–52
    incipient 85
    Iranian 84
    'Little Barrier Reef' 66
    and mangroves 164, 166
    patch 64–6, 70, 79
    production 209
    raised 11, 24–8, 33, 65–7, 228, 248
    ridge 73–5
    simplest 249–50
    Somalian 67–8
    spur and groove systems 70–2
    structure and formation 24–8
    uplift 24–8, 33, 65–7, 228, 248
    see also coral
reptiles 211–14, 292
residential structures 289, 294, 296, 297, 306
Rhabdosargus 137
Rhincalanus 207
Rhizoclonium 187
Rhizophora 162, 163, 168, 170, 237, 311
ridge reefs 73–4, 86
rifting 14–18
rock pools 193–4
rocky shores 190–2
ROPME Sea Area 5

sabkha 175–6, 182–7

Saccostrea 168, 169, 191
salinity
    as environmental control 70–1, 133, 178, 243–7,
        251–5, 301
    and fishes 91–2, 107
    Gulf 49–50, 83–4
    Holocene 56–7
    intertidal areas 178
    and mangroves 162–3, 173
    Red Sea 44–8, 56
salt domes 19, 28, 33, 73, 122, 248
salt flats *see* sabkha
salt marshes 176, 178–82
    and seagrasses 147, 156–7
    and seaweeds 133
Salvadora 178
Salwah *see* Gulf of Salwah
sands
    coral 159
    sandy beaches 67–9, 188–90
Sanganeb 66, 73–5
Sarcophyton 116, 117, 118–19
Sardinella 271
Sargassopsis 135, 136, 137
Sargassum 73, 90
    and fisheries 275
    and fishes 90, 101
    in Red Sea and Gulf 131–4, 214
    and seagrasses 148, 153
    and seasonality 125, 128, 135
    and stresses 244, 253–5
Saudi Arabia
    environmental pressures 288, 290, 294, 296–7,
        299–300, 302, 306
    fisheries 262, 264–5, 269–73, 276–8, 280, 285
    management and conservation 309–11, 313,
        314–16
    mangroves 163–5, 170–1
    map 9
    seagrasses 144–5, 153–6
Saurida 268
Scartelaos 171
Scarus 97, 264
Scomberomorus 271
Scylla 169, 171
sea breezes 40–1, 53
sea level changes 47–8, 52–4, 301
    Holocene 19–24, 56–8
sea snakes 210–12, 235, 292
    biogeography 235–6
seagrasses 141–60, 166, 307
    biogeography 236
    ecology and dynamics 145–53
    along environmental gradients 143–5
    and environmental pressures 293, 299
    and fisheries 263, 277

and fishes 151, 154–5, 156, 157, 159
    productivity 143, 148–50, 152, 156, 158, 159
    role 153–6
seas, formation of 15–19
seasonality
    and algae 121–40
    atmospheric cycles 36–43
        Indian Ocean 37–40
        local wind systems 40–2
    kelp communities in Oman 135–7
    plankton 137–9, 199–203
    tides 52, 192
    turf algae and algal lawns 130–1
seaweeds see algae
sediments
    biota 109–10, 119, 123
    control by 64, 67
    and environmental pressures 296–7, 306
    Gulf 18–19, 25, 29
    production 18–19, 116
    sabkha 182–7
    stabilisation 110, 249–50
Sepia 283
Sesarma 170
sewage 180–2, 202, 289, 294, 295
shading 133–5
Shamal wind 41–2
shark 278
Shatt al Arab 19, 29, 70, 178. 180–2
shipping and ports 289
shoreline terraces: structure and formation 24–8
Shotur 69
shrimps 309
    and seagrasses 152, 153, 157, 159
    and environmental pressures 293, 304
    fisheries 263, 267–8, 273–9, 285, 287
    and mangroves 162, 169
sibling species 232
Siderastrea 73, 225, 243–4, 247, 249, 251
Siganus 137, 151
Sinai
    fisheries 265, 280, 284
    mangroves 162–3, 166–7, 169, 173
    map 7
    seagrasses 145
Sinularia 116, 119
Smargdia 151
snails 169
snakes see sea snakes
snapper 88, 106, 168, 264, 267, 278
soft corals 114, 116–19, 134
soft-bottomed systems
    fishes 267
    mangroves 164–5, 167, 173
    subtidal 156–9
    see also sediments

soils 176–82
solid waste pollution 294–5
Somali current 42–3, 52
Somalia
    fisheries 271
    map 8
    reefs 67–8
South Equatorial current 42–3
Southeast Arabia see Oman; Yemen
Southwest Monsoon 39, 50
speciation
    corals 226–9
    fishes 231–2
Sphacelaria 128
Sphyraena 155
Spirolina 168
Spirydia 166, 168
splash zone 115, 118, 192
sponge 109–10, 116, 157, 168, 169
spreading centre 14
spurs and grooves 70–2, 86, 111
squid 283
starfish 114, 298
Stegastes 115
Stereonephthya 117
Strait of Hormuz 9, 15, 30, 97, 165, 290
Straits of Tiran 7, 18, 32
stress 242–57
    and see environmental stresses
Strombus 151, 166, 169, 281
structure and formations 14–34
    geological history and formation of seas 15–19
    Gulf and Gulf of Suez 28–9
    islands and archipelagoes 31–4
    Oman and Yemen 30–1
    shoreline terraces and layers of reefs 24–8
    substrate precursors 19–24
Stylophora 64, 115, 228, 247, 253, 293, 295
Suaeda 177
Suakin Archipelago 67
subprovinces 225, 237–40
substrates
    and algae 121–40
    non-reefal 75–6, 122, 248–52
    precursors and Quaternary sea level change 19–24
    and seagrasses 147
    see also reefs; corals; sediments
subtidal systems, soft 156–9, 160
Sudan 313
    fisheries 262–3, 265–6, 269, 272–3, 278, 281–2
Suez Canal 45, 233
    fisheries 273, 275, 281, 283
Suez, Gulf of see Gulf of Suez
sulphur spring vegetation 192–3
sunfish 92

supratidal productivity 177
surgeonfish 88–9, 94, 99–101, 104–6
    fisheries 264, 266
sustainable devlopment 308
Synodus 103
Syringodium 144, 145, 149, 151, 159

Talorchestia 189
Tamarix 193
Tapes 281
Tarut Bay 244–6
tectonic activity 14–18, 23, 26–8, 122
    heating 47
Tectus 281
Temnotrema 159
Temoropia 207
temperature
    and corals 51, 69, 133, 243–55
    deep 44, 46–7
    environmental controls 41–2, 49–50, 51–2, 118,
        129–31, 176, 243–55, 301
    and fishes 91–2, 107
    global warming 289, 300–1
    Gulf 41–2, 48–50
    and mangroves 41, 162, 165, 173
    Red Sea 44–8
    and seagrasses 147
    and seaweeds 129–31
Tethys 19
Tetraclita 191
Thais 191
Thalassia 144, 145, 147–9, 159
Thalassodendron 144–9, 159, 236
Thalassoma 101
Thunnus 271
tides
    exposure 53, 192
    ranges 52–4
    red 185, 203–4, 294
    seasonality 52, 192
Tigris river 29, 180–2
Tiran, Strait of 7, 18, 32
tourism *see* recreation and tourism
Trachurus 271
Trachyphyllia 230
transfer coefficients 208–11
trawling 263, 267
Trichodesmium see Oscillatoria
Tridacna 281
Tripneustes 151, 159, 168, 169
tripod fish 278
Trochus 191, 298
trolling 272, 277
tuna 271–2, 285
turbidity 252–5
    and fishes 92, 96, 97, 107

    see also upwelling
Turbinaria algae 125, 130–1
Turbinaria coral 84, 246, 247, 249
Turbo 191
turf algae and algal lawns 111, 113, 121, 127–31
    and coral reef fishes 104, 105, 111, 113
    diversity and distribution 127–9
    productivity and biomass 129–30
    seasonality 130–1
Turritella 157
turtles 137, 151, 155, 212–14
    and environmental pressures 292, 295, 298–9,
        306
Tylodiplax 171
Typha 180, 181

UAE *see* United Arab Emirates
Uca 167, 170, 172, 189
Ulva 299
UNEP Regional Seas Programme 313, 316
United Arab Emirates
    barrier islands 29–34
    corals 243
    fisheries 272, 277
    mangroves 164–5
    map 9
    sabkha 182–3
    seagrasses 145
United Nations 313, 316
uplift
    reefs 24–8, 33, 65–7, 228, 248
    shoreline 23–8, 34
upwelling 40, 51, 68, 96, 117, 134–9, 157
    Holocene 57–9, 238–9
    see also turbidity
urban development *see* residential
urchins 112–14, 130, 234, 298
    and seagrasses 151, 155, 159

Valona 193
Variola 265
vehicles and environmental pressures 289, 298
vertical distribution
    of fishes 98–104, 107
    of mangroves 169
    of seagrasses 142
volcanic islands 16, 31

wadis 20–1
Wahiba sands 68, 188
waste 180–2, 202, 289, 294–5
water
    balance
        Gulf 49–50
        Red Sea 45–8
    exchange in Red Sea 47–8

flows in Bab el Mandeb 47–8, 56
level *see* sea level
movement and turnover 126, 173
   see also upwelling
quality and fishes in Red Sea 91–3
wastewater pollution 294
see also hydrographical influences on climate
wave energy 70–2, 126, 129
   water and climate 52–4
Wedj Bank 66, 67
whales 216
winds 40–2
wrasse 88, 94, 101

Xenia 116–19, 120

xerophytes 176–82

Yemen 8
   environmental pressures 292, 300
   fisheries 262, 267, 269–71, 273, 278–9, 283, 287
   mangroves 163–5
   structure and formation 30–1

Zebrasoma 151
Zoanthus 168
zoogeography *see* biogeography
zooplankton 138–9, 198–9, 203–8, 210
Zostera 150
Zygophyllum 170, 177, 179, 193